PRIMARY INSET SERIES

SCIENCE FOR CURRICULUM LEADERS

PRIMARY INSET SERIES

Series editor David Wray

Already published
Language for Curriculum Leaders
Henry Pearson and David Wray

Forthcoming
Mathematics for Curriculum Leaders
Bill Rawson

TECHNOLOGIES FOR UPGRADING EXISTING OR DESIGNING NEW DRINKING WATER TREATMENT FACILITIES

Office of Drinking Water
Center for Environmental Research Information
U.S. Environmental Protection Agency
Cincinnati, Ohio 45268

CRC Press
Taylor & Francis Group
Boca Raton London New York

CRC Press is an imprint of the
Taylor & Francis Group, an **informa** business

CRC Press
Taylor & Francis Group
6000 Broken Sound Parkway NW, Suite 300
Boca Raton, FL 33487-2742

ISBN-13: 978-1-138-47532-8 (hbk)
ISBN-13: 978-0-87762-824-8 (pbk)

Visit the Taylor & Francis Web site at
http://www.taylorandfrancis.com

and the CRC Press Web site at
http://www.crcpress.com

Contents

Tables

Figures

Acknowledgments

This document is a compilation of the material presented in a series of workshops on "Emerging Technologies for Upgrading Existing or Designing New Drinking Water Treatment Facilities." Some of these workshops were sponsored by local sections of the American Water Works Association (AWWA) and others by the Association of State Drinking Water Administrators (ASDWA). Other sponsors included the United States Environmental Protection Agency's (EPA) Offices of Drinking Water and Research and Development and Regions.

Overall management of the workshop series and preparation of this document was provided by Dr. James E. Smith, Jr., of EPA's Center for Environmental Research Information. He was assisted in providing technical direction by representatives of ASDWA, the local AWWA Sections, EPA Regional Personnel, John Trax and Ken Hay of EPA's Office of Drinking Water, and Walter Feige and other members of EPA's Drinking Water Research Division.

Unless otherwise noted, the technical material in this document was taken from handout materials and presentations provided by the following speakers in the workshops: Dr. Robert M. Clark, Director of EPA's Drinking Water Research Division, Cincinnati, Ohio; John F. Dyksen, Director, Department of Water Supply, Village of Ridgewood, New Jersey; Sigurd Hansen, HDR, Inc., Cameron Park, California; Dr. Gary S. Logsdon, Chief of EPA's Microbiological Treatment Branch, Cincinnati, Ohio; Dr. Rip G. Rice, Rice International Consulting Enterprises, Ashton, Maryland; and Dr. Edward Singley, James Montgomery Consulting Engineers, Gainesville, Florida. Appreciation is also expressed to those individuals around the country who participated in panels and presented case histories of their treatment experiences.

Organization and presentation of the material, technical writing, and editorial work was provided by Lynn Knight, John Reinhardt, and Susan Richmond of Eastern Research Group, Inc., Arlington, Massachusetts. Appreciation is expressed to those individuals in EPA's Offices of Drinking Water and Research and Development who assisted in reviewing drafts of this publication; especially Peter Cook, Ronnie Levin, Michael Schock, and Tom Sorg. The workshop presenters are acknowledged for their assistance in reviewing materials, especially John Dyksen who provided overall technical review.

In addition, EPA gratefully acknowledges the assistance of:

Kalyan Ramen, Environmental Engineer, Malcolm Pirnie, Inc.
Ross Hymen and Paul Anderson, Massachusetts Dept. of Environmental Quality Engineering.
Tom Boshar, Lally Associates
Dr. Carl Nebel, PCI Ozone Corporation
Fred Zinnbauer, Aquionics, Inc.
Gary Swanson, Robert Peccia Associates

Chapter 1

Introduction

This document discusses drinking water treatment technologies that address contaminants and contaminant categories regulated under the Safe Drinking Water Act (SDWA - 42 U.S.C. 300f, et seq.) and its 1986 amendments. The information was distilled from materials used in a series of workshops conducted from 1987 through 1989 in locations throughout the United States. The workshops were sponsored by the Offices of Drinking Water (ODW) and Research and Development (ORD) of the United States Environmental Protection Agency (EPA), and the Association of State Drinking Water Administrators (ASDWA).

The 1986 statutory provisions of the SDWA amendments bring a large number of previously unregulated or minimally regulated water systems under significant regulatory control. This document covers both established and emerging technologies needed to comply with these new regulations. Descriptions of each technology include an overview of the process, performance, design considerations, operating and maintenance aspects, costs, and experiences. This information is meant to assist public water system engineers, operators, and decision-makers faced with the many new regulatory requirements in selecting methods of compliance.

Chapter 2 is an overview of the selection process and the potential technological solutions for each contaminant or contaminant category. This chapter serves as a guide to subsequent chapters that discuss each treatment technology in more detail. It includes many tables that compare relevant information between technologies. Case histories illustrating experience with each technology are provided in appendices.

Chapter 3 covers prefiltration elements of a water treatment system, including rapid mixing, chemical dosage, coagulation, flocculation, and sedimentation. While these elements usually precede filtration, they may be found with other treatment technologies as well. These elements impact the performance of subsequent components in the treatment train, which are described in Chapters 4 through 7.

Chapter 4 describes filtration technologies that address removal of turbidity and microbial contamination. Technologies covered include conventional, direct, slow sand, diatomaceous earth, membrane, and cartridge filtration systems.

Chapter 5 reports on the five major disinfection technologies: chlorine, ozone, ultraviolet (UV) radiation, chlorine dioxide, and chloramines. The problem of disinfection by-products and strategies for their control are also addressed.

Chapters 6 and 7 describe technologies that address organic and inorganic contamination, respectively. Treatment technologies for organics removal include granular activated carbon, packed column aeration, powdered activated carbon, diffused aeration, multiple tray aeration, oxidation, mechanical aeration, catenary grid aeration, Higee aeration, and membrane filtration. Treatment technologies for inorganics removal include corrosion control, reverse osmosis, ion exchange, activated alumina, aeration, and powdered activated carbon.

Chapter 8 reviews the recent research activities of EPA's Drinking Water Research Division.

Finally, Chapter 9 lists the references used in the entire document.

Chapter 2

Selecting and Evaluating Treatment Processes

This chapter serves as a guide to the entire document. It begins in Section 2.1 with a brief summary of the statutory and regulatory framework that applies to public drinking water systems. The new statutory and regulatory provisions drive the treatment objectives for drinking water. Identifying these objectives and selecting treatment alternatives are discussed in Section 2.2. Section 2.3 summarizes the various technological alternatives available to water utilities for complying with the new regulatory provisions. Section 2.4 discusses the final selection process.

2.1 Overview of Federal Drinking Water Regulations

This overview provides a general context for discussing the available treatment technologies presented in this document. It is not intended as a substitute for the Code of Federal Regulations that describes each provision's many regulatory specifications, variances, and exceptions.

The 1986 amendments significantly strengthen and expand the 1974 Safe Drinking Water Act (SDWA). A major focus of the amendments is establishing a strict schedule for promulgating drinking water regulations that use numerical standards, referred to as Maximum Contaminant Levels (MCLs), or treatment technique requirements for 83 contaminants (see Table 2-1). Prior to the 1986 amendments, only 22 MCLs had been set. In addition, EPA must regulate 25 additional contaminants every 3 years beginning in 1991.

The regulatory provisions derived from the 1986 amendments use the following terms to describe controls for water contaminants:

- *Maximum Contaminant Levels (MCLs):* These are the maximum permissible levels of contaminants delivered to a user of public water supplies.
- *Maximum Contaminant Level Goals (MCLG):* These are nonenforceable limits that indicate the level of the contaminant that does not cause any known or anticipated adverse effect on human health. They were formerly termed Recommended Maximum Contaminant levels (RMCLs).
- *Secondary Maximum Contaminant Levels (SMCLs):* These are nonenforceable goals for preserving the aesthetic qualities of drinking water.

The amendments authorize EPA to set treatment technique requirements in lieu of MCLs when it is not economically or technically feasible for water suppliers to determine the level of contaminants. In addition, the amendments require EPA to set best available technology (BAT) for purposes of complying with National Primary Drinking Water Regulations. For instance, the statute lists granular activated carbon (GAC) as BAT for synthetic organic chemical (SOC) removal. For purposes of complying with the regulations, any other technology, treatment technique, or control strategy selected to remove SOCs must be as effective as GAC. BAT is also set for purposes of issuing variances under Section 1415.

In addition to the quantitative provisions noted above, there are several far-reaching provisions on monitoring, filtration, disinfection, and use of lead materials included in the 1986 Amendments to the SDWA, as follows:

- *Monitoring:* The amendments instruct EPA to promulgate regulations requiring monitoring of certain unregulated contaminants. The monitoring schedule is varied based on the number of persons served by the system, the

Table 2-1. Contaminants Required to be Regulated Under the SDWA Amendments of 1986

Volatile Organic Chemicals	
Trichloroethylene	Benzene
Tetrachloroethylene	Chlorobenzene
Carbon tetrachloride	Dichlorobenzene
1,1,1-Trichloroethane	Trichlorobenzene
1,2-Dichloroethane	1,1-Dichloroethylene
Vinyl chloride	trans-1,2 Dichloroethylene
Methylene chloride	cis-1,2-Dichloroethylene
Microbiology and Turbidity	
Total coliforms[a]	Viruses
Turbidity[a]	Standard plate count
Giardia lamblia	*Legionella*
Inorganics	
Arsenic[a]	Molybdenum
Barium[a]	Asbestos
Cadmium[a]	Sulfate
Chromium[a]	Copper
Lead[a]	Vanadium
Mercury[a]	Sodium
Nitrate[a]	Nickel
Selenium[a]	Zinc
Silver[a]	Thallium
Fluoride[a]	Beryllium
Aluminum	Cyanide
Antimony	
Synthetic Organics	
Endrin[a]	1,1,2-Trichloroethane
Lindane[a]	Vydate
Methoxychlor[a]	Simazine
Toxaphene[a]	PAHs
2,4-D[a]	PCBs
2,4,5-TP[a]	Atrazine
Aldicarb	Phthalates
Chlordane	Acrylamide
Dalapon	Dibromochloropropane (DBCP)
Diquat	1,2-Dichloropropane
Endothall	Pentachlorophenol
Glyphosate	Pichloram
Carbofuran	Dinoseb
Alachor	Ethylene dibromide (EDB)
Epichlorohydrin	Dibromomethane
Toluene	Xylene
Adipates	Hexachlorocyclopentadiene
2,3,7,8-Chlorinated dibenzo furans (Dioxin)	Total trihalomethanes[a]
Radionuclides	
Radium 226 and 228[a]	Gross alpha particle activity[a]
Beta particle and photon radioactivity[a]	Radon
Uranium	

[a]Regulated prior to 1986 amendments to SDWA.
Source: U.S. EPA Fact Sheet; February 1989.

source of the supply, and the contaminants likely to be found.

- *Filtration:* EPA was required to specify criteria under which filtration is required for surface water sources and procedures for States to determine which systems must filter.

- *Disinfection:* All public water supplies will be required to disinfect their water.

- *Lead:* The amendments prohibit the use of solders and fluxes containing more than 0.2 percent lead and pipes and pipe fittings containing more than 8 percent lead.

The three primary entities involved with the regulatory effort for the 1986 amendments are:

- U.S. EPA, with the primary roles of national primary and secondary drinking water regulations, designating BATs, and overseeing State programs and enforcement.

- States, with the primary responsibility of implementation, program administration, and enforcement.

- Utilities, which will have to increase monitoring, install new treatment processes, and increase public awareness of contamination problems.

These are very simplified descriptions of each role. The statute and regulations require significant interplay between the three groups.

The regulatory effort is divided into the following five phases:

- Phase I: Volatile organic compounds
- Phase II: Synthetic organic compounds, inorganic compounds, and unregulated contaminant monitoring
- Phase III: Radionuclide contaminants
- Phase IV: Disinfectant and oxidant by-products
- Phase V: Inorganic compounds and synthetic organic compounds

On July 8, 1987, EPA promulgated final regulations for eight VOCs and fluoride. The final MCLGs and MCLs for the VOCs are shown in Table 2-2. The regulations published July 8, 1987, also include monitoring requirements for 51 unregulated contaminants, which are shown in Table 2-3.

In May 1989, EPA issued proposed regulations for 30 SOCs and 8 inorganic chemicals (IOCs), which are listed in Table 2-4. For two contaminants, epichlorohydrin and acrylamide, EPA proposed a treatment technique in lieu of an MCL and monitoring

Table 2-2. Final MCLGs and MCLs for VOCs

	Final MCLG[a] (mg/L)	Final MCL (mg/L)
Trichloroethylene	zero	0.005
Carbon tetrachloride	zero	0.005
Vinyl chloride	zero	0.002
1,2-Dichloroethane	zero	0.005
Benzene	zero	0.005
para-Dichlorobenzene	0.075	0.075
1,1-Dichloroethylene	0.007	0.007
1,1,1-Trichloroethane	0.2	0.2

[a] Final MCLGs were published Nov. 13, 1985. The MCLG and MCL for p-dichlorobenzene were reproposed at zero and 0.005 mg/L on April 17, 1987; comment was requested on levels of 0.075 mg/L and 0.075 mg/L, respectively.
Source: U.S. EPA Fact Sheet; February 1989.

requirements. The proposal also established nine new secondary MCLs, which are listed in Table 2-5.

In June 1988, EPA issued proposed regulations to define MCLs and MCLGs for lead and copper. The MCLG proposed for lead is zero, and 1.3 mg/L for copper. The proposed MCLs applicable to water entering the distribution system were 0.005 mg/L for lead and 1.3 mg/L for copper. EPA also was considering MCLs and/or action levels for lead and copper and other related water quality parameters at the consumers' tap. These proposed regulations are in various stages of finalization.

On June 29, 1989, the Surface Water Treatment Rule (SWTR) and the Coliform Rule were promulgated. According to the SWTR, all public water systems using surface water or ground water under direct influence of surface water, must disinfect and may be required to filter if certain source water quality requirements and site-specific conditions are not met. The MCLGs established in the rule are:

- *Giardia lamblia* - 0
- Viruses - 0
- *Legionella* - 0

No MCLGs were set for turbidity and heterotrophic plate count. Treatment requirements also were established in lieu of MCLs for *Giardia*, viruses, HPC, *Legionella*, and turbidity. Treatment must reliably achieve at least 99.9 percent removal/inactivation of *Giardia lamblia* cysts and 99.99 percent removal/inactivation of viruses.

The Coliform Rule requires all public water systems to meet the coliform MCL and monitor total coliform with frequencies depending on population served, and requires small systems to conduct a sanitary survey. To comply with the coliform MCL, no more than 50 percent of all total coliform samples per month can be total coliform-positive.

Table 2-3. Monitoring for Unregulated VOCs

Required for All Systems:	
Chloroform	1,2-Dichloropropane
Bromodichloromethane	1,1,2,2-Tetrachloroethane
Chlorodibromomethane	Ethylbenzene
Bromoform	1,3-Dichloropropane
trans-1,2-Dichloroethylene	Styrene
Chlorobenzene	Chloromethane
m-Dichlorobenzene	Bromomethane
Dichloromethane	1,2,3-Trichloropropane
cis-1,2-Dichloroethylene	1,1,1,2-Tetrachloroethane
o-Dichlorobenzene	Chloroethane
Dibromomethane	1,1,2-Trichloroethane
1,1-Dichloropropane	2,2-Dichloropropane
Tetrachloroethylene	o-Chlorotoluene
Toluene	p-Chlorotoluene
p-Xylene	Bromobenzene
o-Xylene	1,3-Dichloropropane
m-Xylene	Ethylene dibromide
1,1-Dichloroethane	1,2-Dibromo-3-chloropropane
Required for Vulnerable Systems Only:	
1,2-Dibromo-3-chloropropane (DBCP)	
Ethylenedibromide (EDB)	
At Each State's Discretion:	
1,2,4-Trimethylbenzene	1,3,5-Trimethylbenzene
1,2,4-Trichlorobenzene	p-Isopropyltoluene
1,2,3-Trichlorobenzene	Isopropylbenzene
n-Propylbenzene	tert-butylbenzene
n-Butylbenzene	sec-butylbenzene
Naphthalene	Fluorotrichloromethane
Hexachlorobutadiene	Dichlorodifluoromethane
	Bromochloromethane

Source: U.S. EPA Fact Sheet; February 1989.

2.2 Selecting Treatment Technologies

Defining treatment objectives and selecting control technologies involve eight basic considerations:

- Effluent requirements
- Influent characteristics
- Existing system configuration
- Cost
- Operating requirements
- Pretreatment and posttreatment components
- Waste management
- Future needs of the service area

Table 2-4. Proposed MCLGs and MCLs for SOCs and IOCs

Contaminant	Existing NPDWR[a] (mg/L)	Proposed MCLG (mg/L)	Proposed MCL (mg/L)
Acrylamide	–	zero	TT[b]
Alachlor	–	zero	0.002
Aldicarb	–	0.01	0.01
Aldicarb sulfoxide	–	0.01	0.01
Aldicarb sulfone	–	0.04	0.04
Atrazine	–	0.003	0.003
Carbofuran	–	0.04	0.04
Chlordane	–	zero	0.002
cis-1,2-Dichloroethylene	–	0.07	0.07
Dibromochloropropane (DBCP)	–	zero	0.0002
1,2-Dichloropropane	–	zero	0.005
o-Dichlorobenzene	–	0.6	0.6
2,4-D	0.1	0.07	0.07
Ethylenedibromide (EDB)	–	zero	0.00005
Epichlorphydrin	–	zero	TT[b]
Ethylbenzene	–	0.7	0.7
Heptachlor	–	zero	0.0004
Heptachlor epoxide	–	zero	0.0002
Lindane	0.004	0.0002	0.0002
Methoxychlor	0.1	0.4	0.4
Monochlorobenzene	–	0.1	0.1
PCBs (as decachlorobiphenyl)	–	zero	0.005
Pentachlorophenol		0.2	0.2
Styrene[c]	–	zero/0.1	0.005/0.1
Tetrachloroethylene	–	zero	0.005
Toluene	–	2.0	2.0
2,4,5-TP2,4,5-TP	0.01	0.05	0.05
Toxaphene	0.005	zero	0.005
trans-1,2-Dichloroethylene	–	0.1	0.1
Xylenes (total)	–	10.0	10.0
IOC			
Asbestos	–	7 F/Ld	7 F/L[d]
Barium	1.0	5.0	5.0
Cadmium	0.010	0.005	0.005
Chromium	0.05	0.1	0.1
Mercury	0.002	0.002	0.002
Nitrate (as nitrogen)	10.0	10.0	10.0
Nitrite (as nitrogen)	–	1.0	1.0
Selenium	0.01	0.05	0.05

[a] NPDWR = National Primary Drinking Water Regulations.
[b] TT = Treatment Technique.
[c] EPA proposes MCLs of 0.1 mg/L based on a group C carcinogen classification and 0.005 mg/L based on a B_2 classification.
[d] 7 million fibers/liter (only fibers longer than 10 µm).
Source: U.S. EPA (1989)

Effluent Requirements and Influent Characteristics

Federal drinking water regulations, which include effluent requirements, are the primary factors determining water treatment goals. Supplementary effluent criteria from regulatory recommendations (e.g., U.S. EPA Health Advisories), professional organizations, State regulations, and consumer preferences are also considerations. Comparisons of existing influent characteristics with effluent requirements provide the basis for identifying treatment needs. The influent is properly characterized by an historical profile of the chemical and biological constituents of the water source. Non-contaminant parameters, such as pH levels, are also useful in characterizing influent because of their impact on treatment process efficiencies.

Table 2-5. Proposed SMCLs[a]

Contaminant	Level (mg/L)
Aluminum	0.05
o-Dichlorobenzene	0.01
p-Dichlorobenzene	0.005
Ethylbenzene	0.03
Pentachlorophenol	0.03
Silver	0.09
Styrene	0.01
Toluene	0.04
Xylene	0.02

[a] Other secondary regulations may be proposed in the future, as appropriate.
Source: U.S. EPA (1989).

Existing System Configuration

The adaptability of an existing system configuration to larger or different processes is an important consideration in assessing treatment options. Also, a system's ability to blend treated water with raw water can affect its achievement of regulatory standards, which are generally based on contaminant concentrations.

Cost

Total treatment costs are divided into one-time capital costs and annual operating and maintenance costs. Each treatment option has a different combination of capital and operating and maintenance costs, with higher capital costs often being associated with lower operating and maintenance costs. Smaller systems with limited capital budgets are generally saddled with higher operating and maintenance costs because they cannot afford the appropriate capital equipment.

Operating Requirements

Consistency in the nature and volume of the influent is a crucial operating parameter. As the consistency of the influent decreases, the necessity of monitoring and the operating complexity of most systems increases. Consequently, the level of operator training and attention required, as well as the amount of instrumentation, controls, and automation, also increases as influent consistency decreases. Other important operating considerations include:

- Energy requirements
- Chemical availability and consumption rate
- Instrumentation and automation
- Preventative maintenance
- Noise
- Esthetics
- Backup/Redundant systems
- Startup phase requirements before full removal capacity is achieved
- Cleaning and backwashing requirements

Pretreatment and Posttreatment Processes

All water treatment technologies perform differently with different pretreatment and posttreatment processes. The compatibility of all the processes in the treatment train is key to achieving individual treatment goals. For instance, a lower pH is desirable for maximum efficiency of chlorine disinfection; however, lower pH accelerates corrosion of the water distribution system. For successful implementation of a treatment technology, all elements of the train must interact efficiently and effectively.

Waste Management

Waste management is a concern associated with the removal of contaminants in drinking water. Most treatment processes concentrate contaminants, sometimes hazardous, into residuals that requires special handling. Such wastes are in the form of solutions, gases, sludges, and solids. Extensive regulations cover the management of these wastes, particularly those classified as hazardous.

Future Service Area Needs

The last major factor in selecting a treatment technology is the future of the service and supply areas. Population and economic forecasts of the service area are the basic tools for evaluating future demands. Examination and analyses of the watershed(s) of present and potential water supplies are important to determine the vulnerability of supplies to anthropogenic and natural threats.

2.3 Overview of Available Treatment Alternatives

This section provides an overview of the available treatment technologies associated with all categories of treatment objectives. Summary data are presented for each technology and alternatives are compared on the basis of:

- Performance
- Suitability to treatment plant size
- Degree of acceptance
- Conditions required for effective operation
- Operating and maintenance requirements
- Compatibility with other treatment processes

Table 2-6 lists the treatment options associated with each of five categories of treatment objectives: filtration, disinfection, organic and inorganic contaminant removal, and corrosion control. The table indicates four levels of treatment technology acceptance: experimental, emerging, established, and BAT. Experimental technologies have shown promise in some applications, but have not been extensively tested. Emerging technologies have proven themselves in the laboratory, but not in the field. Established treatments are commonly used in the water industry. BAT is a regulatory designation that indicates the level of contaminant removal achievable through specification of a technology or a technology equivalent, rather than an MCL.

Detailed discussions of all technologies are provided in Sections 3 through 7, as follows:

- Prefiltration and filtration for treatment of primarily turbidity and color – Chapters 3 and 4.

- Disinfection for treatment of pathogenic organisms, including *Giardia* cysts, bacteria, and viruses – Chapter 5.

- Organic contaminant removal, including volatile organic chemicals and other synthetic organic chemicals – Chapter 6.

- Inorganic contaminant removal, including products of distribution system corrosion and radioactive contaminants – Chapter 7.

2.3.1 Filtration

The three basic regulatory requirements associated with filtration systems are control of turbidity, color, and biological contamination (*Giardia* cysts, enteric viruses, and coliform bacteria). Other key considerations in selecting a filtration system include:

- Frequency of the cleaning cycle
- Chemical requirements
- Operational complexity
- Sludge volume and toxicity

Each of these factors is site specific. In addition, climate will determine whether it is necessary to house the filter operation, which in some cases is prohibitively expensive.

In most conventional filtration systems, mixing, flocculation, and sedimentation processes typically precede actual filtration. While these pretreatment elements are always found in conventional filtration

Table 2-6. Overview of Water Treatment Technologies

Treatment Requirements Under the New Regulations	Technological Options to Meet Regulatory Requirements	Stage of Acceptability	Size Suitability	Comments
Filtration of surface water supplies to control turbidity and microbial contamination				
	Conventional filtration	Established	All	Most common; adaptable for adding other processes
	Direct filtration	Established	All	Lower cost alternative to conventional filtration
	Slow sand filtration	Established	Especially small, but all sizes	Operationally simple; low cost, but requires large land areas
	Package plant filtration	Established	Mostly small	Compact; variety of process combinations available
	Diatomaceous earth filtration	Established	Mostly small	Limited applicability; potentially expensive for small systems
	Membrane filtration	Emerging	Mostly small	Experimental, expensive
	Cartridge filtration	Emerging	Small	Experimental, expensive
Disinfection of all public water supplies				
	Chlorine	Established	All	Most widely used method; concerns about health effects of by-products
	Chlorine dioxide	Established	All	Relatively new to the United States; concerns about inorganic by-products
	Monochloramine	Established	All	Secondary disinfectant only; some by-product concerns
	Ozone	Established	All	Very effective and requires a secondary disinfectant
	Ultraviolet radiation	Established	All	Simple, no established harmful by-products and requires secondary disinfectant
	Advanced oxidation (ozone plus H_2O_2 and ozone plus ultraviolet radiation)	Emerging	All	Not much information concerning disinfection aspects of this process
Organic contamination control, including 50 specific compounds				
	Granular activated carbon	BAT	All	Highly effective; potential waste disposal issues
	Packed column aeration	BAT	All	Highly effective for volatile compounds; potential air emissions issues
	Powdered activated carbon	Established	Large	Requires conventional treatment process train for application
	Diffused aeration	Established	All	Variable removal effectiveness
	Multiple tray aeration	Established	All	Variable removal effectiveness
	Oxidation	Experimental	All	By-products concerns
	Reverse osmosis	Emerging	Small to medium	Variable removal effectiveness; expensive
	Mechanical aeration	Experimental	All	Mostly for wastewater treatment; high energy requirements, easy to operate
	Catenary grid	Experimental	All	Performance data are scarce; potential air emissions issues
	Higee aeration	Experimental	Small	Compact, high energy requirements; potential air emissions issues

(Continued)

Table 2-6. (Continued)

Treatment Requirements Under the New Regulations	Technological Options to Meet Regulatory Requirements	Stage of Acceptability	Size Suitability	Comments
	Resins	Experimental	Small	Data scarce
	Ultrafiltration	Emerging	Small	Primarily for turbidity; data for organics removal are scarce
Inorganic contamination control, including 36 specific inorganic contaminants, and 5 radioactive contaminants				
	Reverse osmosis	Established	Small to medium	Highly effective; expensive; potential waste disposal issues
	Ion exchange	Established	Small to medium	Highly effective; expensive; potential waste disposal issues
	Activated alumina	Established	Small	Highly effective; expensive; potential waste disposal issues
	Granular activated carbon	Experimental	Small	Experimental for radionuclide removal; potential waste disposal issues
Corrosion controls	pH control	Established	All	Potential to conflict with other treatments
	Corrosion Inhibitors	Established	All	Variable effectiveness depending on type of inhibitor

Source: Derived from tables and text included in Chapters 3 through 7.

systems, they are also sometimes found in nonconventional systems. In addition to the three prefiltration steps, additives, such as chemicals for pH adjustment, coagulants, coagulant aids, and polymers, are commonly used in conjunction with filtration. All of these elements affect the performance of the filter system. These effects are discussed in detail in Chapter 3.

The seven filtration options reviewed in this document include:

- Conventional filtration
- Direct filtration
- Slow sand filtration
- Package plant filtration
- Diatomaceous earth filtration
- Membrane filtration (reverse osmosis)
- Cartridge filtration

The performance of each filter type depends on the quality of the influent and proper design and operation. The range of influent characteristics for which various filters are effective is provided in Table 2-7.

Conventional filtration, with rapid mix, flocculation, and sedimentation, is clearly the most versatile in its effectiveness in treating variable influents. Coagulation/filtration systems are more difficult to operate compared to slow sand or diatomaceous earth filters because they involve adjusting water chemistry for proper coagulation. Slow sand, diatomaceous earth, membrane, and cartridge filtration units do not have a coagulation step. The complexity of operating a package filtration plant varies with the manufacturer and model. Package plants, slow sand, membrane, and cartridge filtration are most applicable to small systems.

The removal capacities for *Giardia* cysts and viruses of the seven filter systems are presented in Table 2-8.

2.3.2 Disinfection

The recently promulgated SWTR requires water treatment systems to inactivate 99.9 percent of *Giardia* cysts and 99.99 percent of enteric viruses. Disinfection systems alone, or in conjunction with filtration systems, can meet these requirements. At the same time, regulations regarding disinfection by-products in the finished water must be met. Currently, total trihalomethanes are the only disinfection by-products regulated by EPA; however, new disinfection by-product regulations are anticipated.

Primary disinfectants are those used for the inactivation of *Giardia* cysts, viruses, and bacterial contaminants, while secondary disinfectants suppress biological regeneration in the distribution system. Common primary disinfectants are chlorine, chlorine dioxide, ozone, and UV radiation. Most are suitable for both ground water and surface water. An exception is UV radiation, which is only suitable for ground water because it is not effective against

Table 2-7. Typical Influent Characteristics and Capacities for Filtration Technologies

Filtration Options	Turbidity (NTUs)	Color (in color units)	Coliform Count (per 100 mL)	Typical Capacity (MGD)
Conventional	No restrictions	<75	<20,000	>All sizes
Direct	<14	<40	<500	>All sizes
Slow sand	<5	<10	<800	<15
Package Plant		[depends on processes utilized]		<6
Diatomaceous earth	<5	<5	<50	<100
Membrane	<1	[fouling index of <10]		<0.5
Cartridge	<2	NA	NA	<1.0

NA = not available.
1 MGD = 0.044 m^3/sec
Source: See tables and text in Chapters 3 and 4.

Table 2-8. Removal Capacities of Seven Filter Options (percent removal)

Filtration Options	Achievable *Giardia* Cyst Levels	Achievable Virus Levels
Conventional	99.9	99.0
Direct	99.9	99.0
Slow sand	99.99	99.9999
Package plant	varies with	manufacturer
Diatomaceous earth	99.99	>99.95
Membrane	100	Very low
Cartridge	>99	Little data available

Source: See tables and text in Chapters 3 and 4.

Giardia cysts. Secondary disinfectants include chlorine, chlorine dioxide, and chloramines.

Disinfection effectiveness is measured in terms of the residual concentration and length of contact time necessary to achieve the desired inactivations. Four chemical disinfectants listed in descending order of their effectiveness are ozone, chlorine dioxide, chlorine, and chloramines. The effectiveness of a particular disinfectant is also influenced by water quality, temperature, and pH. Lower disinfectant dosages may be used when:

- There is filtration or oxidation prior to the disinfection step.
- The water temperature is high.

The dosage of chlorine required for effective disinfection is reduced as the pH of the water is reduced. A qualitative summary of the advantages and disadvantages of the five disinfectants is provided in Table 2-9.

Some disinfectants are unstable and, therefore, have to be generated on site. Production considerations for the five disinfectants are contained in Table 2-10. The table also provides the number and type of alternative production methods.

The preferred application point for the various disinfectants are described in Table 2-11. Table 2-12 compares basic operational considerations, pH, presence of by-products, operational simplicity, and maintenance required.

One of the most important considerations in assessing disinfectants is balancing inactivation or biocidal effectiveness with by-product production. The by-products of greatest current concern are trihalomethanes and other halogenated organic compounds; chlorine has the greatest potential for generating harmful by-products. The amount of these by-products produced by chlorine is affected by:

- Chlorine dosage
- Types and concentrations of organic material in the influent
- Influent temperature
- Influent pH
- Contact time for free chlorine
- Nature of residual (free chlorine vs. combined chlorine)
- Presence of bromide ion

If chlorine produces an unsatisfactory level of by-products, then other disinfectants are potential alternatives. Chlorine dioxide is effective, but the total levels of chlorine dioxide and its oxidation/reduction products may limit its applicability. Ozone and UV radiation are very effective primary disinfectants, but require the use of secondary disinfectants. Ozone will produce harmful halogenated by-products with influent containing bromide ions. It will also produce harmful oxidation products in the presence of certain synthetic organics such as heptachlor. Chlorine dioxide and monochloramines are effective secondary disinfectants, but require careful dosage and application management to avoid producing harmful by-products in finished water. (Chapter 5 contains a more complete discussion of disinfection by-products and strategies for their control.)

Table 2-9. Advantages and Disadvantages of Five Disinfectants

Disinfectant	Advantages	Disadvantages
Chlorine	Effective. Widely used. Variety of possible application points. Inexpensive. Appropriate as both primary and secondary disinfectant.	Harmful halogenated by-products. Potential conflict with corrosion control pH levels, when used as a secondary disinfectant.
Ozone	Very effective. Minimal harmful by-products identified to date. Enhances slow sand and GAC filters. Provides oxidation and disinfection in the same step.	Requires secondary disinfectant. Relatively high cost. More complex operations because it must be generated on-site.
Ultraviolet radiation	Very effective for viruses and bacteria. Readily available. No harmful residuals. Simple operation and maintenance.	Inappropriate for water with *Giardia* cysts, high suspended solids, high color, high turbidity, or soluble organics. Requires a secondary disinfectant.
Chlorine dioxide	Effective. Relatively low cost. Generally does not produce THMS.	Some harmful by-products. Low dosages currently recommended by U.S. EPA may make it ineffective. Must be generated on-site.
Chloramines	Mildly effective for bacteria. Long-lasting residual. Generally does not produce THMS.	Some harmful by-products. Toxic effects for kidney dialysis patients. Only recommended as a secondary disinfectant. Ineffective against viruses and cysts.

Source: See tables and text in Chapter 5.

Table 2-10. Disinfectant Production Considerations

	Chlorine	Chlorine Dioxide	Mono-chloramine	Ozone	Ultraviolet Radiation
Chemically stable	Yes	Yes	Yes	No	NA
On-site production required	No	Yes	Yes	Yes	Yes
Number of alternative on-site generation techniques	NA	3[a]	2[b]	3[c]	NA

[a]Including:
1. Treating sodium chlorite solution with chlorine gas
2. Treating sodium chlorite solution with sodium hypochlorite and mineral acid
3. Treating sodium chlorite solution with mineral acid

[b]Including:
1. Adding ammonia to a water and chlorine solution
2. Adding chlorine to a water and ammonia solution

[c]Including:
1. Ambient air
2. Pure oxygen
3. Oxygen-enriched air

NA = not applicable.
Source: See text and tables in Chapter 5.

2.3.3 Organic Contaminant Removal

The 1986 Amendments to the SDWA require the establishment of new MCLs for many organic contaminants, including disinfection by-products. The regulations designate BATs as well as MCLs for organic contaminants. The July 1987 regulations specify packed column aeration (PCA) and GAC as the BAT for seven of the eight VOCs. PCA is BAT for vinyl chloride, the eighth VOC. Water utilities with these contaminants will have to provide removals at least equivalent to those achieved by the designated BATs.

Table 2-11. Desired Points of Disinfectant Application[a]

Chlorine	Towards the end of the water treatment process to minimize THM formation and provide secondary disinfection
Ozone	Prior to the rapid mixing step in all treatment processes, except GAC and conventional treatment processes; prior to filtration for GAC; post-sedimentation for conventional treatment. In addition, sufficient time for biodegradation of the oxidation products of the ozonation of organic compounds is recommended prior to secondary disinfection.
Ultraviolet radiation	Towards the end of the water treatment process to minimize presence of other contaminants that interfere with this disinfectant
Chlorine dioxide	Prior to filtration; to assure low levels of ClO_2, ClO_2^-, and ClO_3, treat with GAC after disinfection.
Monochloramines	Best applied towards the end of the process as a secondary disinfectant

[a] In general, disinfectant dosages will be lessened by placing the point of application towards the end of the water treatment process because of the lower levels of contaminants that would interfere with efficient disinfection. However, water plants with short detention times in clear wells and with nearby first customers may be required to move their point of disinfection upstream to attain the appropriate CT value under the Surface Water Treatment Rule.
Source: See text and tables in Chapter 5.

The removal capabilities of selected treatment technologies for several organic contaminants and contaminant classes are provided in Table 2-13. The first listed treatment, coagulation/filtration is widely used primarily for turbidity and microbial contaminant control, but can be somewhat effective in removing certain organic compounds. GAC and PCA are designated as BAT for many of the organic chemicals and are much more effective. Costs for either technology vary depending on the contaminant

Table 2-12. Disinfectant Application Considerations

	Cl_2	ClO_2	Mono-chloramine	O_3	UV
Optimum water pH	7	6-9	7-8	6	N/A
By-products present	Yes	Yes	Yes	Yes	No
Operational simplicity	Yes	No	No	No[a]	Yes
Maintenance required	Low	Low	Low	High	High

[a]Operationally simplified with an automated system.
NA = not applicable.
Source: See text and tables in Chapter 5.

removed and whether waste management is a potential issue. Powdered activated carbon (PAC), the fourth treatment listed in the table, is only suited for application to conventional systems with rapid mix, flocculation, sedimentation, and filtration components. PAC also addresses some taste and odor problems. This form of activated carbon is especially suitable for seasonal organic contaminant problems since it can be added as needed. In some instances, however, such high dosages of PAC are required to achieve organic removal that waste management becomes a problem.

All other listed treatments, including the different aeration configurations, are less established. The aeration treatments, developed for specific applications, are generally equal to or less effective than PCA, but have higher energy requirements. For some, controlling biological growth in the systems is a problem. The remaining treatments, besides the aeration technologies, all tend to be very effective in removing contaminants, but their application to specific organics is still experimental. The efficiencies of all treatments depend on the type and concentration of the contaminants. While pretreatment is not always required, it can increase the effectiveness of some of the treatments. Table 2-14 illustrates the variation in operating conditions for these treatments.

2.3.4 Inorganic Contaminant Removal and Control

Inorganic contaminant treatments are categorized as prevention strategies or removal technologies. Corrosion controls prevent or minimize the presence of corrosion products (inorganic contaminants) at the point of use. Removal technolcgies treat source water that is contaminated with metals or radionuclides.

2.3.4.1 Corrosion Controls

Corrosion controls address the two primary aspects of corrosion:

- Water quality characteristics
- Materials subject to corrosion

The two most significant water quality characteristics that influence corrosion are pH and carbonate/bicarbonate alkalinity. Components of the distribution system subject to corrosion include pipes, valves, meters, plumbing, solder, and flux. The longer the contact time between the water and corrodible materials in the distribution system, the higher the concentration of dissolved metals in the drinking water.

The four general types of corrosion controls are:

- Adjustments to water pH
- Addition of corrosion inhibitors to the water to form protective coatings over the potentially corrodible metal
- Electronic cathodic protection
- Applied coatings and linings

The most commonly used corrosion control is pH adjustment because it is inexpensive and easily applied. Table 2-15 compares corrosion controls.

Corrosion controls involving pH adjustment may conflict with ideal pH conditions for disinfection and control of disinfection by-products. The treatment methods selected for both treatment objectives should be carefully coordinated to avoid diminishing the effectiveness of either process.

2.3.4.2 Inorganic Contaminant Removal

There are 10 treatment processes for addressing the many inorganic contaminants, including radionuclides. Most treatment processes are effective for only a specific set of contaminants under certain circumstances. Fortunately, all of the inorganic contaminants do not often occur simultaneously. The most appropriate applications for each treatment process are shown in Table 2-16. This table distinguishes between ground water and surface water by the generally higher concentrations of suspended solids in most surface water. Table 2-17 provides removal efficiencies for specific contaminants with the same treatments. For radon removal, which is not included in the table, aeration is highly effective.

The most important factors affecting inorganic contaminant removal are:

- Contaminant type and valence
- Influent contaminant concentration
- Influent levels of dissolved solids and pH
- Desired effluent concentration

Table 2-13. **Treatment Technology Removal Effectiveness Reported for Organic Contaminants (percent)**

Contaminant	Coagulation/ Filtration	GAC	PCA	PAC	Diffused Aeration	Oxidation[a]	Reverse Osmosis
Acrylamide	5	NA	0-29	13	NA	NA	0-97
Alachlor	0-49	70-100	70-100	36-100	NA	70-100	70-100
Aldicarb	NA	NA	0-29	NA	NA	NA	94-99
Benzene	0-29	70-100	70-100	NA	NA	70-100	0-29
Carbofuran	54-79	70-100	0-29	45-75	11-20	70-100	70-100
Carbon tetrachloride	0-29	70-100	70-100	0-25	NA	0-29	70-100
Chlordane	NA	70-100	0-29	NA	NA	NA	NA
Chlorobenzene	0-29	70-100	70-100	NA	NA	30-69	70-100
2,4-D	0-29	70-100	70-100	69-100	NA	W	0-65
1,2-Dichloroethane	0-29	70-100	70-100	NA	42-77	0-29	15-70
1,2-Dichloropropane	0-29	70-100	70-100	NA	12-79	0-29	10-100
Dibromochloropropane	0-29	70-100	30-69	NA	NA	0-29	NA
Dichlorobenzene	NA	70-100	NA	NA	NA	NA	NA
o-Dichlorobenzene	0-29	70-100	70-100	38-95	14-72	30-88	30-69
p-Dichlorobenzene	0-29	70-100	70-100	NA	NA	30-69	0-10
1,1-Dichloroethylene	0-29	70-100	70-100	NA	97	70-100	NA
cis-1,2-Dichloroethylene	0-29	70-100	70-100	NA	32-85	70-100	0-30
trans-1,2-Dichloroethylene	0-29	70-100	70-100	NA	37-96	70-100	0-30
Epichlorohydrin	NA	NA	0-29	NA	NA	0-29	NA
Ethylbenzene	0-29	70-100	70-100	33-99	24-89	70-100	0-30
Ethylene dibromide	0-29	70-100	70-100	NA	NA	0-29	37-100
Heptachlor	64	70-100	70-100	53-97	NA	70-100	NA
Heptachlor epoxide	NA	NA	NA	NA	NA	26	NA
High molecular weight hydrocarbons (gasoline, dyes, amines, humics)	NA	W	NA	NA	NA	NA	NA
Lindane	0-29	70-100	0-29	82-97	NA	0-100	50-75
Methoxychlor	NA	70-100	NA	NA	NA	NA	>90
Monochlorobenzene	NA	NA	NA	14-99	14-85	86-98	50-100
Natural organic material	P	P	NA	P	NA	W	P
PCBs	NA	70-100	70-100	NA	NA	NA	95
Phenol and chlorophenols	NA	W	NA	NA	NA	W	NA
Pentachlorophenol	NA	70-100	0	NA	NA	70-100	NA
Styrene	0-29	NA	NA	NA	NA	70-100	NA
Tetrachloroethylene	NA	70-100	NA	NA	73-95	W	70-90
Trichloroethylene	0-29	70-100	70-100	NA	53-95	30-69	0-100
Trichloroethane	NA	70-100	NA	NA	NA	NA	NA
1,1,1-Trichloroethane	0-29	70-100	70-100	40-65	58-90	0-29	15-100
Toluene	0-29	70-100	70-100	0-67	22-89	70-100	NA
2,4,5-TP	63	70-100	NA	82-99	NA	30-69	NA
Toxaphene	0-29	70-100	70-100	40-99	NA	NA	NA
Vinyl chloride	0-29	70-100	70-100	NA	NA	70-100	NA
Xylenes	0-29	70-100	70-100	60-99	18-89	70-100	10-85

W = well removed.
P = poorly removed.
NA = not available.
[a]The specifics of the oxidation processes effective in removing each contaminant are provided in Chapter 6.
Note: Little or no specific performance data were available for:

1. Multiple Tray Aeration
2. Catenary Aeration
3. Higee Aeration
4. Resins
5. Ultrafiltration
6. Mechanical Aeration

Source: See text and tables in Chapter 6.

Table 2-14. Operational Conditions for Organic Treatments

Technology	Level of Operational Skill Required	Level of Maintenance Required	Energy Requirements
Coagulation/ Filtration	High	High	Low
GAC	Medium	Low	Low
PCA	Low	Low	Varies
PAC	Low	Medium	Low
Diffused aeration	Low	Low	Varies
Multiple tray aeration	Low	Low	Low
Oxidation	High	High	Varies
Reverse osmosis	High	High	High
Mechanical aeration	Low	Low	Low
Catenary grid	Low	Low	High
Higee aeration	Low	Medium	High
Resins	Medium	Medium	Low
Ultrafiltration	Medium	High	Medium

Source: See text and tables in Chapter 6.

Table 2-15. Corrosion Control Considerations

Corrosion Controls	Amount of System Covered by the Control	Completely Agreeable with Consumers	Optimum Level of Additive in the Water
Water Treatment			
pH adjustment	Total	Yes	>8.0 pH
Inorganic phosphates	Total	Usually	Varies
Silicates	Total	Yes	2-12 mg/L
Cathodic Protection	Partial	Yes	NA
Coatings and Linings	Partial	Yes	NA

NA = not applicable.
Source: See tables and text in Chapter 7.

The first three factors are site specific. Table 2-18 summarizes the advantages and disadvantages of each of the treatment mechanisms.

2.4 Final Process Selection and Design

When modifying an existing or designing a new treatment system, consideration must be given to all the regulations impacting the system. For surface water, regulations regarding microbiological, disinfection by-product, and lead contaminants are most likely to impact the treatment system. For ground-water systems, regulations regarding organic chemicals (particularly VOCs), radon, lead, and disinfection will have the most impact.

Table 2-16. Most Probable Treatment Application for Inorganic Treatments

Treatment	Most Probable Application: Source Water	Most Probable Application: Contaminants
Conventional with coagulation	Surface	Ag, As, Cd, Cr, or Pb
Lime softening	Surface (hard)	As V, Cd, Cr III, F, or Pb
	Ground water	Ba or Ra
Ion exchange: cation	Ground water	Ba or Ra
anion	Ground water	NO_3, Se
Reverse osmosis	Ground water	All
Powdered activated carbon	Surface water	Organic Hg from spills
Granular activated carbon	Surface water Ground water	Organic Hg Organic Hg
Activated alumina	Ground water	As, F, or Se

Source: See tables and text in Chapter 7.

Balancing the processes selected for microbiological control, disinfection by-product removal, and lead control will be important for both surface and ground-water treatment. One water quality parameter that affects all three processes is pH. A low pH is desirable for disinfection, disinfection by-product removal, and minimizing by-product formation in the distribution system. However, minimizing lead corrosion in the distribution system requires a higher pH. One option to meet all conditions involves use of a corrosion inhibitor, thus permitting a lower pH in the finished water for disinfection and by-product control.

For ground-water supplies, process selection should consider treatments capable of controlling a number of contaminants. For example, PCA can remove both VOCs and radon. It also removes carbon dioxide from the water, which raises the pH to levels desirable for lead corrosion control. As another example, oxidation techniques, such as ozone, may oxidize organic chemicals as well as provide disinfection. Also, reverse osmosis might be considered when both organic and inorganic chemical contaminants are present in the raw water. Use of one treatment technique to meet multiple regulatory requirements will help reduce operating complexity and costs.

Evaluation of potential treatment technologies proceeds from literature searches to bench-scale or field tests. Sludge volume and composition are also important considerations in the final design selection. Construction costs for the final alternatives are used to select the appropriate option before developing the final design.

Table 2-17. Removal Effectiveness for Nine Processes by Inorganic Contaminant

	Contaminant																				
Treatment	Ag	As	As^{III}	As^{V}	Ba	Cd	Cr	Cr^{III}	Cr^{VI}	F	Hg	$Hg^{(O)}$	$Hg^{(I)}$	NO_3	Pb	Ra	Rn	Se	$Sel^{(V)}$	$Se^{(II)}$	U
Conventional treatment	H	–	M	H	L	H	–	H	H	L	–	M	M	L	H	L	–	–	M	L	M
Coagulation - aluminum	H	–	–	H	–	M	–	H	–	–	M	–	–	–	H	–	–	–	–	–	–
Coagulation - iron	M	–	–	H	–	–	–	H	H	–	–	–	–	–	–	–	–	–	–	–	–
Lime softening	–	–	M	H	H	H	–	H	L	M	–	L	M	L	H	H	–	–	M	L	H
Reverse osmosis and electrodialysis	H	–	M	H	H	H	H	–	–	H	H	–	–	M	H	H	–	H	–	–	H
Cation exchange	–	L	–	–	H	H	–	H	L	L	–	–	–	L	H	H	–	L	–	–	H
Anion exchange	–	–	–	–	M	M	–	M	H	–	–	–	–	H	M	M	–	H	–	–	H
Activated alumina	–	–	H	–	L	L	–	–	–	H	–	–	–	–	–	L	–	H	–	–	–
Powdered activated carbon	L	–	–	–	L	M	–	L	–	L	–	M	M	L	–	L	–	–	–	–	–
Granular activated carbon	–	–	–	–	L	M	–	L	–	L	–	H	H	L	–	L	H	–	–	–	–

H = High = >80% removal.
M = Medium = 20-80% removal.
L = Low = <20% removal.
"–" = indicate no data were provided.
Source: See text and tables in Chapter 7.

Table 2-18. Advantages and Disadvantages of Inorganic Contaminant Removal Processes

Precipitation and Coprecipitation Used in Coagulation/Filtration

Advantages

- Low cost for high volume
- Often improved by high ionic strength
- Reliable process well suited to automatic control

Disadvantages

- Stoichiometric chemical additions required
- High-water-content sludge must be disposed of
- Parts-per-billion effluent contaminant levels may require two-stage precipitation
- Not readily applied to small, intermittent flows
- Coprecipitation efficiency depends on initial contaminant concentration and surface area of primary floc

Ion Exchange

Advantages

- Operates on demand
- Relatively insensitive to flow variations
- Essentially zero level of effluent contamination possible
- Large variety of specific resins available
- Beneficial selectivity reversal commonly occurs upon regeneration

Disadvantages

- Potential for chromatographic effluent peaking
- Spent regenerant must be disposed of
- Variable effluent quality with respect to background ions
- Usually not feasible at high levels of total dissolved solids

Activated Alumina

Advantages

- Operates on demand
- Insensitive to flow and total dissolved solids background
- Low effluent contaminant level possible
- Highly selective for fluoride and arsenic

Disadvantages

- Both acid and base are required for regeneration
- Media tend to dissolve, producing fine particles
- Slow adsorption kinetics
- Spent regenerant must be disposed of

Membranes (Reverse Osmosis and Electrodialysis)

Advantages

- All contaminant ions and most dissolved non-ions are removed
- Relatively insensitive to flow and total dissolved solids level
- Low effluent concentration possible
- In reverse osmosis, bacteria and particles are removed as well

Disadvantages

- High capital and operating costs
- High level of pretreatment required
- Membranes are prone to fouling
- Reject stream is 20-90% of feed flow

Source: Clifford (1986).

The chapters that follow elaborate on each group of treatment techniques. Filtration technologies are discussed in Chapters 3 and 4, disinfection in Chapter 5, organic contaminant treatments in Chapter 6, and technologies that control corrosion and remove inorganic compounds and radionuclides in Chapter 7.

Chapter 3

Prefiltration Treatment Elements

Conventional treatment is the most widely used train of processes to control microbial and turbidity levels in surface supplied drinking water. The precise order and composition of conventional treatment elements are not invariably the same, but commonly include chemical feed, rapid mix, flocculation, sedimentation, filtration, and disinfection.

The performance of a filtration technology is greatly impacted by the processes that precede it. Prefiltration elements, including chemical feed, rapid mix, flocculation, and sedimentation may need upgrading to improve overall system performance or accommodate system expansions. This chapter presents typical problems and recommended solutions associated with prefiltration processes.

It is often necessary to modify existing conventional treatment plants due to changes in population, financial constraints, raw water quality, and regulatory atmosphere. These changes may upgrade the present system to:

- Improve the water quality
- Increase plant capacity
- Improve reliability
- Reduce maintenance
- Reduce costs

Upgrading conventional treatment plants can effect significant improvements. For example, upgrading can increase the capacity of most soundly constructed plants by 100 to 200 percent, provided traditional design parameters were used in the original plant installation.

Any of the five basic components of conventional treatment can be upgraded, alone or in combination. A deficiency, or a set of deficiencies, in one or more components of conventional treatment will typically prompt an evaluation of the potential for upgrading. Table 3-1 summarizes the causes of potential deficiencies for each treatment element. Techniques for upgrading fall into five basic categories, according to treatment element:

- Modifying chemical treatment or dosage
- Modifying or adding rapid coagulant mixing
- Improving flocculation
- Improving sedimentation
- Modifying filtration processes

Table 3-1. Summary of Conventional Treatment Operating Issues

Treatment Element	Potential Causes of Problems
Chemical feed	–Choice of chemical(s) –Choice of chemical dose and pH –Control of chemical addition; performance of chemical pumping equipment –Maintenance of chemical feed lines –Flexibility in feed system to allow for changing the point of addition, adding chemicals at more than one point, etc. –Sequence of adding different chemicals –Degree of dilution of chemicals before injection
Rapid mix	–Type of rapid mix; in-line versus mechanical mix –Number of rapid mixers –Method of chemical addition –Mixing speed/detention time
Flocculation	–Optimum detention time –Optimum mixing intensity –Number of stages –Adequate baffling to approximate plug flow conditions
Sedimentation	–Surface loading rates –Short circuiting due to wind, temperature, density differences, inlet and outlet design –Amount and rate of accumulation of sludge –Sludge removal
Filtration	–Filter rate and rate control –Hydraulics –Chemical pretreatment of water reaching the filter –Inadequate backwashing

The remainder of this chapter discusses the first four categories of modification. The last category, modifying filtration processes, is discussed in Chapter 4.

3.1 Modifying Chemical Feed

Proper chemical feed is critical to optimal performance of flocculation, sedimentation, filtration, and disinfection systems. The three basic aspects of chemical feed systems are chemical type, dosage management, and method of chemical application. This chapter addresses these topics and the costs associated with modifying chemical feed systems.

3.1.1 Chemical Type

The three types of chemicals usually applied to the raw water at the beginning of the conventional treatment train are coagulants, coagulant aids, and pH control substances. Some polymer coagulants are used to coat filters during backwash cycles to increase their effectiveness. However, this application is discussed in the chapter on filtration.

Coagulants

Coagulants are chemicals used to remove turbidity and organic substances from raw water by precipitation. Coagulation also removes bacteria, algae, color, iron oxide, magnesium oxide, calcium carbonate, and clay. Coagulation's effectiveness in removing humic acids is important because humic acids combine with chlorine to form trihalomethanes. Coagulants act by overcoming the charges of suspended particles thus allowing larger particle groupings to form. They are introduced during the initial stage of water treatment just before the rapid mixing step. Coagulants are effective by themselves and in conjunction with coagulant aids and pH modifiers. The two most common coagulant types are metallic salts and polymers; the most common metallic salt coagulants are aluminum sulfate (alum) and ferric chloride.

The selection of a particular coagulant depends on the required level of effectiveness. A standard jar test is the recommended method to determine the relative effectiveness of various coagulants for a particular raw water supply. Jar test procedures are described later in this chapter. If different coagulants are equally effective, then the factors that should be considered are cost; availability; overall safety; and ease of storage, handling, and application. Metallic salts, for example, are much less expensive than polymers, as shown in Table 3-2, which lists a variety of chemical costs.

Alum is the most widely used coagulant because of its availability, low cost, ease of use, and ease of storage.

Table 3-2. Chemical Costs for Conventional Treatment ($1986)

Chemical	Small (<1 MGD) Systems, $/ton	Large (>1 MGD) Systems, $/ton
Alum (dry)	$ 500	$ 250
Alum (liquid)	300	125
Lime (quick)	100	75
Lime (hydrated)	150	100
Ferric chloride	500	275
Ferrous sulfate	277	250
Ferric sulfate	200	155
Soda ash	250	200
Sodium hydroxide	595	316
Chlorine	500	300
Sodium hypochlorite	190	150
Liquid carbon dioxide	350	100
Sodium hexametaphosphate	1,160	1,100
Zinc orthophosphate	1,520	1,000
Ammonia, aqua	230	200
Ammonia, anhydrous	410	350
Sulfuric acid	140	100
Hydrochloric acid	171	166
Activated carbon powdered	950	800
Activated carbon granular	1,900	1,600
Activated alumina	1,694	1,156
Potassium permanganate	2,800	2,500
Sodium bisulfate (anhydrous)	909	673
Sodium silicate	400	200
Sodium chloride	105	85
Polyelectrolyte	1,500	1,000
Diatomaceous earth	680	310
Magnesium	650	582
Sodium chlorite	3,200	2,800
Sodium hydroxide 76%	590	316
Sodium bicarbonate	490	380
Calcium hypochlorite	2,700	1,540

1 MGD = 0.044 m^3/sec

Ferric chloride, other metallic salts, and polymers are less widely used. Alum's performance, however, is greatly affected by the pH of the influent. The recommended dosage of alum ranges from 5 to 150 mg/L, but the problem of sludge disposal increases with higher alum dosages.

Due to health concerns about aluminum, some water utilities are considering switching to ferric chloride. Although ferric chloride is not always as effective as alum in reducing trihalomethane formation potential (THMFP) and total organic carbon (TOC), it is more effective than alum for water with high dissolved color, low turbidity, and a moderate pH.

Polymers are effective coagulants, coagulant aids, and filter aids. They consist of chains of monomers and are classified according to their charge or lack of charge. A polymer with a charge is an ionized polymer, or a polyelectrolyte. If a polymer has a positive charge, it is cationic; if it has a negative charge, it is anionic; and if it has no charge, it is non-ionic.

Cationic polymers are effective for coagulating negatively charged clay particles, while anionic polymers are not very effective for this purpose. Anionic polymers are generally more effective for removing certain positively charged colloids and when used in conjunction with alum or other metallic coagulants. Nonionic polymers are effective filter aids. Polymers are not effective for coagulating some organic colors. Polymer selection requires frequent use of the standard jar test to adjust and compensate for seasonal changes in the influent characteristics.

In applications where polymers are effective, dosages are generally lower than alum dosages for the same effect. Typical polymer dosages range from 1.5 to 10 mg/L. Consequently, polymer coagulants produce less residual sludge than alum. To avoid large sludge volumes, there is a trend towards replacing alum with either cationic polymers, cationic polymers plus alum, or sodium aluminate. The use of cationic polymers instead of alum is most applicable to raw water with turbidity of less than 5 NTU and where direct filtration is feasible.

Coagulant Aids

Coagulant aids are added to the influent after or simultaneously with the primary coagulant to improve particle capture during flocculation, poor flocculation basin performance,clarification efficiency, and retention in filters of the coagulated particles that agglomerate (floc). Coagulant aids, through formation of faster settling floc, will improve the sedimentation process and reduce turbidity loadings on filters, thus potentially extending filter operating cycles. Nonionic and anionic polymers are used as coagulant aids. Nonionic polymers strengthen floc.

Coagulant aid dosages are stated in terms of their ratio to the amount of alum added to the influent. Dosages are determined by raw water characteristics and operational factors. The ratio of alum to polymer dosages ranges from 100:1 to 50:1. Again, standard jar tests are required to determine precise coagulant aid dosage.

pH and Alkalinity Controls

The optimum pH range makes metallic coagulants insoluble and improves the strength of the floc, and stronger floc enhances turbidity removal through sedimentation and filtration. The water must contain sufficient alkalinity for aluminum or iron floc formation. When pH is too low, adding soda ash, lime, sodium hydroxide, or sodium bicarbonate will increase the water's pH and alkalinity. Without sufficient alkalinity, the coagulation process will not proceed to completion. Very low pH levels will also lead to high aluminum ion residual in the finished water, which is undesirable.

When the pH level is too high due to factors such as algae and aquatic plant activity, adding acid is necessary. Typical acids used for this application are sulfuric acid and hydrochloric acid.

3.1.2 Chemical Dosage Management

Coagulant dosage management is a universal problem that is critical for achieving low turbidities in any plant, both well equipped and poorly equipped. Too little coagulant will result in excessive turbidity; too much coagulant could also result in excessive turbidity and wasted coagulant. Coagulant aid and pH control dosages are determined using jar test procedures.

While coagulant type is largely determined by influent characteristics, coagulant dosage is chiefly determined by effluent testing. The tests are performed either in a batch reactor or on a continuous basis. Batch tests provide indications of proper coagulant dosage selection at a single point in the influent stream, but continuous tests are preferable because they can be used to constantly monitor the coagulation process and adjust operations. The three basic tests are the jar test, particle charge observation technique, and pilot filters. Turbidity monitoring provides an indication of coagulation effectiveness.

Jar Tests

The most common technique, the standard jar test, involves laboratory tests of samples, the results of which are extrapolated to full-scale operations. Different coagulant dosages are applied to influent samples, with each dosage level in a different jar. For prescribed periods of time, the resultant turbidity levels are observed to derive the optimal dosage.

Jar tests are simple, but certain aspects require attention. Because jar tests are not very accurate for low turbidity water, coagulant dosages for influent turbidities less than 5 NTU are difficult to ascertain with only jar test results based on settled water turbidity. In such cases, it may be appropriate to filter settled water through filter paper before measuring turbidity. Careful attention is necessary to ensure that the jar test water temperature is the same as that of the plant influent. In addition, jar tests are inappropriate for establishing expected performance of direct filtration plants because the sedimentation process simulated in the jar test is not part of a direct filtration system.

Zeta Potential Observation Procedures

The zeta potential observation technique is a batch test method that measures particle charge with electrodes. However, it can be adapted to a continuous test method using a streaming current detector (SCD), which provides a continuous indication of particle charge.

The particle charge test relates electrode output to the zeta potential, a measure of the electrical potential between the water and the counter ions surrounding the colloid particles. Colloids are usually negatively charged and resist coagulation. Different colloids require different coagulants to adjust this zeta potential and thus enhance their removal from the water stream. Zeta potential after coagulation ranges from +5 to -10 minivolts (mv). Organic colloids require near 0 zeta potentials for coagulation, while clay colloids require negative zeta potentials. Particle charge observation tests are calibrated with jar tests or filtrate monitoring.

The results of an SCD, which continuously measures indicators of zeta potential, can be used to continually control the chemical feed pumps. Streaming current detectors provide information on coagulant dose more rapidly than jar tests.[1] Therefore, they enable operators to optimize chemical doses and cut chemical costs, especially in systems with variable influent quality. Use of an SCD improves plant operations, but is not a substitute for an effective operator. Figure 3-1 is a schematic diagram of an SCD.

The matching of the zeta potential with specific plant output must be carefully observed because, while there is a high correlation between zeta potential and finished water quality, there is no causal relationship. The lack of careful scrutiny of the relationship between particle charge observation test results and plant effluent will lead to false conclusions about the finished water's clarity. Bench-scale particle charge tests are best suited for water with small annual variations in quality and are also useful in research. They are not practical for plant operations, because they require a high level of skill and patience to perform and interpret the results.

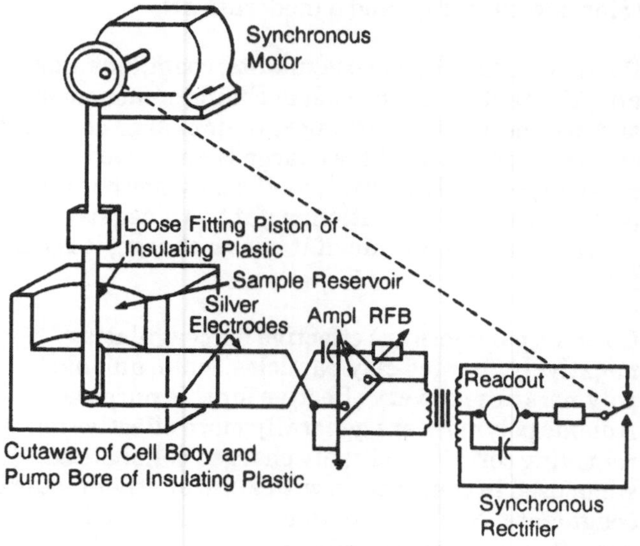

Figure 3-1. Simplified diagram of streaming current detector.

Pilot Filters

Pilot filters measure the water quality on a continuous basis. They provide an accurate and direct measure of the expected plant effluent turbidity, avoiding the intermediate step of extrapolating lab tests to predict plant-scale performance. Pilot filters determine the adequacy of the coagulant dosage and are used with manual or automated coagulant adjustment mechanisms. Their accuracy in monitoring effluent turbidity can achieve significant cost savings through improved efficiency of coagulant usage. Pilot filters have the advantage of providing indications of floc strength and filter headloss rate. This information is useful for filter management and minimizing backwash water.

In general, pilot filters are used to determine whether an applied coagulant dosage is adequate to produce acceptable quality water from the plant filters. A small sample of water, taken immediately after the addition of coagulants, is processed through a small (usually an 11.43-cm [4.5-in] inside diameter plexiglass column) pilot filter. The turbidity of the pilot filter is monitored continuously, and the information provides advance warning of improper coagulant dosages, which may lead to excessively high turbidity levels from the plant filter. Improper coagulant dosages can be detected with pilot filters within 10 to 15 minutes of coagulant application.

Floc breakthrough in the pilot filter can contribute to effluent turbidities and interfere with accurate coagulant dosage determination. To prevent floc

1 For a thorough discussion of SCDs, see Steven K. Dertel and Christine M. Kingery; November 1988; "An Evaluation of Streaming Current Detectors," American Water Works Association; Denver, Colorado.

breakthrough, excessive doses of polymer filter aids are applied to pilot filter influent.

Automated coagulant control systems can be used with the pilot filter systems. Some units are equipped with plant flow pace meters to account for flow variation. Override controls can also be placed throughout the system.

Turbidity Monitoring

Turbidity is a measure of suspended particles. Monitoring filtered water turbidity is important because it reflects the effectiveness of the coagulation process. In addition, turbidity monitoring is:

- Required of all surface water systems by the recently passed Surface Water Treatment Rule
- Related to probable level of potentially pathogenic organisms in the finished water
- Reflective of the overall treatment effectiveness
- A primary aesthetic indicator of the finished water quality

The current availability of inexpensive and accurate turbidimeters makes monitoring the turbidity of finished water very cost effective. Continuous monitoring of each filter unit is especially useful for optimizing the quality of effluent blended from different filters.

Particle counting is another method of measuring turbidity that provides a profile description of the particle sizes. Particles as small as 1 micron can be measured by this method, but since many particles are less than a micron in size, the method is of limited applicability. However, particle counting is very useful for pilot plants and research applications, especially for asbestiform fiber detection.

3.1.3 Chemical Application Methods and Considerations

The following discussion covers the equipment and methods used to apply treatment chemicals, including operation and maintenance considerations. The following four chemical feed systems are reviewed: liquid alum, polymer, sodium hydroxide, and sulfuric acid.

Chemical feed system options change as plants increase in size. Plants with a capacity of 0.12 m^3/sec (2.5 MGD) or greater generally are able to justify bulk purchases and storage of chemical stock and install the requisite storage and handling equipment. Special storage facilities for chemicals are also usually needed for larger plants. For example, liquid alum is available from 15,140-L (4,000-gal) tank trucks for larger plants, while smaller plants often use dry bagged chemicals that are mixed with water to produce desired treatment-strength solutions.

The point of application of any chemical will have a great impact on its effectiveness. Chemicals are added at seven basic points in the water treatment process:

- To raw water before rapid mixing
- During rapid mixing
- Before flocculation
- Before sedimentation
- Before filtration
- After filtration
- Prior to filter backwash

The optimal application point depends on site-specific factors. The best situation is to have the capability to add chemicals at several points, so that the plant operator can perform tests to determine the ideal application point.

Most chemical feed systems commonly include several basic elements such as mixing tanks, mixers, and metering pumps. In the chemical feed process, operators manually add chemicals either wet or dry, in specific concentrations to the mixing tank. A metering pump ensures accurate feeding of known concentrations of coagulants, polymers, coagulant aids, or pH adjustment chemicals into the water stream.

Liquid Alum Feed

Specialized liquid alum feed systems become appropriate for plants larger than about 0.12 m^3/sec (2.5 MGD). Design elements for these facilities include:

- Storage facilities for a 15-day supply (recommended)
- Fiberglass reinforced or polyethylene tanks
- Indoor uncovered tanks for smaller systems
- Outdoor, covered, and heated tanks for larger systems, because crystallization occurs at under 10°C (50°F)

Polymer Feed

In plants of less than about 0.12 m^3/sec (2.5 MGD), polymer application can be performed using a basic chemical feed system. In small plants, a polymer feed may include two tanks of equal size, a mixer, and a metering pump. One tank is used for mixing and one for storage of the polymer feed solution. The polymer mixing tank is elevated above the storage tank to permit gravity feed of the polymer solution. Optionally, a single tank suitable for water treatment operations can be interrupted to prepare the polymer solutions. A preparation of 0.25 percent stock polymer solution is generally used for all plant sizes.

Plants greater than 0.12 m^3/sec (2.5 MGD) generally require two tanks of equal size and a mixer, but, in addition, require a dry chemical storage hopper and a dry chemical feeder.

Sodium Hydroxide Feed

Plants less than about 0.12 m^3/sec (2.5 MGD) are able to use basic chemical feed systems to apply sodium hydroxide solution prepared from up to 90.72 kg (200 lb)/day of dry sodium hydroxide. These plants require:

- Piping and valves to convey water for mixing
- 98.9 percent pure dry sodium hydroxide
- A volumetric feeder for mixing
- Two mixing tanks because of the slow mixing rate necessary to accommodate the high heat of the process
- Application of the sodium hydroxide in a 10 percent solution
- A dual headed metered pump to deliver the sodium hydroxide solution to the point of application

Plants greater than 0.12 m^3/sec (2.5 MGD) are designed to use liquid sodium hydroxide, which requires equipment over and above the basic feed system. Plants using liquid sodium hydroxide feed a premixed 50 percent solution containing 6.38 pounds of sodium hydroxide per gallon. To store a 15-day supply of this stock solution, plants need fiberglass reinforced or steel tanks. The sodium hydroxide must be kept indoors or stored in heated tanks to avoid crystallization of the solution that occurs at or below 12.2°C (54°F).

Sulfuric Acid Feed

Sulfuric acid feed systems are designed to directly apply a 93 percent solution to the water through a metering system that includes standby metering pumps.

Most plants will purchase acid in bulk, which must be stored in fiberglass reinforced tanks. Only very small plants will use sulfuric acid delivered in drums that will require indoor storage facilities. All plants should have storage capacity for a 15-day supply.

Capital and operating and maintenance costs for alum, polymer, sodium hydroxide, and sulfuric acid feed systems are presented in Tables 3-3 through 3-6.

3.2 Modifying or Adding Rapid Coagulant Mixing

Rapid mixing is generally the first stage of the treatment process. This step is essential for quick and complete dispersion and mixing of the coagulant; the absence of complete mixing could result in excessive effluent turbidity. While some plants only have one rapid mixer to add coagulants, other plants have multiple rapid mixers to achieve optimum performance.

Increasing the plant flow rates often necessitates improvements to existing rapid mixing facilities to ensure complete and thorough mixing of coagulants with incoming water. Effective rapid mixing also can reduce coagulant dosage requirements.

Improvements to rapid mixing involve adjusting both the means of mixing the coagulant with the influent and the coagulant's point of application. Refinements to the rapid mixing process usually result in significant improvements to the finished water. The savings in coagulant usage justify the low capital cost of the improvements.

Rapid mixer effectiveness is usually gauged by contact time and the velocity gradient imparted to the water stream. Contact time refers to the detention time of the water in the rapid mixer. The velocity gradient is the difference in water speeds in the rapid mixing chamber (where water with the highest speed is in proximity to the mixing device and the slowest water is near the chamber wall). Velocity gradient is related to the amount of energy imparted to the water during mixing.

Complete coagulant dispersion is accomplished in very little time and with a wide range of velocity gradients. The reaction times for coagulation range from microseconds up to 30 seconds, and the velocity gradients of various mixers range from 700 to 5,000 sec^{-1}. Table 3-7 presents contact times and velocity gradients for four types of rapid mixers.

The contact time required for successful coagulation depends on which of the two primary coagulation processes - adsorption or sweep-floc (enmeshment) - is necessary for the particular raw. Adsorption takes place very quickly; sweep-floc coagulation takes longer. In adsorption, if polymers are not formed during the application process, then coagulation can occur in microseconds. If polymers are formed, coagulation will still take less than a second. Sweep floc, a process that is triggered by large doses of alum, ferric chloride, and other precipitative coagulants, takes from 1 to 7 seconds to complete.

The types of devices used for rapid mixing include mechanical mixers, static mixers, jet injection mixers, and coagulant diffusers. Table 3-8 presents

Table 3-3. Estimated Costs for Supplementing Surface Water Treatment by Adding Alum Feed facilities (10 mg/l) ($1978)

Category	Plant Capacity, MGD	Average Flow, MGD	Capital Cost, $1,000	Operation and Maintenance Costs $1,000/yr	Operation and Maintenance Costs ¢/1,000 gal	Total Cost, ¢/1,000 gal
1	0.026	0.013	12.2	2.3	47.8	87.1
2	0.068	0.45	15.9	2.6	16.0	26.2
3	0.166	0.13	20.3	5.4	11.5	16.5
4	0.50	0.40	28.0	7.7	5.3	7.5
5	2.50	1.30	40.7	4.1	0.9	1.9
6	5.85	3.25	47.1	7.9	0.7	1.1
7	11.59	6.75	58.1	14.9	0.6	0.9
8	22.86	11.50	70.4	24.4	0.6	0.8
9	39.68	20.00	81.6	41.3	0.6	0.7
10	109.90	55.50	128	112	0.55	0.63
11	404	205	250	402	0.54	0.58
12	1,275	650	575	1,245	0.52	0.55

Source: U.S. EPA (1979).

Table 3-4. Estimated Costs for Supplementing Surface Water Treatment by Adding Polymer Feed Facilities (0.3 mg/l) ($1978)

Category	Plant Capacity, MGD	Average Flow, MGD	Capital Cost, $1,000	Operation and Maintenance Costs $1,000/yr[a]	Operation and Maintenance Costs ¢/1,000 gal	Total Cost, ¢/1,000 gal
1	0.026	0.013	2.8	1.3	28.3	35.3
2	0.068	0.045	3.3	1.4	8.6	11.0
3	0.166	0.13	3.8	3.5	7.3	8.2
4	0.50	0.40	4.6	3.7	2.5	2.9
5	2.50	1.30	66.1	5.8	1.2	2.9
6	5.85	3.25	66.7	6.8	0.6	1.2
7	11.59	6.75	68.8	8.4	0.3	0.7
8	22.86	11.50	72.7	10.7	0.2	0.5
9	39.68	20.00	78.6	14.6	0.2	0.3
10	109.90	55.50	107.2	31.1	0.2	0.2
11	404	205	159.9	100	0.1	0.2
12	1,275	650	227.3	299	0.1	0.1

[a]Polymer cost for Categories 1 through 4 is 75¢/lb, and for Categories 5 through 12 it is $1,000/ton
Source: U.S. EPA (1979).

Table 3-5. Estimated Costs for Supplementing Surface Water Treatment by Adding Sodium Hydroxide Facilities ($1978)

Category	Plant Capacity, MGD	Average Flow, MGD	Capital Cost, $1,000	Operation and Maintenance Costs $1,000/yr[a]	Operation and Maintenance Costs ¢/1,000 gal	Total Cost, ¢/1,000 gal
1	0.026	0.013	2.4	1.2	24.3	30.2
2	0.068	0.045	2.7	1.4	8.8	10.7
3	0.166	0.13	3.3	3.3	7.6	8.4
4	0.50	0.40	5.4	6.2	4.2	4.7
5	2.50	1.30	33.3	8.8	1.8	2.7
6	5.85	3.25	36.9	18.4	1.6	1.9
7	11.59	6.75	43.0	35.8	1.4	1.6
8	22.86	11.50	56.3	59.4	1.4	1.6
9	39.68	20.00	76.4	101	1.4	1.5
10	109.90	55.50	159	275	1.4	1.4
11	404	205	353	1,015	1.4	1.4
12	1,275	650	697	3,210	1.4	1.4

[a] Costs include storage and feed facilities to add NaOH at a concentration of 10 mg/L. Dry sodium hydroxide is used for Categories 1 through 4, while a liquid solution is used for bulk delivery for Categories 5 through 12.
Source: U.S. EPA (1979).

Table 3-6. Estimated Costs for Supplementing Surface Water Treatment by Adding Sulfuric Acid Feed Facilities ($1978)

Category	Plant Capacity, MGD	Average Flow, MGD	Capital Cost, $1,000	Operation and Maintenance Costs		Total Cost, ¢/1,000 gal
				$1,000/yr[a]	¢/1,000 gal	
1	0.026	0.013	2.8	0.9	20.0	27.0
2	0.068	0.045	3.2	1.3	7.6	9.9
3	0.166	0.13	4.0	3.2	6.7	7.6
4	0.50	0.40	5.7	5.6	3.9	4.3
5	2.50	1.30	21.6	2.4	0.5	1.0
6	5.85	3.25	25.6	3.6	0.3	0.6
7	11.59	6.75	29.7	5.8	0.2	0.4
8	22.86	11.50	37.9	8.8	0.2	0.3
9	39.68	20.00	49.9	14.1	0.2	0.3
10	109.90	55.50	86.9	36.3	0.2	0.2
11	404	205	210.3	128.3	0.2	0.2
12	1,275	650	431.0	404.7	0.2	0.2

[a] Costs include storage (15 days), feed, and metering facilities for delivering concentrated acid directly from storage to application point. Categories 1 through 6 include delivery in drums, stored indoors. Categories 7 through 12 include bulk delivery and outdoor storage in FRP tanks. Application rate is 2.5 mg/L.
Source: U.S. EPA (1979).

Table 3-7. Typical Rapid Mixer Contact Times and Velocity Gradients

Mixer Type	Typical Contact Time Range (in seconds)	Typical Velocity Gradients (in sec^{-1})
Mechanical mixers	10-30	700-1,000
In-line mechanical and static blenders	0.5	3,000-5,000
In-line jet injectors	0.5	1,000-2,000
Hydraulic mixers	2	800

Source: Williams and Culp (1986).

the estimated capital and operating costs associated with adding rapid mixing.

Mechanical Mixers

The most common mixing devices are mechanical mixing tanks, also termed completely mixed or back-mixed units. They use turbines or impellers to mix coagulant with water. The three advantages of mechanical mixers are that they (1) are effective, (2) impart little headloss (headloss being loss of pressure or the reduction of the water velocity within the plant), and (3) are unaffected by the volume of water and flow variations.

A variable speed drive can alter mixer speeds to achieve different velocity gradients. Lower velocity gradients are used following the application of polyelectrolyte coagulant aids. Labor is required for daily jar testing operations and routine inspections, as well as annual inspections, cleaning, tank drainage, and oil changes.

In-Line Static Mixers

In-line static mixers consist of a series of baffle elements placed in a pipe section to impart alternating changes in flow direction and intense mixing action as water flows through the device. Headloss ranging from 0.3 to 1.8 m (1 to 6 ft) occurs as the flow passes through the static mixer. Static mixers achieve virtually instantaneous mixing, are relatively maintenance free, and are less expensive than other rapid mixing processes. The only disadvantage of in-line static mixers is that mixing intensity depends upon the plant's water flow rate. As rates decrease, the mixing intensity slows.

In-Line Mechanical Blenders

In-line mechanical blenders provide rapid mixing of treatment chemicals with water flowing in a pipeline under pressure. These devices consist of a housed propeller driven by an electrical motor. They have the advantage over static mixers of imparting considerably less headloss and not being affected by change in flow. In addition, in-line blenders offer the following advantages:

- Virtually instantaneous mixing with a minimum of short-circuiting
- Minimal headlosses
- Less expense than more conventional rapid mix units

Jet Injection Blending

Jet injection mixers, a type of in-line blender, are used for attaining nearly instantaneous dispersion of coagulant with raw water, usually at larger treatment plants. Jet injectors can avoid the backmixing inefficiencies of turbine or impeller mixers, and the recommended detention times are

Table 3-8. Estimated Costs for Supplementing Surface Water Treatment by Adding Rapid Mix ($1978)

Category	Plant Capacity, MGD	Average Flow, MGD	Capital Cost, $1,000	Operation and Maintenance Costs $1,000/yr	Operation and Maintenance Costs ¢/1,000 gal	Total Cost, ¢/1,000 gal
1	0.026	0.013	13.2	2.8	58.6	91.3
2	0.068	0.045	17.5	2.9	17.6	30.1
3	0.166	0.133	22.5	7.0	14.7	20.3
4	0.50	0.40	30.9	7.9	5.4	7.9
5	2.50	1.30	47.7	13.3	2.8	4.0
6	5.85	3.25	63.7	22.4	1.9	2.5
7	11.59	6.75	88.2	38.2	1.6	2.0
8	22.86	11.50	139	69.2	1.6	2.0
9	39.68	20.00	218	116	1.6	1.9
10	109.90	55.50	587	313	1.5	1.9
11	404	205	2,100	1,130	1.5	1.8
12	1,275	650	6,670	3,540	1.5	1.8

Source: U.S. EPA (1979).

shorter than for mechanical mixers. The primary disadvantages are that the orifices in the injection pipe tend to become plugged and the mixing intensity cannot be varied. Figure 3-2 illustrates a jet injection mixer.

Coagulant Diffusers

Coagulant diffusers are also used to improve the rapid mixing process. Figure 3-3 is a schematic diagram of a coagulant diffuser. Coagulant diffusers are similar to jet-injection mixers in design, except that jet injectors are usually used in a pipe setting and coagulant diffusers are typically used in basin settings. Coagulant diffusers are designed to apply the coagulant at the point of maximum turbulence. In some plants, multiple application points spaced uniformly across the flow cross section permit rapid and thorough contact of chemical solution with the entire incoming water flow. This rapid and uniform dispersion of coagulant prevents its hydrolysis, which is a common problem in systems with single application points. Where turbulence in incoming channels or pipelines provides good mixing, the simple addition of a coagulant diffuser could improve coagulation and lead to improved filtered water quality.

3.3 Improving Flocculation

Flocculation usually follows the rapid mixing process in conventional treatment plants. Flocculation is a time-dependent process that directly affects clarification efficiency by providing multiple opportunities for particles suspended in water to collide through gentle and prolonged agitation. The process takes place in a basin equipped with a mixer that provides agitation. This agitation must be thorough enough to encourage interparticle contact, but gentle enough to prevent disintegration of existing flocculated particles.

Effective flocculation is important for the successful operation of the sedimentation process. Expanding plant capacity will require flocculation improvement to maintain plant performance. For example, a doubling of plant capacity resulting from a change in filter media from rapid sand to mixed- or dual-media can increase flow capacity from 1.34 to 2.7 L/sec/m^2 (2 to 4 GPM/sq ft). The increased flow will cut the flocculation detention time in half, thereby preventing completion of the flocculation process, and reducing sedimentation effectiveness.

Three basic methods of improving the flocculation process are:

- Installing new mixing equipment to improve agitation
- Improving inlet and outlet conditions to minimize destabilizing turbulence
- Adding baffling to reduce basin short-circuiting

3.3.1 Improving Mixing

Different mixing equipment can improve turbulence patterns in the basin to maximize the formation of flocculated particles and minimize the destruction of previously formed floc. There are several types of mixers used for flocculation, with the mechanical mixer being the most common. Mechanical mixers for flocculation basins are differentiated by overall design and type of agitator.

The most important goal in flocculator design is the efficient removal of floc during sedimentation and filtration. Many flocculator units, especially low-energy mixers, tend to maximize floc size rather than floc density, which affects the speed with which floc

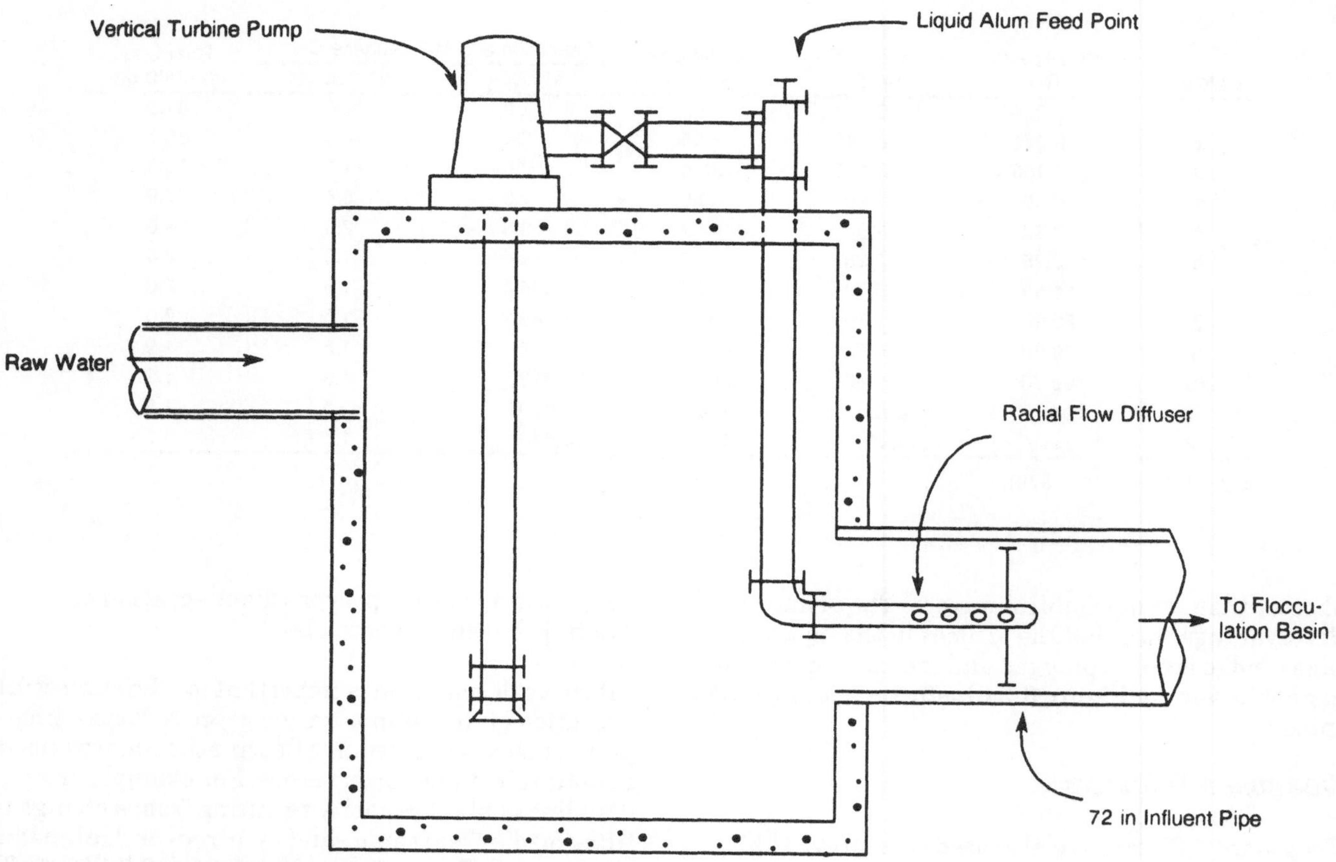

Figure 3-2. In-line jet injection mixer.

settles. High-energy mixers create smaller and denser floc which settles faster and occupies less volume in the filter bed than the larger floc created by low-energy mixers.

Paddle, walking-beam, and flat blade turbine mixers are known as low-energy mechanical mixers. The paddle or reel mixer operates with low tip speeds of 2 to 15 revolutions per minute (rpm) to prevent floc destruction. The flat blade turbine typically operates at 10 to 15 rpm. Flat blade turbines can produce excessively high velocity gradients of over 45 rpm.

Axial flow propellers or turbines, known as high-energy mechanical flocculators, typically operate at 150 to 1,500 rpm with no limit on blade tip speed. Axial flow propellers are favored in some situations because they produce uniform turbulence and are simple to install and maintain.

Careful evaluation is needed to determine whether a plant should change to high-energy flocculators. Sometimes, higher energy gradients can be achieved through modifying paddle-style units by adding more paddles or installing a new higher speed mechanical drive.

3.3.2 Improving Flocculator Inlet and Outlet Conditions

The desired detention time (i.e., residence time) determines the size and occasionally the design configuration of the flocculation basin. The basin's internal design then determines detention time effectiveness. The design of the basin inlet, outlet, and internal circulation patterns all affect floc formation and destruction.

Typical detention times range from 15 to 45 minutes. These are influenced by the influent water conditions, type of coagulant used, and requirements of downstream processes. Cold, low turbidity water could require 30 minutes of detention, while the same water undergoing direct filtration at higher temperatures could require only 15 minutes of detention. In small plants with high efficiency volumetric mixing, 10 to 15 minutes of detention might be acceptable.

If water passes through the flocculation basin in much less than the volumetric residence time, the influent stream is said to have "short-circuited." To achieve effective flocculation and minimize short-circuiting, designers must pay particular attention to inlets, outlets, and internal baffling. Inlet and outlet turbulence is the major source of destructive energy

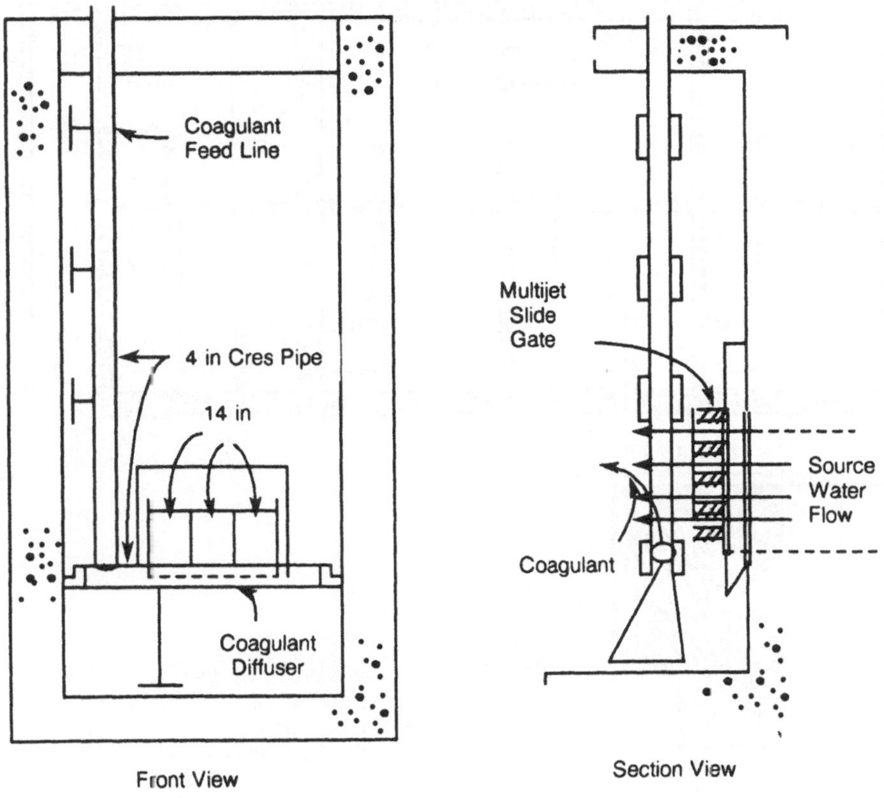

Figure 3-3. Coagulant diffuser.

in the flocculation basin contributing to short-circuiting.

Effective baffle, inlet, and outlet design reduces short-circuit problems. Improvements that are often overlooked can include:

- Adding inlet diffusers to improve the uniformity of the distribution of incoming water
- Enlarging connecting conduits to reduce floc-disrupting turbulence
- Adding a secondary entry baffle across the inlet to the flocculation basin to impart increased headloss for uniform water entry

Improving inlet and outlet conditions will reduce basin turbulence and thus lessen floc breakup, increase detention times, and allow more efficient coagulant usage.

3.3.3 Improving Basin Circulation with Baffles

Flocculation basins can be baffled to direct the movement of the water through the basin. Baffling, usually near the basin entrances and exits, can improve basin circulation and achieve more uniform flocculation. Baffles, commonly made of wood, plastic, concrete, or steel are either "over-under" or "around-the-end" in design. "Over-under" designs direct water flow either over or under the baffle, while "around-the-end" designs direct the water around either end of the baffle. Baffles should be designed to direct the water flow such that the velocity gradients are less than those produced in the flocculation process but greater than those in the flow moving laterally across the inlets. An example of a basin divided with baffling is illustrated in Figure 3-4.

Estimated costs for adding flocculation are presented in Table 3-9.

3.4 Improving Sedimentation

Sedimentation is the step in conventional water treatment systems that follows flocculation and precedes filtration. Its purpose is to enhance the filtration process by removing particulates. Sedimentation requires that the water flow through the basin at a slow enough velocity to permit the particles to settle to the bottom before the water exits the basin. The equipment required for this process includes a settling basin of either rectangular,

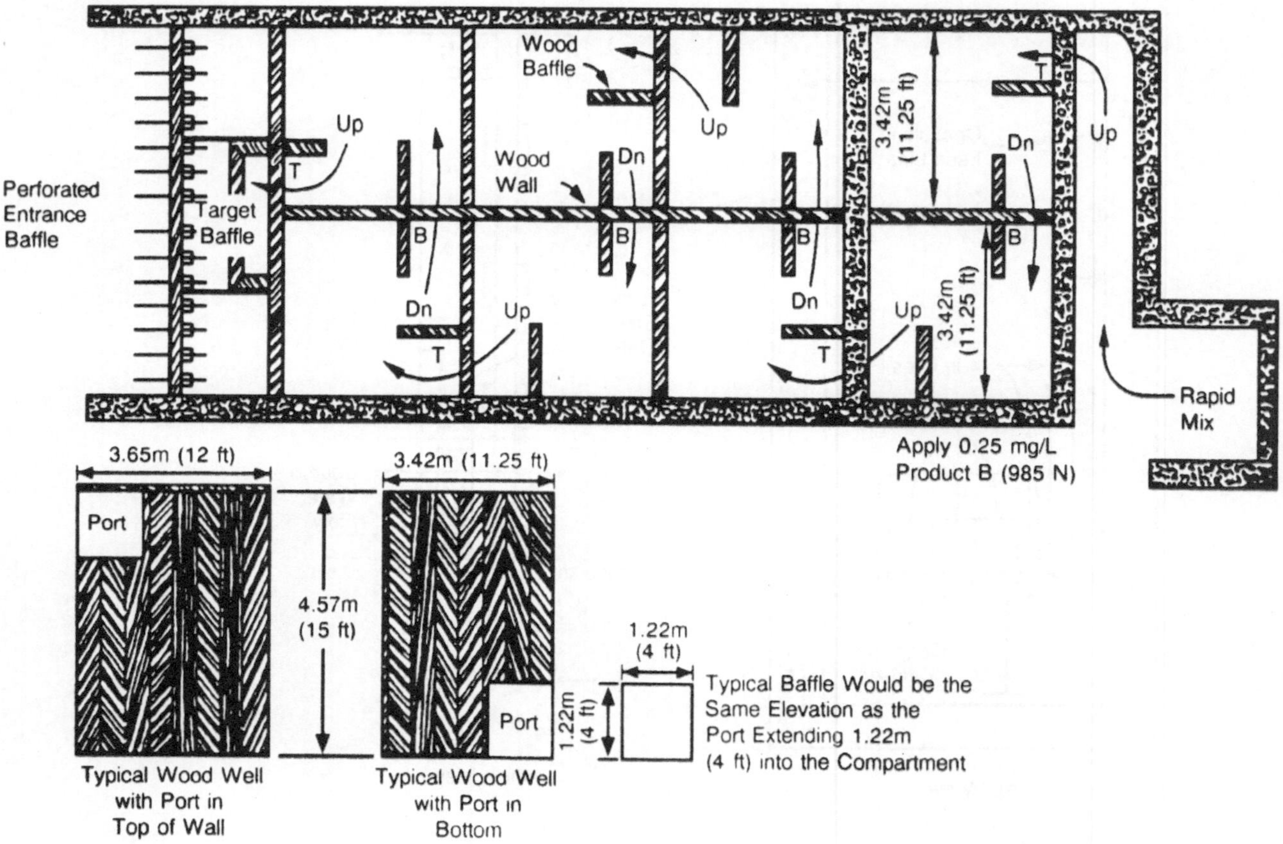

Figure 3-4. Divided flocculation basin.

Table 3-9. Estimated Costs for Supplementing Surface Water Treatment by Adding Flocculation ($1978)

Category	Plant Capacity, MGD	Average Flow, MGD	Capital Cost, $1,000	Operation and Maintenance Costs $1,000/yr	Operation and Maintenance Costs ¢/1,000 gal	Total Cost, ¢/1,000 gal
1	0.026	0.013	10	1.0	21.7	45.2
2	0.068	0.045	18	1.1	6.9	20.1
3	0.166	0.133	34	2.3	4.9	13.3
4	0.50	0.40	73	2.7	1.8	7.7
5	2.50	1.30	217	3.8	0.8	6.2
6	5.85	3.25	325	5.6	0.5	3.7
7	11.59	6.75	418	8.7	0.4	2.3
8	22.86	11.50	587	14.5	0.4	2.0
9	39.68	20.00	840	22.9	0.3	1.7
10	109.90	55.50	1,830	53.9	0.3	1.3
11	404	205	6,060	182	0.2	1.2
12	1,275	650	19,200	569	0.2	1.2

Source: U.S. EPA (1979).

square, or circular configuration. The basin includes provisions for inlet and outlet structures, and a sludge collection system. In addition, sedimentation systems are optionally equipped with tube or plate settlers to improve performance.

Improvements in other elements of the treatment process often require modification of the settling basins to maintain effluent water quality. Typical system improvements that necessitate sedimentation modification involve expanded flow rates, resulting in decreased detention times and increased clarification rates.

Problems with settling basins that might require modification include:

- Poor entry and/or exit conditions
- Destructive basin turbulence
- Excessive clarification rate
- Inadequate sludge collection and removal

Settling basin inlets are often responsible for creating turbulence that results in breaking floc. One way to avoid such destructive turbulence is to eliminate conduits and pipes between flocculation and settling basins by conducting both flocculation and sedimentation in one basin separated by perforated baffles. Baffles will minimize the breaking of floc and evenly distribute the flocculated water to the settling basin to maximize settling efficiency.

Improperly designed outlets are often responsible for sedimentation basin short-circuiting. One method of reducing short-circuiting in rectangular basins is to use finger launders extending inward from the exit wall in a uniform pattern. Finger launders are small troughs with V-notch weir openings that collect water uniformly over a large area of the basin. Weir loading rates ranging from 126.2 to 252.3 m^3/day/m (10,000 to 20,000 gpd/ft) of weir length are optimal. For circular or radial flow basins, clarified water is collected in a continuous peripheral weir trough. These collection troughs should be designed with the same loading rate as weir troughs used in rectangular basins.

Tube or plate settlers are often added to settling basins to improve their efficiency, especially if flows are to be increased beyond original design conditions. This established technology is widely used to reduce floc carryover in existing basins. Tube settlers can be installed in most conventionally designed settling basins and, in many cases, result in twice the basin's original settling capacity. Also, the use of tube and plate settlers in new plant construction minimizes settling basin costs by minimizing the basin size necessary to attain a desired level of treatment. Tube settlers can also be used with vacuum sludge removal systems.

Tube settlers typically use 0.6 to 0.9 m (2 to 3 ft) long tubes or plates placed from 5.1 to 10.2 cm (2 to 4 in) apart and installed over part or all of a basin. They are generally designed to accept flow rates ranging from 0.68 to 2.04 L/sec/m^2 (1 to 3 GPM/ft^2). The shallow settling depths and the large surface area provided by tube settlers permit effective clarification at detention times of several minutes versus several hours in conventional sedimentation basins.

There are several different types of commonly used tube settling basin designs. The two most common choices are horizontal flow basins (either rectangular or circular) and upflow solids contact clarifiers. Either design uses tube settling modules or plates to apply the principles of shallow depth sedimentation, permitting operation at higher clarification rates than conventional clarifiers.

The major by-product of sedimentation is a volume of sludge, which must be removed on a continuous basis. Sludge collection systems are either manual or automated. Automated mechanical sludge collection equipment is suitable for larger plants to attain continuous and complete sludge removal.

3.4.1 Horizontal Flow Sedimentation Basins

As the name implies, the water generally follows a horizontal flow path in this type of basin. Water is introduced at one end of the basin and suspended particles settle to the bottom as the water moves through to the exit. The particular basin configuration will affect the design requirements for installing tube or plate settlers to improve performance or increase capacity.

The terms overflow rate, clarification rate, and loading rate all refer to the amount of water that can be adequately processed by sedimentation (i.e., the velocity of water through the tubes expressed in GPM/sq ft). The clarification rate at which tubes or plate settlers can be operated depends on the type of clarifier and water characteristics, including temperature and the desired effluent quality. For example, the tube clarification rate must be reduced when influent turbidity increases or water temperature decreases. The specific relationships between tube loading rates and influent and effluent turbidity for cold and warm water are presented in Table 3-10.

The preferred location for tube modules in horizontal flow basins is away from areas of possible turbulence such as basin entrances. For example, in a horizontal basin, often as much as one third of the basin length at the inlet end may be left uncovered by the tubes so that it may be used as a "stilling" zone, where water turbulence is diminished in preparation for sedimentation. This is permissible in most basins because the required number of tube settlers will cover only a portion of the basin. Tube modules are usually placed at least 0.6 m (2 ft) below the surface in standard depth (3 to 3.9 m [10 to 13 ft]) settling basins. They can be submerged as much as 1.2 m (4 ft) for deeper basins (4.8 to 6 m [16 to 20 ft]). In all cases sufficient clearance must be maintained beneath the tube settling module for movement of sludge collection equipment.

Table 3-10. Horizontal Flow Basins Loading Rates

Overflow Rate, Based on Total Clarifier Area (GPM/ft^2)	Raw Water Turbidity	Overflow Rate, Portion of Basin Covered by Tubes (GPM/ft^2)	Probable Effluent Turbidity, NTU
For water temperatures 40°F or less:			
2.0	0-100	2.5	1-5
2.0	0-100	3.0	3-7
2.0	100-1,000	2.5	3-7
2.0	100-1,000	3.0	5-10
For water temperatures 50°F or above:			
2.0	0-100	2.5	1-3
2.0	0-100	3.0	1-5
2.0	0-100	4.0	3-7
3.0	0-100	3.5	1-5
2.0	100-1,000	2.5	1-5
2.0	100-1,000	3.0	3-7

1 GPM = 0.06308 L/sec; 1 ft^2 = 0.0929 m^2; °C = (°F - 32) × .556.

Figure 3-5 shows a typical tube settler installation in a rectangular basin illustrating the arrangement of tube settling modules and the utilization of collection troughs in the covered area to ensure uniform flow collection. In radial flow basins (see Figure 3-6), the required quantity of modules can be placed in a ring around the basin periphery, leaving an inner-ring open area between the modules and the center well to dissipate inlet turbulence.

3.4.2 Upflow Solids Contact Clarifiers

The design of upflow solids contact clarifiers is based upon maintenance of a layer or blanket of flocculated material through which water flows in a vertical direction in the clarifier. The purpose of the layer, known as a sludge blanket, is to entrap slowly settling small particles and achieve a high level of clarification. The sludge blanket is maintained at a certain level and concentration by the controlled removal of sludge. The precise height is determined by the clarification rate. When the flow is increased, the clarification rate is greater and the level of the blanket rises.

In solids contact clarifiers, the clarification efficiency of tube settlers depends upon both the clarification rate and the concentration of floc in the sludge blanket. The allowable loading rate or velocity of flow through tube settlers is dependent upon:

- Average settling velocity of the floc layer or sludge blanket
- Ability of the clarifier to concentrate solids
- Ability of the sludge removal system to maintain an equilibrium solids concentration; if the sludge is not removed quickly enough, then the solids layer will pass through the tubes and into the effluent
- Temperature of the water

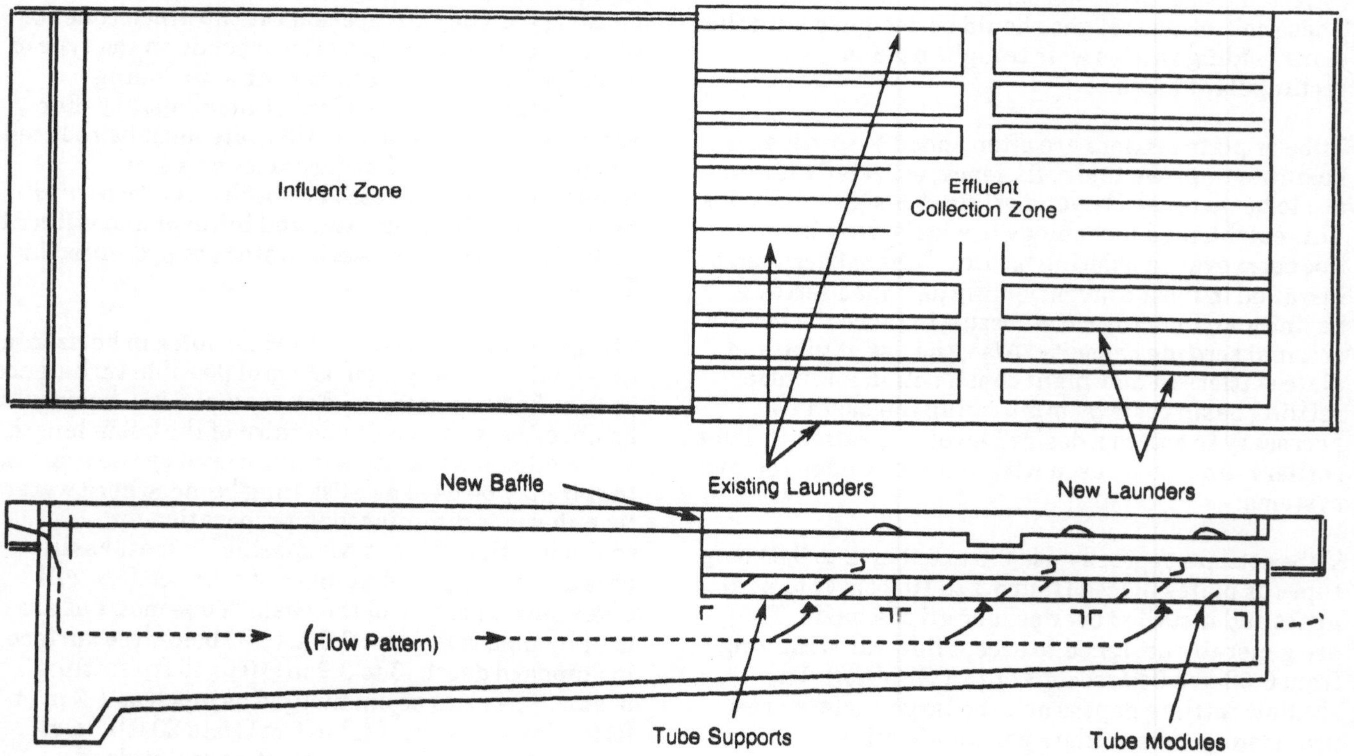

Figure 3-5. Typical tube settler installation in rectangular basin.

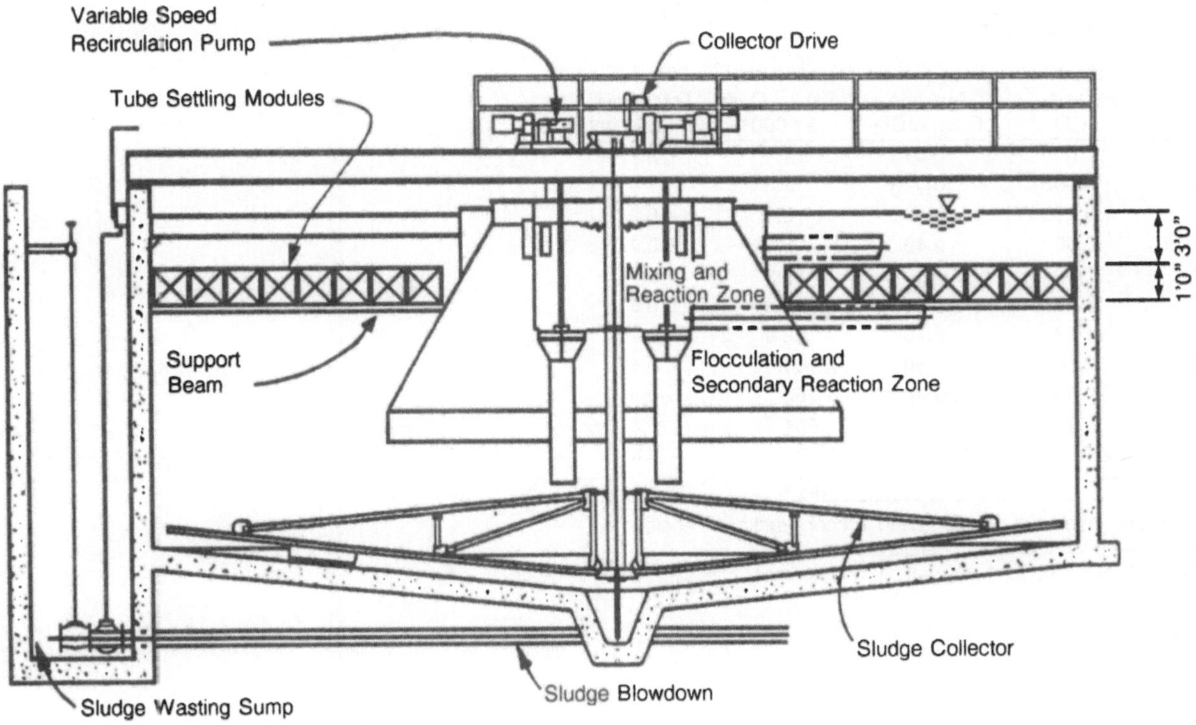

Figure 3-6. Radial solids contact clarifier with tube settlers.

Recommended loading rates for upflow clarifiers are provided in Table 3-11.

Table 3-11. Upflow Clarifier Loading Rates for Cold Water

Overflow Rate, Based on Total Clarifier Area (GPM/ft²)	Overflow Rate, Portion of Basin Covered by Tubes (GPM/ft²)	Probable Effluent Turbidity, NTU
For water temperatures 40°F or less:		
1.5	2.0	1-3
1.5	3.0	1-5
1.5	4.0	3-7
2.0	2.0	1-5
2.0	3.0	3-7
For water temperatures 50°F or above:		
2.0	2.0	1-3
2.0	3.0	1-5
2.0	4.0	3-7
2.5	2.5	3-7
2.5	3.0	5-10

1 GPM = 0.06308 L/sec; 1 ft^2 = 0.0929 m^2;
°C = (°F - 32) × .556.

Expansion of the capacity of this type of clarifier is limited by its ability to remove solids from the system. Generally, the maximum expansion from original design capacities is limited to 50 to 100 percent.

Tube module placement is governed by the same general rule as that for horizontal flow basins; that is, tube modules are best located away from influent turbulence. In terms of placement depth and general proximity, the recommendations regarding distance from the inlet and depth from the surface are also the same. Lastly, collection launders should be placed at 3- to 4.5-m (10- to 15-ft) intervals to ensure uniform collection of flow over the area covered by the tube settling modules.

Estimates of the costs of adding tube modules to conventional water systems are provided in Table 3-12. These costs include modules, supports, anchor braces, transition baffles, effluent launders, and installation.

Table 3-12. Estimated Costs for Supplementing Surface Water Treatment by Adding Tube Settling Modules ($1978)

Category	Plant Capacity, MGD	Average Flow, MGD	Capital Cost, $1,000[a]	Total Cost, ¢/1,000 gal
1	0.026	0.013	1.1	2.7
2	0.068	0.045	2.3	1.6
3	0.166	0.133	3.9	0.9
4	0.50	0.40	8.2	0.7
5	2.50	1.30	25.2	0.6
6	5.85	3.25	46.8	0.5
7	11.59	6.75	92.0	0.4
8	22.86	11.50	163.0	0.4
9	39.68	20.00	250	0.4
10	109.90	55.50	684	0.4
11	404	205	2,500	0.4
12	1,275	650	7,850	0.4

[a] Costs include modules, supports and anchor brackets, transition baffle, effluent launders, and installation.
1 MGD = 0.044 m^3/sec; 1,000 gallon = 3.785 m^3
Source: U.S. EPA (1979).

Chapter 4

Filtration Technologies

This section describes several available filtration technologies ranging from commonly used conventional systems to new and emerging technologies. Filtration is one of the most important elements in traditional water treatment systems and also plays an important role in controlling some organic contaminants.

The filtration process itself may need upgrading due to expansions or the need to meet stricter effluent quality limits. Section 4.1 addresses typical concerns and describes beneficial modifications. Many water supply systems will have to add filtration to comply with the 1986 Safe Drinking Water Act (SDWA) Amendments and the resultant Surface Water Treatment Rule. Sections 4.2 through 4.6 provide overviews of the following filtration technologies:

- Conventional treatment and direct filtration
- Direct filtration (gravity and pressure filters)
- Slow sand filtration
- Package plants
- Diatomaceous earth filtration
- Membrane filters
- Cartridge filters

Conventional treatment and direct filtration (Section 4.2) are the most widely used systems. Slow sand filters (Section 4.3), package plants (Section 4.4), and diatomaceous earth filtration (Section 4.5) are considered newer technologies, but have broad applicability. While slow sand filters have been used for many decades, they do not have an established record of performance with a large number of water systems in this country. Membrane and cartridge filtration systems (Section 4.6) are considered emerging technologies because they show promise but have not generally been used for treatment of drinking water. Package plants, slow sand, diatomaceous earth, membrane, and cartridge filters are considered best suited for small systems (less than 0.44 m^3/sec [10 MGD]).

The sections describing filtration technologies provide information on each type of system covering the following aspects:

- Process description
- System design
- Operation and maintenance
- System performance
- System costs

The last subsection, Section 4.8, provides some guidance in selecting the appropriate filtration technology for particular applications. Appendix A includes descriptions of experiences with the various filtration technologies.

4.1 Modifying Filtration Systems

Filtration is usually the final step in conventional treatment trains, although disinfection frequently follows filtration. This section addresses improvements that can be made to plants that use rapid sand, mixed-media, or dual-media filters.

Filtration systems are regarded as effective for removal of turbidity and microbial contaminants. Microbial contaminants of special concern include coliform bacteria, *Giardia lamblia*, enterovirus, and *Legionella*.

The Surface Water Treatment Rule (SWTR) requires that filters achieve turbidities of less than 0.5 NTU in 95 percent of the finished water samples, and never exceed 5 NTU. Turbidity is a measure of suspended particles, which can include organic solids, viruses, bacteria, and other substances. Turbidity particles range in size from less than 1 micron to 100 microns.

4.1.1 General Effectiveness of Filtration Systems

The effectiveness of filtration systems is determined by their ability to remove microorganisms, turbidity, and color. Color is imparted to water supplies by organic material and can be removed by chemical coagulation. Once the color is coagulated and combined with the floc particles, color removal can be related to turbidity removal. The measurement and control of microbes and turbidity are distinct procedures. The effectiveness of filter systems in removing microbial contaminants, however, is heavily related to influent turbidity. The general rule of thumb is to minimize turbidity to maximize microbial removal.

The filtration systems discussed in this document are appropriate for raw water with varying characteristics. Table 4-1 contains the recommended upper limits for several influent variables, including coliform bacteria, turbidity, and color, for four filtration technologies. Conventional treatment is clearly the most accommodating of the filtration systems in the table because it includes flocculation and sedimentation, which reduce turbidity before the water is actually processed by the filter. Diatomaceous earth filtration systems, which include little pretreatment, require high-quality influent.

Giardia lamblia is of particular concern in drinking water supplies because it forms a cyst that cannot be effectively killed by traditional chlorine disinfection. Effective removal can be attained only by filtration. Conversely, viral and bacterial pathogens are relatively easy to destroy with disinfection.

Figure 4-1 is a graph of the relationship between filter effluent turbidity and cyst removal efficiency after filtration. The graph clearly shows that filtered water with very low levels of turbidity, ranging below 0.1 NTU, contained almost no cysts (DeWalle et al., 1984). The four filtration systems shown in Table 4-2 are very effective in removing *Giardia*; the only exception is direct filtration without coagulation, which does not effectively reduce turbidity. If water supplies containing *Giardia* are not effectively coagulated, the cysts will pass through the entire treatment process, including the filters.

The viral removal efficiencies of four filter technologies found to be very effective are shown in Table 4-3. Of these processes, direct filtration provides the poorest removal of viruses, ranging from 90 to 99 percent.

4.1.2 Filtration System Improvements

Improving filtration systems can increase plant capacity and improve effluent quality. Increasing the capacity and effectiveness of rapid sand filters, originally designed to operate at 1.36 L/sec/m^2 (2 GPM/sq ft), is usually achieved by either changing the filter media to dual- or mixed-media or replacing the top layer of sand with anthracite coal (also termed "capping" the filter). With these modifications, filter rates often can be increased to 2.7 to 3.4 L/sec/m^2 (4 to 5 GPM/sq ft), doubling plant capacity and producing higher quality effluent.

A thorough hydraulic study is required to determine the feasibility of improving a filtration system. For example, if a plant is going to be expanded, sedimentation basins and interprocess transfer systems have to be evaluated in terms of their ability to accommodate the increased flow from larger filters. Assessing the maximum carrying capacities for all components of the system is a prerequisite to evaluating the expansion potential of a treatment plant. Field testing of the entire system is the preferred method for conducting hydraulic evaluations.

If a hydraulic evaluation reveals that it is feasible to increase plant capacity, a specific evaluation of filter modification is appropriate. Such an evaluation will include consideration of filter box design, underdrain type, surface wash system, air scour system, flow-limiting devices and piping, filtration rate controls, filter aid application, and backwash additives. Figure 4-2 illustrates the basic components of a sand filter system that need to be evaluated.

Filter Boxes

Filter boxes, the structures that contain the filter media, should be examined first to see if they can accommodate a filter that will operate at higher flows. Most rapid sand filters are easily converted to higher volume dual- or mixed-media filters. Backwash rates, the primary operational consideration, are about the same for these two filter types.

When flow capacities are increased using either dual- or mixed-media filters, wash troughs and certain filter cleaning technologies will require specific consideration. Filter wash troughs, structures at the top of filter boxes that collect waste backwash water during filter cleaning, may need to be deepened to prevent loss of the lighter anthracite coal media. Wash troughs that are shallow and wide and spaced closely together can create overly high rise velocities that wash out anthracite coal during backwash. If filter troughs cover more than half the area of the filter box, narrower and deeper troughs covering less surface area may be needed to prevent excessive coal loss during filter backwashing.

Table 4-1. Generalized Capability of Filtration Systems To Accommodate Raw Water Quality Conditions

	Recommended Upper Limits For Influent		
Treatment Technology	Total Coliforms, #/100 mL	Turbidity, NTU	Color, CU
Conventional treatment	< 20,000	No Restrictions	75
- with no predisinfection	< 5,000	No Restrictions	75
Direct Filtration	< 500	< 7-14	< 40
Slow Sand Filtration	< 800	< 10	< 5
Diatomaceous Earth Filtration	< 50	< 5	< 5

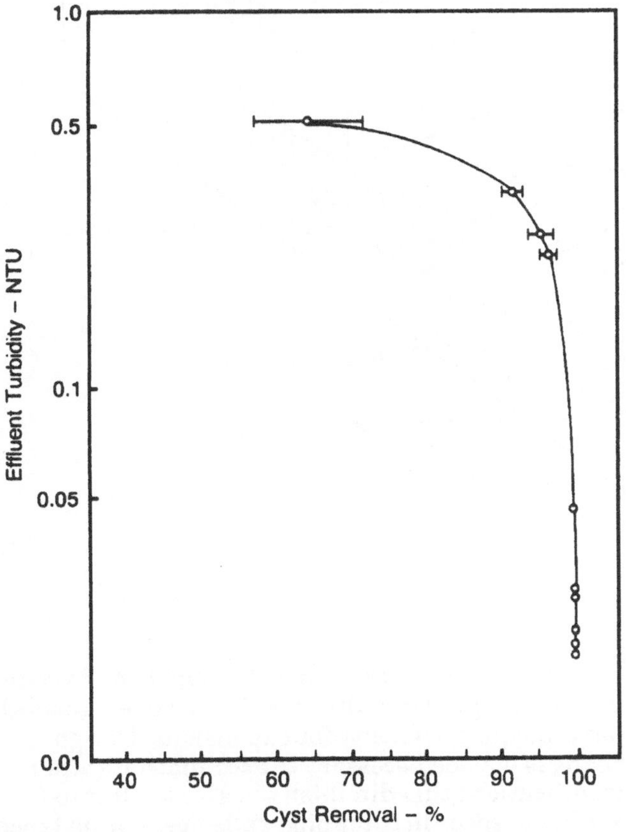

Figure 4-1. Relationship between cyst removal and filtered water turbidity.

Table 4-2. Removal Efficiencies Of *Giardia Lamblia* By Water Treatment Processes

Unit Process	Percent Removal
Rapid Filtration with Coagulation, sedimentation	96.9-99.9
Direct Filtration	
- with coagulation	95.9-99.9
- without coagulation	10-70
- with flocculation	95-99
Diatomaceous Earth Filtration	99-99.99[a]
Slow Sand Filtration	98-100[b]

[a] Aided by coagulation.
[b] Fully-ripened filter.

Table 4-3. Removal Efficiencies Of Viruses By Water Treatment Processes

Unit Process	Percent Removal
Slow Sand Filtration	99.9999
Diatomaceous Earth Filtration (with filter aid)	>99.95[a]
Direct Filtration	90-99
Conventional Treatment	99

[a] No viruses recovered.

Underdrains

The condition of the underdrain gravel, the coarse layer underneath the filter media, is difficult to assess in a filter evaluation. Removing filter media and visually inspecting the condition of the underdrain is the only reliable method of ascertaining whether it is satisfactory. However, indirect indications of underdrain failure include boils incurred during the backwash process or an uneven mounding of the filter surface. As a general practice, underdrain gravel is removed and replaced with new material along with the filter media. Frequently, the existing backwash water distribution system, which commonly consists of pipe laterals, is also replaced at the same time.

Surface Wash Systems

Surface wash systems are used to scour the upper layers of the filter during or prior to backwashing. These systems are more important for dual- and mixed-media filters than for sand filters. Surface washers prevent the formation of "mud-balls," which form within the filter media, most frequently at the sand-coal interface. Also, surface washers improve the process of cleaning dual- and mixed-media filters when polyelectrolytes are added, a practice used at filter flow rates of 2.7 to 4.1 L/sec/m^2 (4 to 6 GPM/sq ft) to control floc breakthrough.

Surface wash systems may be designed with one or two rotary arms. The single-rotary-arm systems position the arm at the filter surface. The two-arm system has one arm on the surface and one arm within the filter bed at the sand and coal interface in

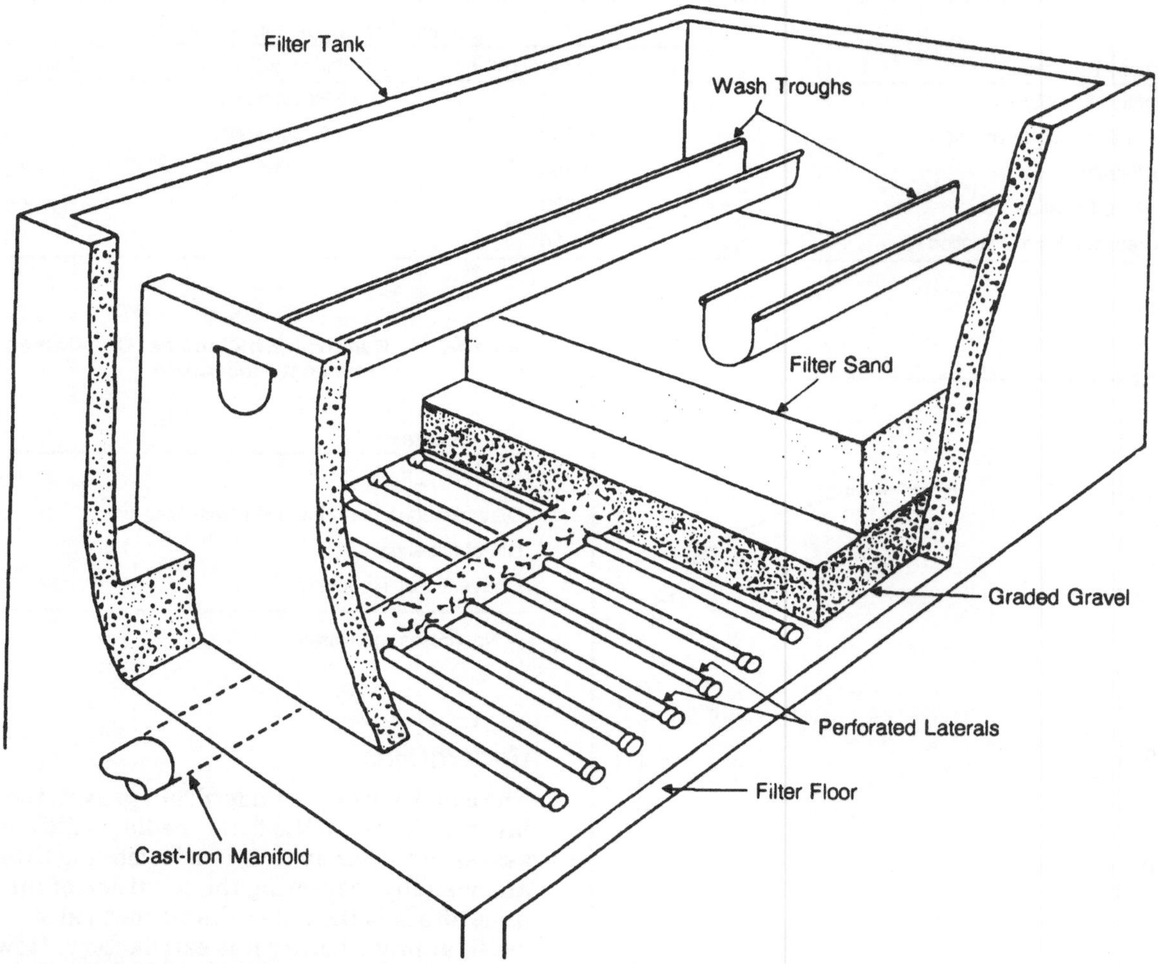

Figure 4-2. Cutaway view of typical rapid sand filter.

a dual media filter. The arms deliver a high velocity stream of water that violently agitates the filter media as it turns.

Air Scour Systems

Air scour systems can be used to clean rapid sand or dual-media filters, eliminating the need for surface wash systems. They require specially designed wash water distribution systems and generally do not use a coarse grade gravel support layer, because the violent mixing action caused by introducing air beneath a filter would upset a graded gravel underdrain system and destroy the filter bed. Instead, specially designed nozzles with retaining screens placed on a false floor in the filter box under the filter media are commonly used. The air introduced beneath either dual- or mixed-media filters destratifies the media. To return a destratified filter to service involves backwashing at 10.2 L/sec/m^2 (15 GPM/sq ft) to reclassify the filter media.

Piping and Controls

Filter effluent piping and filter flow rate controllers may be too small to accommodate expansion of a current system. Backwash water supply and waste piping in rapid sand filter plants, however, generally have adequate capacity for expansions. Design changes may be necessary to accommodate other modifications that diminish the already limited available space in filter pipe galleries. For instance, cast iron piping and gate valves can be replaced by more compact fabricated steel piping and less bulky butterfly valves.

Method of Filter Rate Control

There are two methods for controlling filter rates: constant rate filtration and variable declining rate filtration. Constant rate filtration is the most common method. While constant rate filters have provisions to control both the influent and effluent flow of each filter, variable declining control systems do not use filter rate-of-flow controllers and have a common filter effluent collection flume or pipe. In variable declining rate filtration, filtration rate declines as headloss builds up during the filtering process. In plants designed with multiple filters, filter influent flow frequently is directed automatically to another cleaner filter.

Variable declining rate filtration can be used where the capacity of an existing rapid sand filter is to be increased. Its major advantages over traditional constant rate filtration methods are its relatively low cost and its elimination of rate-of-flow controllers which require maintenance. Also, the filters are never operated at negative heads with this type of control, thus eliminating the potential for air binding. The major limitation is that the filtration rate of individual filters is uncontrolled at the beginning of a filter cycle.

Filtration Aid Application

Filtration aids prevent premature turbidity breakthrough by controlling floc penetration into the filter. Chemically, they are usually nonionic polyelectrolytes, and are very useful in maximizing the performance of high-rate filters and systems with cold influent water. These aids should be applied in systems that upgrade from sand to dual- or mixed-media filters. An optimal dosage of filtration aid minimizes the effluent turbidity until the filter achieves its maximum headloss. Typical filtration aid dosages are from 0.02 to 0.1 mg/L.

Addition of Polymer to Backwash

The addition of polymers to the backwash water can reduce the initial turbidity peaks during filter ripening following backwash and extend filter operation before breakthrough occurs. Polymer enhances the ability of the filter to retain turbidity particles. Adding polymer also improves the settling and thickening rate of the backwash solids. In addition, savings in backwash water can be achieved by precoating filters with polymer.

Polymer should be added during the first 5 to 7 minutes of the 10- to 15-minute backwash period. Polymer additions ranging from 0.1 to 0.15 mg/L have reduced the initial turbidity breakthrough by one half.

4.1.3 System Design Checklist

The following steps are recommended to maximize the efficiency of any filter modification:

- Use pilot test data to select filter size and medium; however, if pilot test data are not available, use data from analogous applications.
- Provide for the addition of disinfectant directly to filter influent.
- Provide for the addition of polyelectrolytes directly to filter influent for dual- or mixed-media filters.
- Provide for the continuous monitoring and recording of turbidity from each filter.
- Provide for the continuous monitoring and recording of flow and headloss from each filter.
- Provide for the housing of filter controls and pipe galleries.
- Use color-coded filter piping.
- Provide for the easy removal of pumps and valves for maintenance.
- Equip each filter with a filter-to-waste cycle for unsatisfactory water.
- Provide for filter cleansing with either surface wash or air scouring.
- Equip with an automatic filter backwash system.
- Select backwash rate based on filter media and expected wash water temperatures.
- Select backwash supply storage capacity to accommodate a minimum of two filter backwashes.
- Provide backup capacity to the largest single pump for backwashing and surface washing.
- Equip pressure filters with pressure and vacuum release valves.
- Equip backwash supply lines with air release valves.

Appendix A presents brief case histories of plant upgrades, discussing some of those steps in more detail.

4.2 Direct Filtration

Direct filtration systems are similar to conventional systems, but omit sedimentation. This section describes direct filtration technology.

4.2.1 Process Description

Direct filtration is an established technology that was developed because dual- and mixed-media filters are able to effectively process higher influent turbidities without the use of sedimentation. The direct filtration process is expected to be more widely used on water supplies that, up until now, only performed chlorination.

Direct filtration is only applicable for systems with high-quality and seasonally consistent influent supplies. The influent generally should have turbidity of less than 5 to 10 NTU and color of less than 20 to 30 units.

Direct filtration consists of several combinations of treatment processes. It always includes coagulation and filtration, and sometimes includes flocculation or a contact basin after coagulation addition. Typical coagulant dosages range from less than 1 to 30 mg/L. Cationic polymers often successfully coagulate water supplies and assist direct filtration. Nonionic polymers sometimes are added to the filtration step to increase filter efficiency. A flow diagram of a typical direct filtration system is shown in Figure 4-3.

In-line direct filtration, the simplest version of direct filtration, is commonly used in pressure filtration

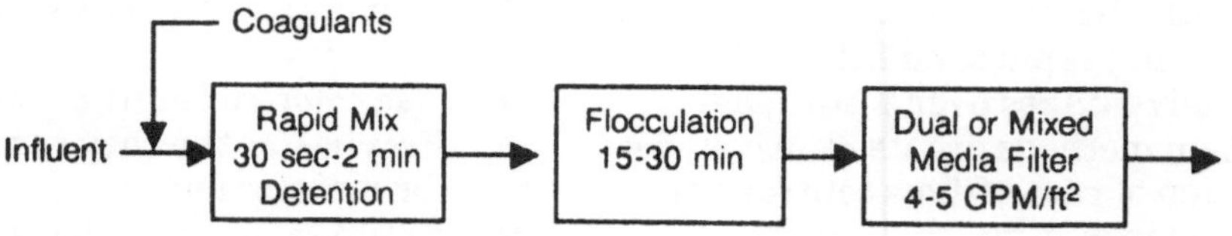

Figure 4-3. Flow diagram of typical direct filtration system.

systems (see Figure 4-4). In this process, chemical coagulant application and mixing are followed by addition of a nonionic polymer aid and filtration. There is no separate flocculation step. This treatment is effective and commonly used in the removal of iron and manganese from ground water when accompanied by appropriate coagulant doses. In-line direct filtration, however, is only applicable for surface waters with seasonal average turbidities of less than 2 to 3 NTU and free of contamination by wastewater sources (when permitted by regulatory agencies).

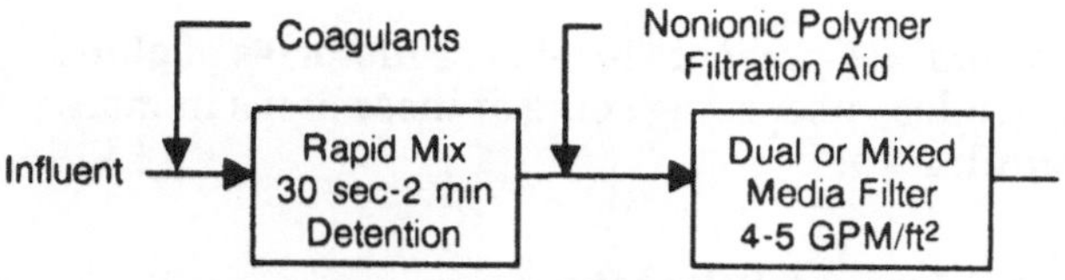

Figure 4-4. Flow diagram of in-line direct filtration system.

Another version of a direct filtration system is the "modified" system, which substitutes a contact basin for the flocculation basin found in the typical direct filtration process. Figure 4-5 is a flow diagram of modified direct filtration. The 1-hour detention contact basin serves primarily to condition the floc prior to filtration. Contact basins also provide pretreatment by decreasing turbidity peaks in the influent, providing silt and sand removal, and allowing for more equal distribution of coagulant.

4.2.2 System Performance

Direct filtration units can perform effectively within the following general influent parameters:

- Less than 500 total coliforms per 100 mL
- Less than 14 NTU of turbidity
- Less than 40 color units

While the direct filtration process is able to operate satisfactorily with influent turbidities as high as 14 NTU, optimally, influent turbidity should be less than 5 NTU. Effective direct filtration performance ranges from 90 to 99 percent and from 10 to 99.99 percent for virus and *Giardia* removal, respectively. The wide variation in direct filtration's *Giardia* removal efficiencies is due to the wide range of available system configurations. The most effective direct filtration configurations for *Giardia* removal must include coagulation.

4.3 Slow Sand Filtration

Slow sand filtration systems have a long history, having been used without disinfection at least since the 1850s in London, England. Slow sand filters were also commonly used in the United States in smaller facilities. Rapid sand filters have replaced many of them to accommodate the increase in water demand; consequently, slow sand filters are now not very common. According to the Surface Water Treatment Rule under the SDWA, many small systems could meet their regulatory filtration obligations with the simple slow sand filters. With the requirement for the multiple barrier disinfection approach in water treatment, slow sand filters are almost always accompanied by disinfection. Because slow sand filters with disinfection have not been used extensively, they are classified as "new" technology in the current literature.

Slow sand filters are similar to single media rapid rate filters (conventional systems) except that they:

- Are 50 to 100 times slower than normal filtration rates
- Use biological processes, which may enhance chemical/physical removal processes
- Require a ripening period before operation
- Use smaller pores between sand (i.e., smaller sand particles)
- Do not require backwashing
- Have longer run times between cleanings

Slow sand filters are most attractive for smaller systems with high quality raw water. Specifically, water which comes from a protected surface water supply, has previously received only chlorination as a treatment, contains less than 10 NTU, and has no color problems is the most suitable for slow sand filtration. While their operational simplicity makes them very suitable for small plants, slow sand filters are also applicable for medium to large plants.

The advantages of slow sand filtration include its simplicity and reliability, low cost, and ability to achieve greater than 99.9 percent *Giardia* cyst removal. In addition, these systems do not require continuous turbidity monitoring since they are applied to water supplies with relatively low turbidity.

Slow sand filters have several limitations, however. Due to a low filtration rate, these filters require

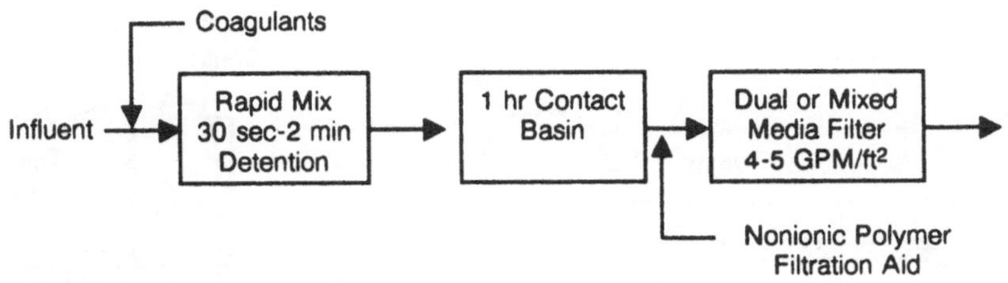

Figure 4-5. Flow diagram of modified direct filtration system.

relatively extensive land area. Without pretreatment (particularly coagulation), limitations exist on the quality of raw water that is suitable for treatment using slow sand filtration. Also because pretreatment is minimal or nonexistent at slow sand filter plants, other contaminants such as synthetic organic chemicals, disinfection by-product precursors, or inorganic chemicals are not readily removed. Systems with raw water containing high color or algae content are probably not appropriate for slow sand filtration, because these contaminants are not removed by slow sand filtration and the raw water will likely contain precursors for chlorination by-products.

4.3.1 System Design Considerations

Slow sand filters require influent that does not exceed the following parameters:

- Turbidity of <10 or up to 20 NTU depending on other operating characteristics
- Color of less than 30 units as prescribed by the Ten State Standards
- Algae of less than 5 mg per cubic meter of chlorophyll A

These are maximum limits, not typical operational parameters. Design flow rates range from 0.94 to 9.35 m^3/m^2 (1 to 10 million gallons per acre)per day, with the usual range from 2.8 to 5.6 m^3/m^2 (3 to 6 million gallons per acre)per day. Slow sand filters require sand with particle sizes ranging in effective diameters from 0.25 to 0.35 mm, with a uniformity coefficient of 2 to 3. A higher uniformity coefficient is acceptable for filters with pilot test verifications of the performance.

Typical underdrains for slow sand filters are made from split tile with laterals placed in coarse stone. These drains routinely discharge to a tile or concrete main drain. Recently constructed slow sand systems are equipped with perforated polyvinyl chloride pipe as laterals. Manually adjusted weirs, outlets, and valves are adequate for these systems. Inlet structures may be located at the end or side of the filter.

Slow sand filters perform poorly during the first 1 to 2 days of operation, called the "ripening period." The ripening period is the time required by the filter after a cleaning cycle to become a functioning biological filter. Although *Giardia* removal is not usually affected by the ripening period, the overall poor water quality during this period requires provision of a filter-to-waste cycle.

These filters require continuous operation under submerged conditions, ranging in depth from 0.9 to 1.5 m (3 to 5 ft). Hydraulic filter outlet controls keep the filters submerged at all times. The difference between the level of water in the collection gallery and the level of water above the filter media is called the headloss through the media. The initial headloss is about 0.06 m (0.2 ft). Maximum headloss should be less than the submerged depth of 1.2 to 1.5 m (4 to 5 ft) to avoid air binding and the uneven flow of water through the filter medium. The buildup of the maximum headloss is slow, taking up to 6 months.

Redundant, or stand-by, systems are required to accommodate the extended cleaning periods associated with slow sand systems.

In climates subject to below freezing temperatures, slow sand filters usually must be housed. Unhoused filters in harsh climates develop an ice layer that prevents cleaning during the winter months. Uncovered slow sand filters will operate effectively, however, if influent turbidities are low enough to permit the filter to operate through the winter months without cleaning. An illustration of an unhoused system is provided in Figure 4-6; Figure 4-7 shows an example of a housed system.

Because of filter housing costs, slow sand filters are most appropriate for small systems. Due to this expense, large systems can usually only utilize slow sand filters when they are located in moderate climates and therefore do not require housing. The slow sand filter in Salem, Oregon, is an example of a large system that is able to use unhoused slow sand filters because of climate.

4.3.2 Operation and Maintenance

The primary operational consideration for a slow sand filtration system is maintaining a clean filter. Cleaning becomes necessary when headloss reaches about 1.2 m (4 ft). The normal length of time between cleanings ranges from 20 to 90 days, but will vary depending on raw water quality, sand size, and

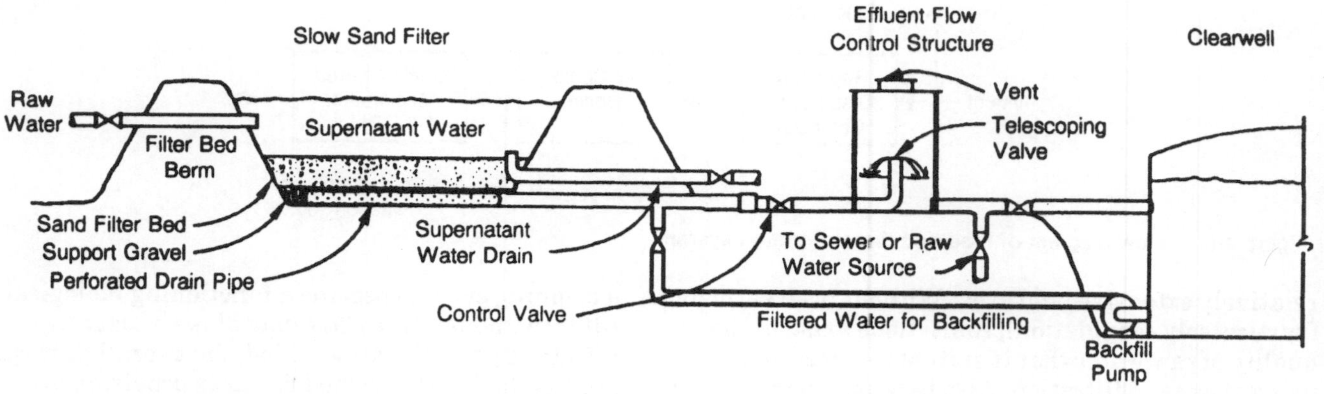

Figure 4-6. Typical unhoused slow sand filter installation.

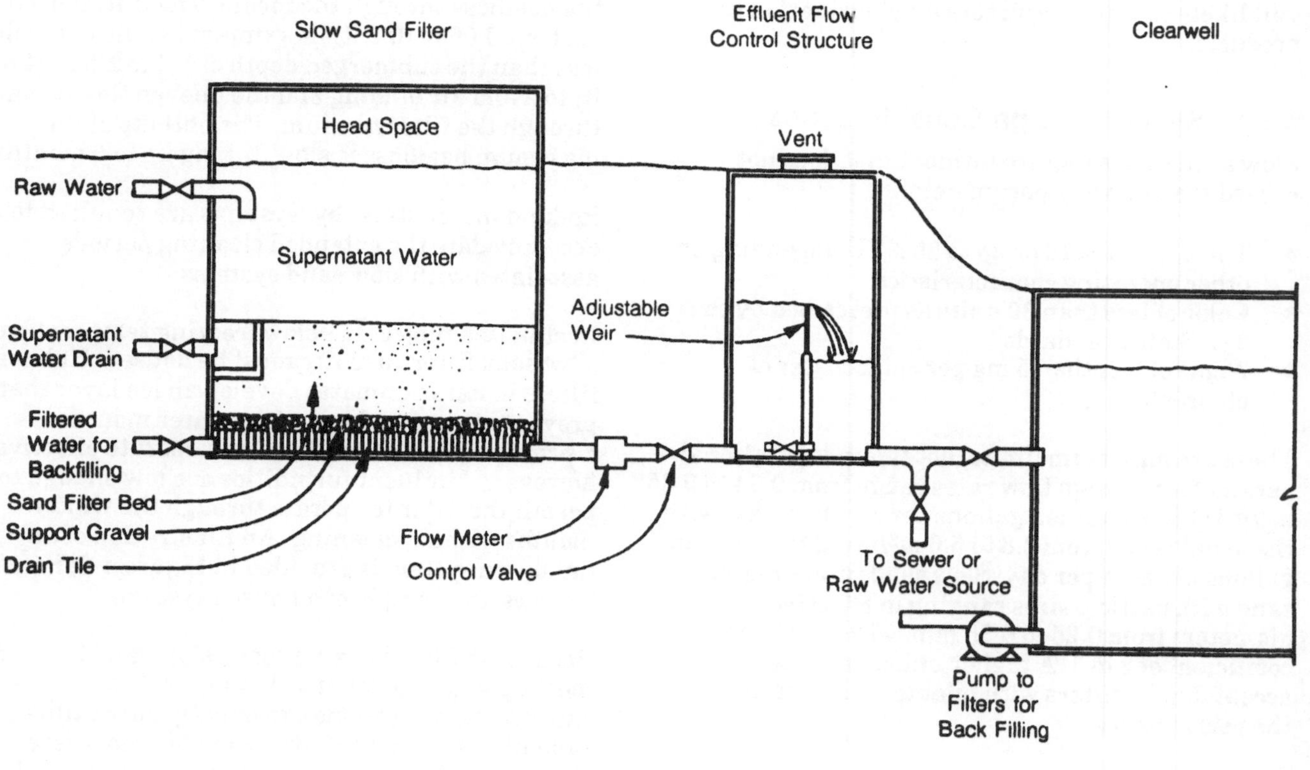

Figure 4-7. Typical housed slow sand filter installation.

filtration rate. Filters should be cleaned every 1 to 2 months.

Cleaning involves manually scraping off the top 2.5 to 5 cm (1 to 2 in) of the filter media. The removed sand is either discarded or treated separately. Most small systems use manual cleaning techniques, but very large systems may use mechanical scrapers.

A number of precautions must be observed in regard to cleaning operations. Cleaning filters necessitates removing the filter from service, after which a ripening period is required to bring the filter back into operation. Operators must minimize intermittent operation by coordinating maintenance tasks. While cleaning, operators should take care to avoid disturbing the filter bed during backwashing, shock chlorination (if practiced), and raking the surface of the filter bed. Scouring and subsequent erosion of the filter media from refilling after cleaning must also be avoided. Gradual filling from the bottom of the filter during the startup phase partially addresses this problem. Excessive drying of the filter beds also should be avoided during cleanings. During the postcleaning ripening period, filtered water should be directed to waste. Operators should periodically remove algae from the filter and, to accommodate seasonal changes, reduce filtration rates during the winter months.

If sand depth drops below 61 cm (24 in), new sand needs to be added. Bed depths of only 30.5 to 50.8 cm (12 to 20 in) have had poor performance records. Replacing sand, however, is not considered a normal operational or maintenance task because with careful cleaning, resanding may be necessary only once every 10 years.

In addition to the above maintenance considerations, slow sand filters require daily inspection, control valve adjustment, and turbidity monitoring. The filter housing structure, if present, also requires routine maintenance.

4.3.3 System Performance

Slow sand filtration units perform best within the following general influent parameters:

- Less than 800 total coliforms per 100 mL
- Less than 10 NTU of turbidity
- Less than 5 color units

Effective slow sand filtration performance, with respect to virus and *Giardia* removal, ranges from 91 to 99.9999 percent and 100 percent, respectively.

4.3.4 System Costs

Slow sand filter cost data are presented in Table 4-4. Smaller plants are typically designed with cast-in-place concrete structures with wood or concrete slab covers. The piping is either cast iron or steel and below grade. Flow meters are used to monitor the output for each filter. A below-grade concrete clear-well is another cost element included in Table 4-4.

Larger plants, on the other hand, use uncovered earthen berm structures. These on-grade plastic membrane-lined structures have all below-grade piping, and some plants use above-grade steel tanks. Flow meters are also used for each filter in the effluent control structure.

The process energy requirements of slow sand filters are negligible assuming they have gravity-fed source water. The housing structure energy requirements include heating, lighting, and ventilation, and are directly related to the size of the structure. Maintenance supply requirements are also negligible since the filters are assumed to operate virtually unattended. Replacement sand is not included because it is needed so infrequently.

4.4 Package Plant Filtration

Package plants are categorized as "new" treatment technology. They are not altogether different processes from other treatments mentioned in this section, although several models contain treatment elements that are innovative, such as adsorptive clarifiers. The primary distinction, however, between package plants and custom-designed plants is that package plants are built in a factory, skid mounted, and transported virtually assembled to the operation site.

Package plants are designed to effectively remove turbidity and bacteria from surface water with generally consistent low-to-moderate turbidity levels. While package plants can treat influent streams with highly variable characteristics, they would require more skilled operators and closer attention.

There are about 650 to 700 package plants operating in the United States with capacities ranging from 27.3 m^3/day to 0.18 m^3/sec (7,200 GPD to 4 MGD). Many are built to conventional design standards. Others, using tube settlers, have reduced size and larger capacities. More recently, package plants have been built with efficient and compact adsorptive clarifiers. Package plants with adsorptive clarifiers have capacities ranging from 54.5 m^3/day to 0.26 m^3/sec (14,400 GPD to 6 MGD).

The four major advantages of package plants are their compact size, cost effectiveness, relative ease of operation, and design for unattended operation. Typically, these types of filtration plants are used to treat small community water supplies and for a variety of special applications, including:

- Emergency supplies
- Recreational areas
- State parks
- Construction sites
- Ski areas
- Military installations
- Other areas not served by municipal supplies

Package plants can differ widely with regard to design criteria, and operating and maintenance considerations.

Table 4-4. Estimated Costs For Supplementing Surface Water Treatment By Slow Sand Filtration ($1978)

Category	Plant Capacity, MGD	Average Flow, MGD	Capital Cost, $1,000	Operation and Maintenance Costs		Total Cost, ¢/1,000 gal
				$1,000/yr[a]	¢/1,000 gal	
1	0.026	0.013	145	0.8	17.9	377.8
2	0.068	0.045	273	1.6	9.9	205.1
3	0.166	0.133	508	5.1	10.5	133.4
4	0.5	0.40	603	9.0	6.2	54.7
5	2.5	1.30	1,213	20.5	4.3	34.3
6	5.85	3.25	2,573	38.0	3.2	28.7
7	11.59	6.75	4,782	62.3	2.5	25.3

[a] Processes include slow sand filters and clearwell storage. Sand filters in Categories 1 through 3 are constructed of concrete and are covered. Sand filters in Categories 4 through 7 are constructed of membrane-lined earthern berms and are uncovered.
1 MGD = 0.044 m^3/s; 1,000 gal = 3.78 m^3.
Source: U.S. EPA (1979).

4.4.1 Selecting a Package Plant System

Package plant systems are most appropriate for plant sizes ranging from 94.6 to 22,710 m^3/day (25,000 to 6,000,000 GPD). Common treatment elements in a package plant are chemical coagulation, flocculation, settling, and filtration.

Some package plant manufacturers have used new treatment technologies to improve their product's performance, for example, tube settlers, adsorptive clarifiers, and high flow rate mixed- or dual-media filters. Package plants employing innovative technologies sometimes encounter regulatory barriers, because many existing State design standards recognize only conventional treatment processes.

Influent water quality is the most important consideration in determining the suitability of a package plant application. Complete influent water quality records should be examined to establish turbidity levels, seasonal temperature fluctuations, and color level expectations. Both high turbidity and color may require coagulant dosages beyond many package plant design specifications. In cases of consistently high levels of turbidity and color, the package plant capacity must be down-rated or a larger model selected. Where turbidity exceeds 100 to 200 NTU, presedimentation may be required as a pretreatment. Pilot tests may be necessary to select a package plant for more innovative designs using high flow rates and shorter detention time unit processes.

4.4.2 System Description and Design Considerations

The three basic types of package plant filter systems are conventional package plants, tube-type clarification package plants, and adsorption clarifier package plants.

Conventional Package Plants

Conventional package plants are manufactured by several firms to a variety of specifications. As their name indicates, they contain the conventional processes of coagulation, flocculation, sedimentation, and filtration. Typical design standards for these units are:

- 20- to 30-minute flocculation detention time
- 2-hour sedimentation detention time
- Rapid sand filters rated at 1.34 L/sec/m^2 (2 GPM/ft^2)

Tube-Type Clarifier Package Plants

A flow diagram of a tube-type clarifier package plant is illustrated in Figure 4-8. This type of plant has two versions with different capacity ranges; one version can treat from 0.63 to 6.3 L/sec (10 to 100 GPM) and the other, equipped with dual units, can treat from 12.6 to 88.3 L/sec (200 to 1,400 GPM).

In these package systems, the disinfectant, primary coagulant, and polyelectrolyte coagulant aid are added before the influent enters the flash mixer. After the flash mixer, the water enters the flocculation chamber where mechanical mixers gently agitate the water for 10 to 20 minutes depending on the flow.

The flocculated water then enters the tube settlers, which consist of many 2.5-cm (1-in) deep, 99-cm (39-in) long split-hexagonal-shaped passageways. The large surface area of the many 2.5-cm (1-in) deep tube settlers achieves an effective clarification overflow rate of less than 6.1 m^3/day/m^2 (150 GPD/ft^2). Adequate clarification is attained with less than 15-minute detention times.

The clarified water then enters a gravity flow mixed-media filter. A constant filtration rate is maintained

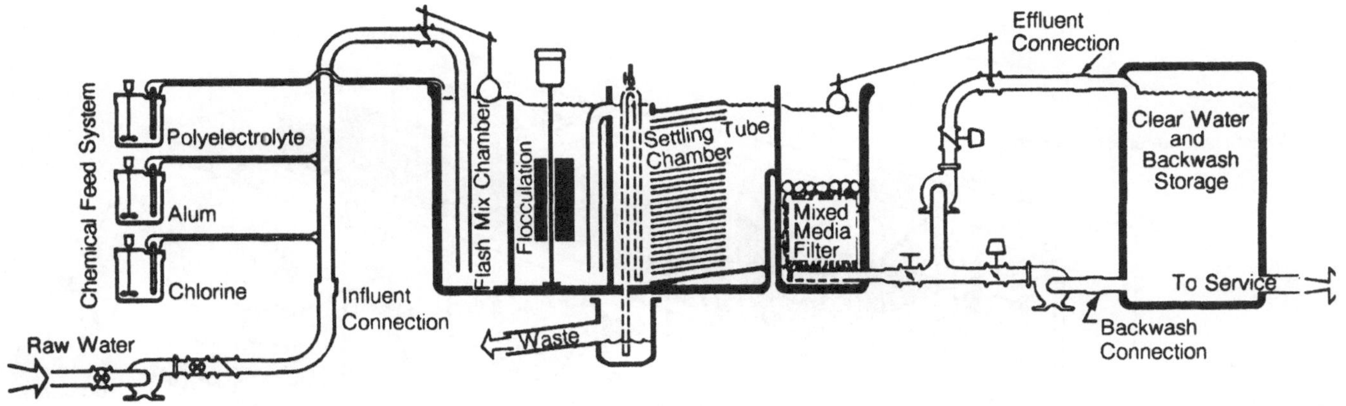

Figure 4-8. Flow diagram of a package plant.

by a low-head filter effluent pump discharging through a float operated, level control valve. After a preset headloss is reached, backwashing of the filter is initiated automatically. A manual backwash cycle can be initiated any time (if desired). Settled sludge from the tube settlers is flushed during the backwashing cycle. Combining backwashing and tube settler flushing simplifies operations and reduces operator skill requirements.

Adsorption Clarifier Package Plant

The adsorption clarifier package plant is a new concept developed in the early 1980s. It uses an upflow filter with low density plastic bead media (called the adsorption clarifier) followed by a mixed-media filter to complete the water treatment. The adsorption clarifier replaces the flocculation and the sedimentation basins, combining flocculation and sedimentation into one step. A typical example is contained in Figure 4-9.

While passing through the adsorption media, the coagulant and water are mixed, contact flocculated, and clarified. The mixing intensity, as measured by the mean temporal velocity, ranges from 150 to 300 per second. Flocculation is accomplished through turbulence as the water passes through the adsorption media. In addition, flocculation is enhanced by contact between the flocculated materials and the floc-coated media. Turbidity is reduced through adsorption of the coagulated and flocculated solids onto the adsorption media and the previously adsorbed materials. The adsorption clarifier can achieve 95 percent or greater removal at 6.8 L/sec/m^2 (10 GPM/ft^2). This highly efficient clarification method results in extremely compact designs.

Adsorption clarifiers are cleaned by a combination of air scouring followed by water flushing. The air scouring starts the cleaning process for the plastic media used in the clarifier. Adsorption clarifier cleaning is initiated more frequently than filter backwashing because more solids are removed by the clarifier. The clarifier cleaning process is automatically initiated either by a timer or a pressure switch that continuously monitors headloss across the adsorption media.

The air introduced under the adsorption media causes a vigorous scrubbing action. The scrubbing action dislodges solids, which are washed away by the flow of the incoming water. Flushing is generally timed to occur between every fourth and eighth hour. Complete cleaning of the adsorption media is not desired because performance is enhanced by some residual solids. Diagrams illustrating the various cycles of an adsorption clarifier package plant are contained in Figure 4-10.

4.4.3 Operation and Maintenance

Package plant operation is simplified by automated features, and maintenance requirements are well documented in manuals. However, the operator needs to be well acquainted with water treatment principles and the plant manual, and should have attended a comprehensive training session.

Common automated devices found in package plants are effluent turbidimeters and chemical feed controls. The effluent turbidimeters and fail-safe controls are built into many plants to ensure that the finished water does not exceed set turbidity levels. Automated chemical feed systems are especially appropriate for plants without full-time operators or with highly variable influent characteristics.

Typical plant operator and maintenance manuals contain operating principles, methods of establishing proper chemical dosages, operating instructions, and troubleshooting guides.

Periodic visits by the manufacturer to make adjustments to the plant and inspect the equipment

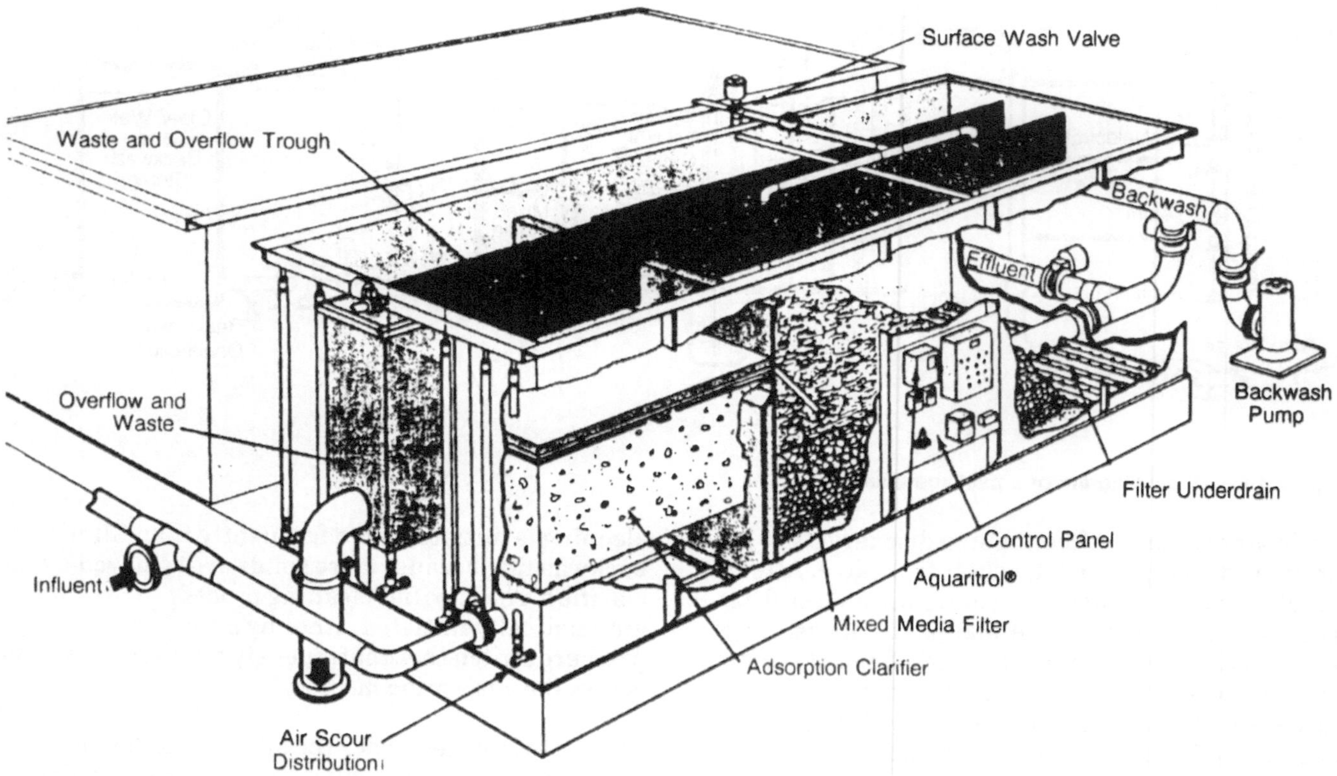

Figure 4-9. Adsorption clarifier package plant (courtesy of Microfloc products).

operations and performance are recommended. The first visit should be no more than 6 months after initial operation; the next should follow in another 6 months. Subsequently, annual visits should be sufficient.

4.4.4 System Performance

According to extensive manufacturer evaluations, system performance, in general, is improving because of better equipment, more highly skilled operators, and greater surveillance by regulators. Table 4-5 contains summaries of evaluations at three plants in Oregon. All finished water reported in the table had less than 0.53 NTU turbidity.

4.4.5 System Costs

Table 4-6 presents capital, operating, and maintenance costs for package plants. Costs range from \$0.14 to \$2.50/m^3 (\$0.52 to \$9.44/1,000 gal).

4.5 Diatomaceous Earth Filtration

Diatomaceous earth filtration, also known as precoat or diatomite filtration, relies on a layer of diatomaceous earth about 0.3-cm (1/8-in) thick placed on a septum or filter element. The septums may be placed in pressure vessels or operated under a vacuum in open vessels. A schematic diagram of a typical pressure system is shown in Figure 4-11.

Diatomaceous earth filters are effective in removing cysts, algae, and asbestos. For water supplies with low amounts of suspended solids, they have lower initial costs than conventional rapid sand filtration systems. Diatomaceous earth filters are especially effective against *Giardia* cysts.

The difficulties in maintaining a perfect film of diatomaceous earth as the filtering layer have discouraged wide usage of diatomaceous earth filters for potable water treatment, except in favorable circumstances (i.e., waters with low turbidity and low bacterial counts). Consequently, while this technology has been used extensively in specialized applications, such as swimming pools, it is a "new" technology for water supply treatment.

Diatomaceous earth filter plants have been chosen for projects with limited initial capital, and for emergency or standby capacity to service large seasonal increases in demand. Since these systems are most suitable for applications where influent is low in turbidity and bacterial counts, water supplies presently receiving just chlorination may consider using diatomaceous earth to meet the filtration requirements of the SWTR.[1]

[1] "Ten State Standards" refers to the publication: *Recommended Standards for Water Works: Policies for the Review and Approval of Plans and Specifications for Public Water Supplies*, 1982, by Committee of the Great Lakes/Upper Mississippi River Board of Sanitary Engineers, published by Health Education Service, P.O. Box 7283, Albany, New York 12224.

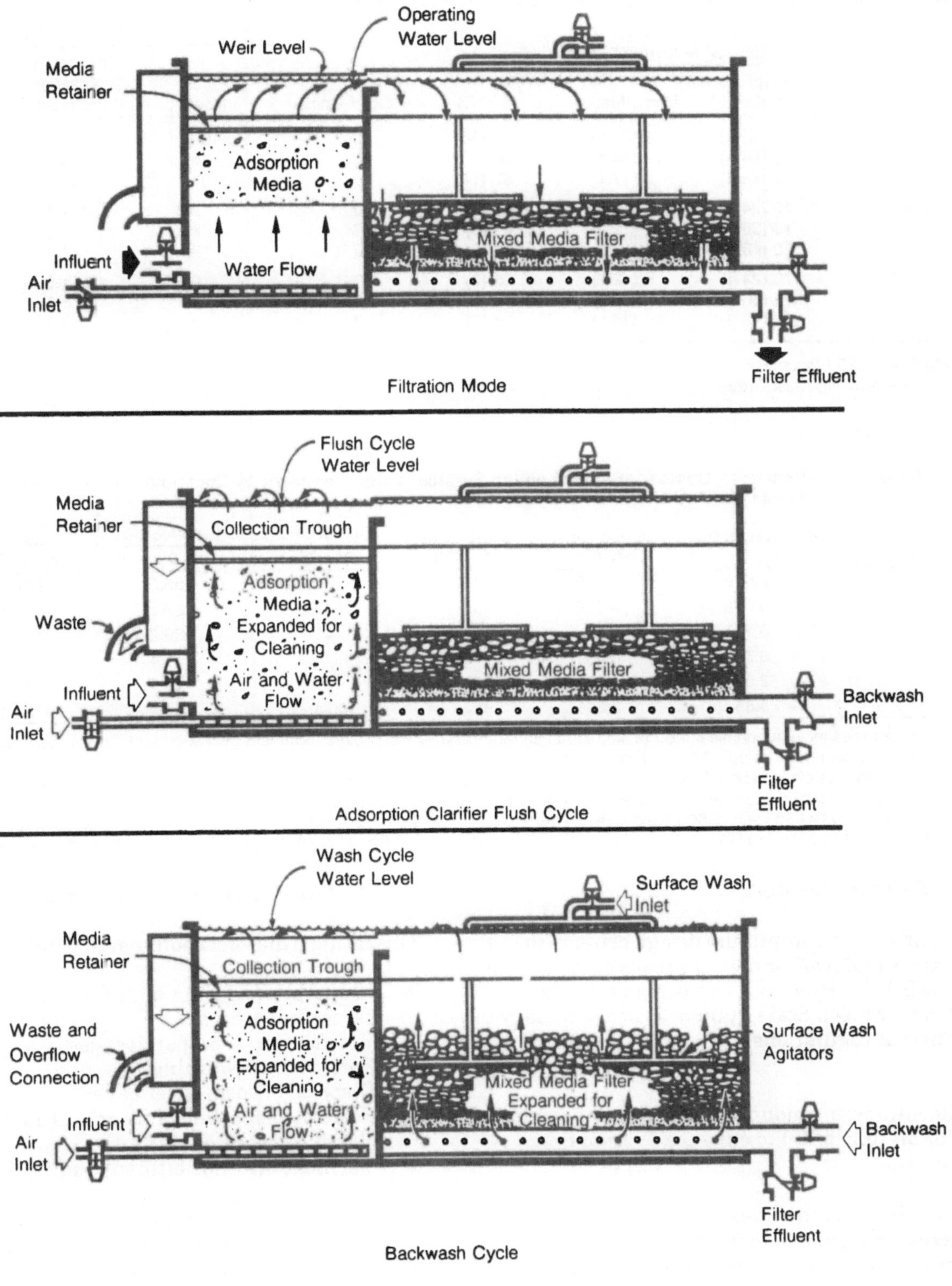

Figure 4-10. Operating cycles of package plant (courtesy of Microfloc products).

Table 4-5. Summary Of Results Of Adsorption Clarification Package Plants

Location	Rates, GPM/ft² (m/h) Adsorption Clarifier	Mixed-Media	Water Temp., °F (°C)	Color Units	Turbidity, NTU Influent	Adsorption Clarifier Effluent	Mixed-Media Effluent
Corvalis, OR	5 (12.2)	9 (22.0)	68 (20)	0	100	8	0.27
	10 (24.4)	10 (24.4)	68 (20)	0	46	10	0.52
	8 (19.5)	7 (17.1)	57 (14)	0	103	21	0.30
Rainier, OR	10 (24.4)	5 (12.2)	41 (5)	30	9.3	1.4	0.24
	15 (36.6)	5 (12.2)	41 (5)	30	8.2	1.6	0.22
	20 (48.8)	5 (12.2)	41 (5)	30	8.1	1.7	0.20
Newport, OR	10 (24.4)	5 (12.2)	45 (7)	8	19	4.3	0.13
	15 (36.6)	7.5 (18.3)	45 (7)	8	15	3.7	0.11
	20 (48.8)	10 (24.4)	45 (7)	9	9	3.8	0.23

1 GPM/ft² = 0.679 L/sec/m².
Source: Microflow Products Group.

Table 4-6. Estimated Costs for Supplementing Surface Water Treatment by Complete Treatment Package Plants ($1978)

Category	Plant Capacity, MGD	Average Flow, MGD	Capital Cost, $1,000	Operation and Maintenance Costs $1,000/yr[a]	¢/1,000 gal	Total Cost, ¢/1,000 gal
1	0.026	0.013	278	12.2	255.2	944.5
2	0.068	0.045	295	15.9	87.5	277.4
3	0.166	0.13	428	42.4	89.2	195.1
4	0.50	0.40	773	75.1	51.4	113.6
5	2.50	1.30	1,770	137	29.0	72.8
6	5.85	3.25	2,952	274	23.1	52.4

[a] Processes include chemical feed (alum, soda ash, and polymer); complete treatment package plant (flocculation, tube settling, and mixed media filtration); backwash storage/clearwell basins; and sludge dewatering lagoons. A separate pumping station is used to transmit unthickened sludge to the sludge dewatering lagoons in Categories 5 and 6. Sludge pumping is included in the cost of the package plant in Categories 1 through 4.
1 MGD = 0.044 m³/sec; 1,000 gallons = 3.78 m³
Source: U.S. EPA (1979).

4.5.1 System Design

The majority of the minimum design criteria in the Ten State Standards for diatomaceous earth systems meet the SWTR. However, two design criteria in addition to the Ten State Standards are necessary to meet current regulations:

1. The minimum amount of filter precoat should be 1 kg/m² (0.2 lb/ft²) to enhance *Giardia* cyst removal.

2. The minimum thickness of the precoat should be increased from 0.3- to 0.5-cm (1/8- to 1/5-inch) (found to be more important than the size graduation of the diatomaceous earth), also to enhance *Giardia* cyst removal.

An additional recommendation is to use coagulant (alum or a suitable polymer) to coat the body feed to improve removal rates for viruses, bacteria, and turbidity. Adding these chemicals to the coating does not improve *Giardia* removal rates.

4.5.2 Operation and Maintenance

Operating a diatomaceous earth filter requires:

- Preparation of filter body feed (diatomaceous earth) and precoat
- Verification of proper dosages
- Periodic backwashing
- Disposal of spent filter cake
- Periodic inspection of filter septum for cleanliness and damage
- Verification of the effluent quality

A common operating difficulty is maintaining a complete and uniform thickness of diatomaceous earth on the filter septum.

In some cases, alum precoating of the diatomaceous earth can improve performance. Typical alum doses range from 1 to 2 percent by weight (1 to 2 kg/100 kg of diatomaceous earth applied). Typical precoats of diatomaceous earth of 0.49 to 0.98 kg/m² (0.1 to 0.2 lb/ft²) are applied to prepare the filter. These filters also require a continuous supplemental body feed of diatomite because the filter is subject to cracking. If

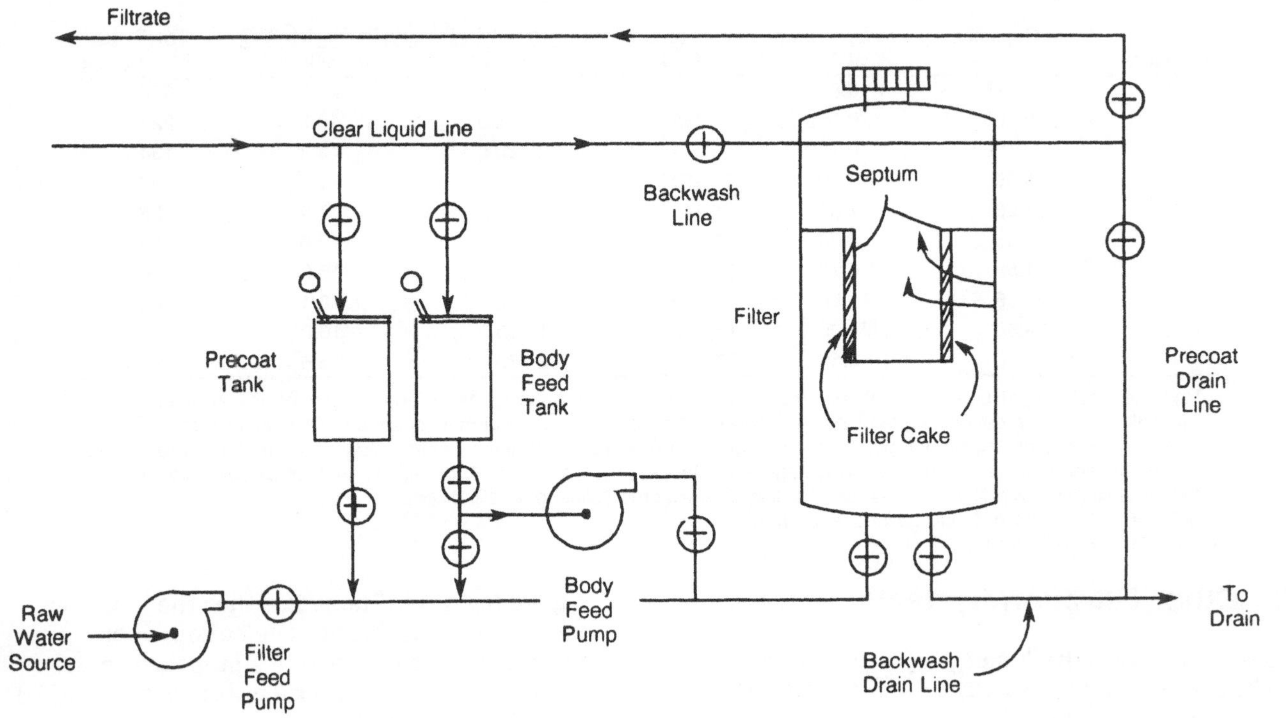

Figure 4-11. Typical pressure diatomaceous earth filtration system.

the filter has no body feed, there will be rapid increases in headloss due to buildup on the surface of the filter cake. Body feed rates must be adjusted for effective turbidity removal. Diatomaceous earth filters do not need a filter-to-waste cycle, because of the precoating process.

Regular cleaning will maintain up to 95 percent of the filter septum area available for filtration after 100 filter runs. The filter cake drops off the septum during an interruption in flow, such as occurs during cleaning. During operating interruptions, clean diatomaceous earth and filter water should be used to recoat the filter to reduce the potential for passage of pathogens.

Filter runs typically range from 2 to 4 days. The rate of the body feed and the diatomite media size are critical in determining the filter run length. Shorter filter runs will minimize filtered water odor and taste problems stemming from the decomposition of organic matter trapped in the filter.

Vacuum diatomaceous earth filters are a variation of this technology that offer the advantages of visibility during backwashing and of not requiring pressure vessels. Their primary disadvantage is that they run an increased risk of the release of gases in the filter cake that shorten filter runs.

4.5.3 System Performance

Diatomaceous earth filtration units perform well on waters with low influent turbidity and bacteria levels. Effective removals of viruses and *Giardia* range up to 99.95 percent and from 99 to 99.99 percent, respectively.

Some researchers have found that diatomaceous earth filters, with added operational steps, are effective in removing polioviruses. The additional steps include coating the filter surface with filter aid or mixing the filter influent with a cationic polymer. The limited research found:

- No viruses were detected in 11 effluent samples from a 12-hour run of a filter coated with 1 mg of cationic polymer per gram of diatomaceous earth.
- Only 1 of 12 effluent samples contained viruses during the operation of another filter coated with 1 mg of cationic polymer per gram of diatomaceous earth.
- No viruses were detected in the effluent in 12 samples from another 12-hour run of an uncoated filter, and the influent was mixed with 0.14 mg of cationic polymer per liter of water.

4.5.4 System Costs

Costs for diatomaceous earth filters are provided in Table 4-7. Costs range from \$0.09 to \$1.78/m^3 (\$0.35 to \$6.73/1,000 gal) depending on plant size.

Table 4-7. Estimated Costs for Supplementing Surface Water Treatment by Direct Filtration Using Diatomaceous Earth ($1978)

Category	Plant Capacity, MGD	Average Flow, MGD	Capital Cost, $1,000	Operation and Maintenance Costs		Total Cost, ¢/1,000 gal
				$1,000/yr[a]	¢/1,000 gal	
1	0.026	0.013	221	6.0	127.0	672.9
2	0.068	0.045	285	8.0	43.7	227.2
3	0.166	0.130	374	20.0	42.2	134.7
4	0.50	0.40	570	30.4	20.8	66.6
5	2.50	1.30	1,573	128	27.0	66.0
6	5.85	3.25	2,538	214	18.0	43.1
7	11.59	6.75	4,433	369	15.0	36.1
8	22.86	11.50	10,713	762	18.1	48.1
9	39.68	20.00	15,982	1,165	16.0	41.7
10	109.90	55.50	37,733	2,730	13.5	35.4

[a] Processes include pressure diatomaceous earth filtration units, diatomaceous earth feed equipment; filtered water storage clearwell; and sludge dewatering lagoons. A separate administration, lab, and maintenance building is included in Categories 5 through 10. Sludge pumps are included in the package facilities used in Categories 1 through 4, but separate sludge pumping stations are included in Categories 5 through 10. Categories 8 through 10 include sludge holding tanks, sludge dewatering with filter presses, and hauling of dewatered solids to landfill disposal.

1 MGD = 0.044 m^3/sec; 1,000 gallons = 3.78 m^3

Source: U.S. EPA (1979).

4.6 Other Filtration Systems

This section covers the "emerging technologies" of membrane and cartridge filtration. Since there are limited operational data concerning these technologies, pilot and case studies assume more importance.

4.6.1 Membrane Filtration

Membrane filtration, also known as ultrafiltration, is extremely compact and does not require coagulation. Membrane filters use hollow fiber membranes to remove undissolved, suspended, and emulsified solids. The membranes are typically able to exclude all particles greater than 0.2 microns from the water stream.

Membrane filters are typically used for specialized applications that require highly purified water, and often serve as:

- Pretreatment processes for reverse osmosis
- Pretreatment to remove colloidal silica from boiler feed water
- Treatment for drinking water supplies with influent turbidities of 1 NTU or less, and fouling indexes of less than 10

Typically, ground water and surface water of high clarity have fouling indexes of less than 10. Fouling of the hollow fibers by turbidity is the major problem preventing widespread application of this technology.

Traditional membrane filters introduce feed water to the inside of the hollow fiber membrane, with the permeate or filtrate emerging on the outside of the membrane. State-of-the-art membrane filters are designed to pass influent to either the inside or outside of the membrane. The hollow fiber membranes are contained in a pressure vessel or cartridge and operate over a pressure range of 10 to 100 psi. Contaminants collect on the end of the hollow fiber and are discharged to waste by reversing the water flow. A sample membrane system is shown in Figure 4-12.

Periodic backflushing and occasional chemical cleaning is necessary to maintain the membrane fibers. Treatment of the flush water containing solids requires separate coagulation and clarification. The clarified flush water is either recycled or discharged after treatment. The sludge collected from this type of system is typically dried and disposed of in a landfill.

Customarily, 90 percent of the feed water is collected as permeate; the other 10 percent is discharged along with the contaminants. These filters can be designed to exclude particles larger than 0.01 μm. Unlike reverse osmosis, this process does not exclude inorganic salts and electrolytes. Hollow fiber filters with the finest membranes remove bacteria, *Giardia*, and some viruses.

The hollow fiber membranes vary in size and porosity and in their corresponding effectiveness, yet all membrane filter fibers are sensitive to influent concentrations of suspended and colloidal solids. Specifically, the flux level (the volume of permeate produced per unit area of membrane filter per day) and the flux stability are affected by:

- Quality of the influent
- Filter cycle duration
- Quality of the backwashing water

In general, influent water with a fouling index of less than 10 will permit filter cycles of 8 hours with a 10

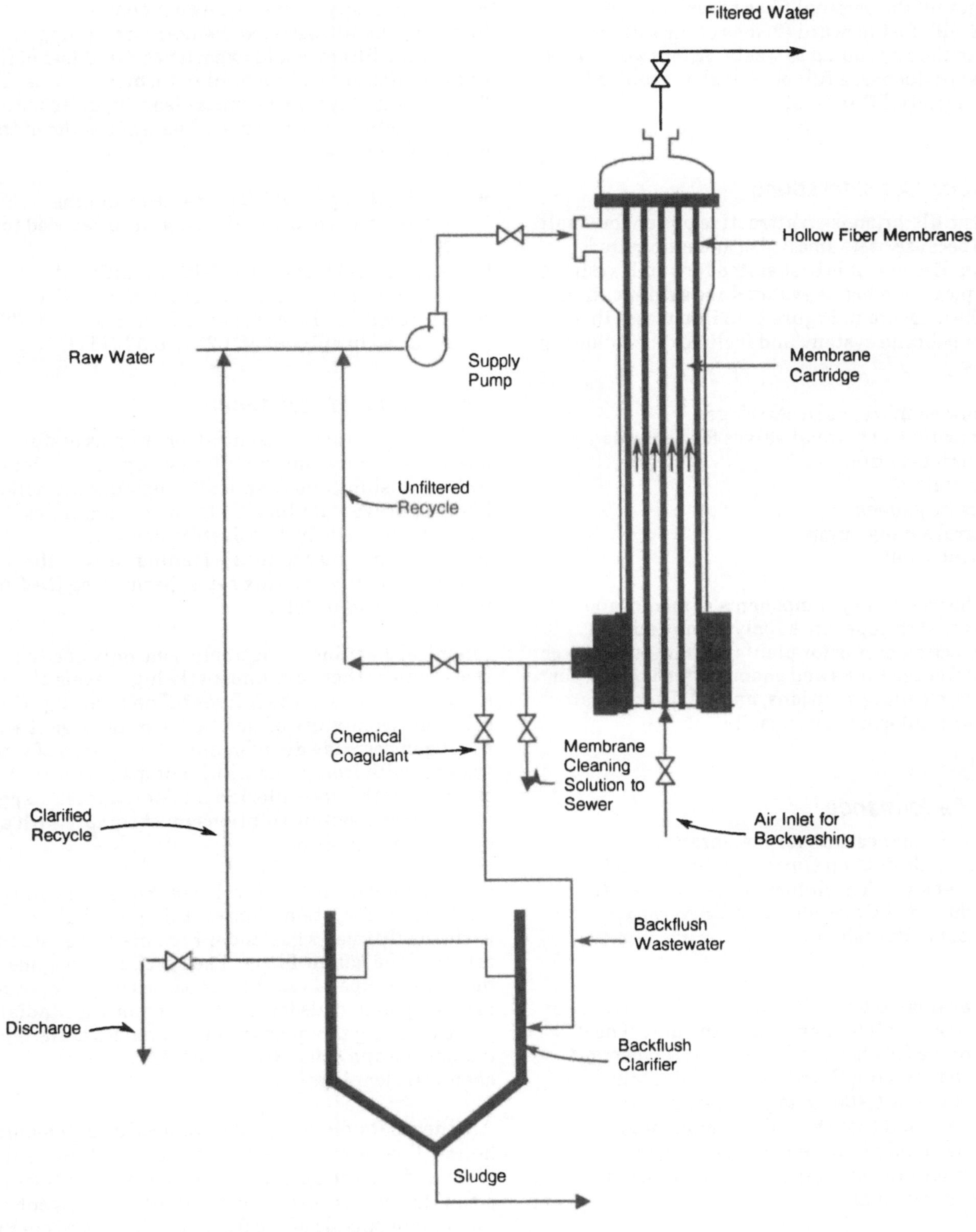

Figure 4-12. Flow diagram of membrane filtration system.

percent reduction in flux. In between backflushes, a high degree of the original fiber porosity can be retained with fast forward flushes of the influent, which can then be routed to waste. After two to three fast forward flushes, a full backflush is required to restore the initial flux level.

Application Considerations

Membrane filtration is an attractive option for small systems because of its small size and automated operation. Because it is best suited for small systems, many typical membrane systems are skid-mounted units. The diagram in Figure 4-13 illustrates this type of membrane system, and includes the following elements:

- Hollow membranes in cartridges
- Automatic and manual valves for backwashing and unit isolation
- Flow meters
- Pressure gauges
- Integral supply pump
- Control panel

Three other necessary components of membrane systems are (1) a separate supply pump and interconnecting piping for plants with multiple filter units, (2) storage tanks and chemical feed pumps for membrane cleaning solutions, and (3) filtered water storage with chlorination capacity.

System Performance

System performance data for membrane filter systems include data on *Giardia*, coliform, and turbidity removal. The following results were from tests conducted at Colorado State University's Department of Pathology.

One test evaluated *Giardia* removal effectiveness for one manufacturer's hollow membrane unit. The 0.2 μm membrane filters were found to be 100 percent effective in removing *Giardia* cysts at influent concentrations of 1,100 cysts per liter. In addition, researchers found that the unit's "radial pulse" cleaning mechanism was effective in preventing membrane fouling and caused no reductions in *Giardia* cyst removal.

Coliform removal was evaluated for a manufacturer's hollow membrane unit. This evaluation used water seeded with *Escherichia coli* bacteria ranging from 20 to 30 million organisms per 100 mL. During a 130-minute test run, the effluent contained less than 1 coliform bacteria per 100 mL. No membrane breakthrough was experienced during three test runs.

Turbidity removal was also evaluated by Colorado State University. Prior to development of intermittent, air-assisted membrane cleaning, membrane filters would experience rapid loss of flux or flow rates at medium or high turbidity levels. The "radial pulse" system permits cleaning units without such reductions in flux rates. The test results of this system found that:

- Turbidity up to 30 NTU created from the introduction of bentonite clay were reduced to 0.2 NTU.
- Actual field tests with 190 NTU influent turbidity resulted in effluent with 0.6 NTU.
- Tests with effluent ranging from 2.4 to 3.0 NTU resulted in effluent of 0.25 to 0.57 NTU.

Operation and Maintenance

Effective cleaning is essential for the successful operation of a membrane filter system. It is achieved by backflushing and chemically cleaning the hollow fibers to prevent fouling. Chlorine is sometimes added to the backflush to destroy bacteria. Backflushing and chemical cleaning restore the original porosity and flux rates, permitting the filter to operate indefinitely.

Chemical cleaning is routinely done only once a week, unless there are unusually high levels of suspended solids in the influent. The cleaning solution is a mixture of caustic-based detergent and hydrogen peroxide disinfectant. Membranes of water systems with iron in their influent may require cleaning with hydrochloric acid. Methods for disposal of the spent cleaning solutions must comply with all applicable regulations.

One manufacturer has developed the self-cleaning "radial pulse" system. This cleaning innovation periodically injects gas under high pressure into the center of the hollow fibers. The specially designed membrane expands as the gas passes through, thus removing materials fouling the membrane. Radial pulse cleaning prolongs the effectiveness of the membrane and thus extends the time between chemical cleanings.

Another manufacturer has developed a self-cleaning hollow-fiber membrane filter system that uses a flow of water from the outside to the inside of the fibers to clean the membranes. A water stream is tangentially introduced outside of the fiber. The filter system then collects filtrate in the interior of the fiber, and the rejected stream of concentrate is either recycled or discharged. Continuous recycling permits extremely high filtered water recovery rates of up to 99 percent. This system uses a more porous fiber than the more traditional inside-to-outside design that is only about 70 percent porous with a 0.2 μm nominal pore size. The higher porosity of these membranes allow for

Figure 4-13. Skid-mounted membrane filtration assembly.

treatment of influent with higher suspended solids concentrations. These filters will meet turbidity standards, exclude all *Giardia* cysts, and remove coliform and other bacteria, but will not remove viruses.

System Costs

Cost data for membrane filter units for small systems are contained in Table 4-8. The total costs range from \$0.37/m^3 for the largest systems to \$1.21/m^3 for the smallest systems (\$1.38 to \$4.56/1,000 gal).

Potential Problems

The primary application concerns with membrane filters are membrane failure and organics removal effectiveness. Pilot testing is needed to qualify each application, because of the lack of experience with the process. Membrane failure is a major concern because the membrane is the only barrier between potentially pathogenic microbial contaminants and the finished water. Most other treatment systems have multiple barriers to pathogenic breakthrough. To guard against this potential problem, some membrane systems include features that trigger an operational shutdown or activate an alarm. One manufacturer includes a manual or automatic device to verify the integrity of each membrane.

The second concern is that membrane filters may not be effective in removing certain organic components. Larger membranes of about 0.2 µm will not effectively remove organic materials contributing color. Smaller inside-to-outside membranes of about 0.1 µm will remove smaller particles, but when influent contains organic materials that contribute to color, such as humic or fulvic acids, supplemental treatments may be necessary.

Pilot testing of membrane systems is generally necessary to establish design criteria and operating parameters. The paucity of available data necessitates pilot testing. The only known membrane filter used for water supply treatment is in Keystone, Colorado. Most membrane experience has been in the medical and electronic fields where tap water is treated for special high purity applications.

4.6.2 Cartridge Filtration

Cartridge filters are considered an emerging technology suitable for removing microbes and turbidity in small systems. These filters are mechanically simple but manually operated, so they could be expensive to operate. They also require low turbidity influent. Cartridge filters use ceramic or polypropylene microporous filter elements that are packed into pressurized housings. They operate by the physical process of straining the water through porous membranes and can exclude particles down to 0.2 µm. The pore sizes that are suitable for producing potable water range from 0.2 to 1.0 µm. The ease of operation and maintenance of cartridge filters makes them very attractive for small systems.

Table 4-8. Estimated Costs for Supplementing Surface Water Treatment by Package Membrane Filtration Plants ($1978)

Category	Plant Capacity, MGD	Average Flow, MGD	Capital Cost, $1,000	Operation and Maintenance Costs		Total Cost, ¢/1,000 gal
				$1,000/yr[a]	¢/1,000 gal	
1	0.026	0.013	142	5.0	105.2	455.6
2	0.068	0.045	269	9.8	53.7	226.8
3	0.166	0.130	503	26.0	54.7	179.2
4	0.50	0.40	1,144	67.7	46.4	138.4

[a] Processes include a complete package membrane filtration unit, clearwell storage, and sludge dewatering lagoons with liquid sludge hauling.
1 MGD = 0.044 m^3/sec; 1,000 gallons = 3.78 m^3
Source: U.S. EPA (1979).

One manufacturer uses single microporous ceramic filter elements packaged together in a cartridge housing with flow capacities of up to 1.5 L/sec (24 GPM). This filter has pore sizes as small as 0.2 µm and is suitable for influent with moderate levels of turbidity, algae, and microbial contaminants. The clean filter pressure drop is about 3.2 kg/cm^2 (45 psi), while the pressure drop during cleaning reaches about 6.2 kg/cm^2 (88 psi).

Application Considerations

Roughing filters, for pretreatment prior to cartridge filtration, are sometimes necessary to remove larger suspended solids and prevent the rapid fouling of the cartridges. Roughing filters can be rapid sand filters, multimedia filters, or fine mesh screens.

Prechlorination is recommended to prevent surface-fouling microbial growth on the cartridge filters and reduce microbial pass-through. Except for chlorine, no chemical additions are necessary. There is a lack of data concerning the effectiveness of cartridge filters for viral removal.

Operation and Maintenance

These systems are operationally simple, apart from cleaning and membrane replacement. There is no need for skilled personnel; personnel are necessary only for general maintenance. Ceramic membranes may be cleaned and used for repeated filter cycles. Polypropylene cartridges become fouled relatively quickly and must be replaced with new units.

In one manufacturer's unit, cleaning the ceramic cartridge filters entails cleaning each vertical filter element with a hand-operated hydraulically driven brush that fits over the element. Some manufacturers use disposable polypropylene filter elements in multi-cartridge stainless steel housing to avoid the cleaning procedures. This type of unit is available with capacities ranging from 0.13 to 45.4 L/sec (2 to 720 GPM). The primary disadvantage of the disposable polypropylene membrane is that it can only be cleaned once before disposal. Manufacturers' guidelines state that these filters have service periods ranging from 5 to 20 days with influent turbidities of 2 NTU or less, depending on the pore size of the filter. Another manufacturer incorporates particles of silver into the ceramic filters to prevent bacterial growth.

4.7 Selecting the Appropriate Filtration Treatment System

This section discusses considerations in selecting an appropriate filtration technology. First, it discusses the steps involved in selecting any filtration system. Next, it discusses the role and objectives of pilot studies, flocculation and sedimentation studies, and filtration studies in selecting the specific operating characteristics for selected filtration technologies.

4.7.1 Steps in an Evaluation

The first step in selecting a new water treatment technology is to review all raw water quality data to establish the requirements for the potential alternatives. Next, a list of alternative technologies should be compiled. The considered alternatives must be able, in theory or as proven under similar conditions, to solve the problems identified with the current filtration system. Once the potential alternatives are selected, one must determine the necessity of pilot or bench-scale tests. If the desired performance of one or more of the alternatives is in doubt, testing is appropriate. (Testing is always useful if time and budget allow.) Otherwise, literature surveys, bench-scale studies, or pilot test results can be used to derive performance characteristics and design considerations for each alternative. For small systems, the alternatives for a particular application may include slow sand filters, package plants, diatomaceous earth filtration, or membrane filters.

Following this initial selection, the basic process concerns for the various alternatives should be identified, including the following:

- Turbidity removal performance
- Color removal performance
- *Giardia* removal performance
- Cleaning cycle frequency
- Necessary chemicals

- Chemical dosages
- Requisite operational skill
- Applicable regulatory standards
- Necessary sludge management

In the next stage, conceptual designs and preliminary layouts for selected alternatives are prepared. Also, construction costs for the alternatives should be developed. One should compare all alternatives for reliability, simplicity, flexibility, and ease of implementation, to select the appropriate application. Finally, one can proceed with the final design.

4.7.2 Need for Pilot Studies

A pilot study is a broad term used for small-scale testing of either complete water treatment processes or merely individual processes. Pilot studies are used to evaluate alternate treatment options and operating techniques. Pilot tests can be relatively short or very long in duration. For example, time requirements for pilot tests of rapid filtration are about 2 weeks; for slow sand filtration, 2 to 4 months; and for corrosion inhibitors, up to 6 months. These tests may be continuous or intermittent. The longer, more involved tests are more expensive. To avoid unnecessary costs, pilot tests should have clearly delineated objectives to prevent including extraneous evaluations. Yet they should be performed long enough to encounter the full range of raw water conditions and process design parameters.

Simple jar tests are usually sufficient to evaluate procedures for traditional coagulation, flocculation, sedimentation, and filtration processes. However, new or innovative technologies usually require the more extensive pilot tests.

Pilot tests have been endorsed by field experts for many years to assess precise design specifications, operational recommendations, and chemical applications procedures. In addition, pilot tests are required for some technologies to adjust individual processes to specific local water conditions. Pilot tests are also used to identify unforeseen design and operating problems, demonstrate operation to regulatory authorities, and develop better information concerning capital and operating costs. Pilot studies may also be a prerequisite to obtaining conditional regulatory approval. They are especially needed for new and emerging technologies and for accepted technologies with innovative elements, such as tube settlers or mixed-media filters.

Pilot studies are important to ensure the suitability of a small-scale prototypical plant for a particular application. This is especially true for raw water with difficult treatment aspects or poor quality, such as highly polluted water with high concentrations of organics, iron, manganese, and algae. Pilot tests are also particularly necessary for plants that use new short-detention and high-rate processes or where direct filtration is being considered.

4.7.3 Flocculation and Sedimentation Studies

Full-scale flocculation and sedimentation studies are necessary because it is difficult to extrapolate the performance of pilot-scale flocculation and sedimentation tanks to full-scale plants. Since flocculator efficiency is directly related to volume, smaller flocculators perform more efficiently than their full-scale counterparts. Consequently, pilot flocculation studies do not provide results applicable to full-scale facilities.

Traditional sedimentation basins present even greater extrapolation problems. Since small-scale versions cannot duplicate the process of traditional basins, which are generally 2.7 to 4.5 m (9 to 15 ft) deep, full-scale sedimentation studies are also necessary. However, tube settlers can be evaluated on a pilot scale. Figure 4-14 shows a sample test module for a settling tube. These pilot-scale units can be suspended from existing basin walls or collection launders and operated at various flow rates.

4.7.4 Filtration Studies

Filtration studies can successfully employ pilot tests. They are necessary for plants considering direct filtration and very useful for plants with unusual raw water characteristics. One of the problems that can be identified and evaluated with pilot tests is the presence of diatoms (filter clogging algae) that do not show up as high turbidity, yet can clog filters. Another such problem involves the presence of paper fiber, a common situation for water intakes below paper plant effluent outfalls. These fibers also may not show up as high turbidity but present filter clogging problems. Filtration pilot tests establish whether the presence of paper fiber or diatoms will cause problems.

Side-by-side pilot filters will provide useful comparisons of different filter media designs being considered for a particular application. A schematic of a pilot filter is shown in Figure 4-15. Each of the three filters can be operated at the same flow rate and is designed to maintain a constant flow as headloss increases. Pilot filters also include provisions to measure headloss. Measurements of the filtered water turbidity and filter headloss are two of the most useful results of side-by-side pilot tests to predict filtration efficiency and filter run length.

Pilot tests are especially recommended when high-rate clarification/filtration processes are being considered. When such systems have short contact times, a significant risk of poor performance is

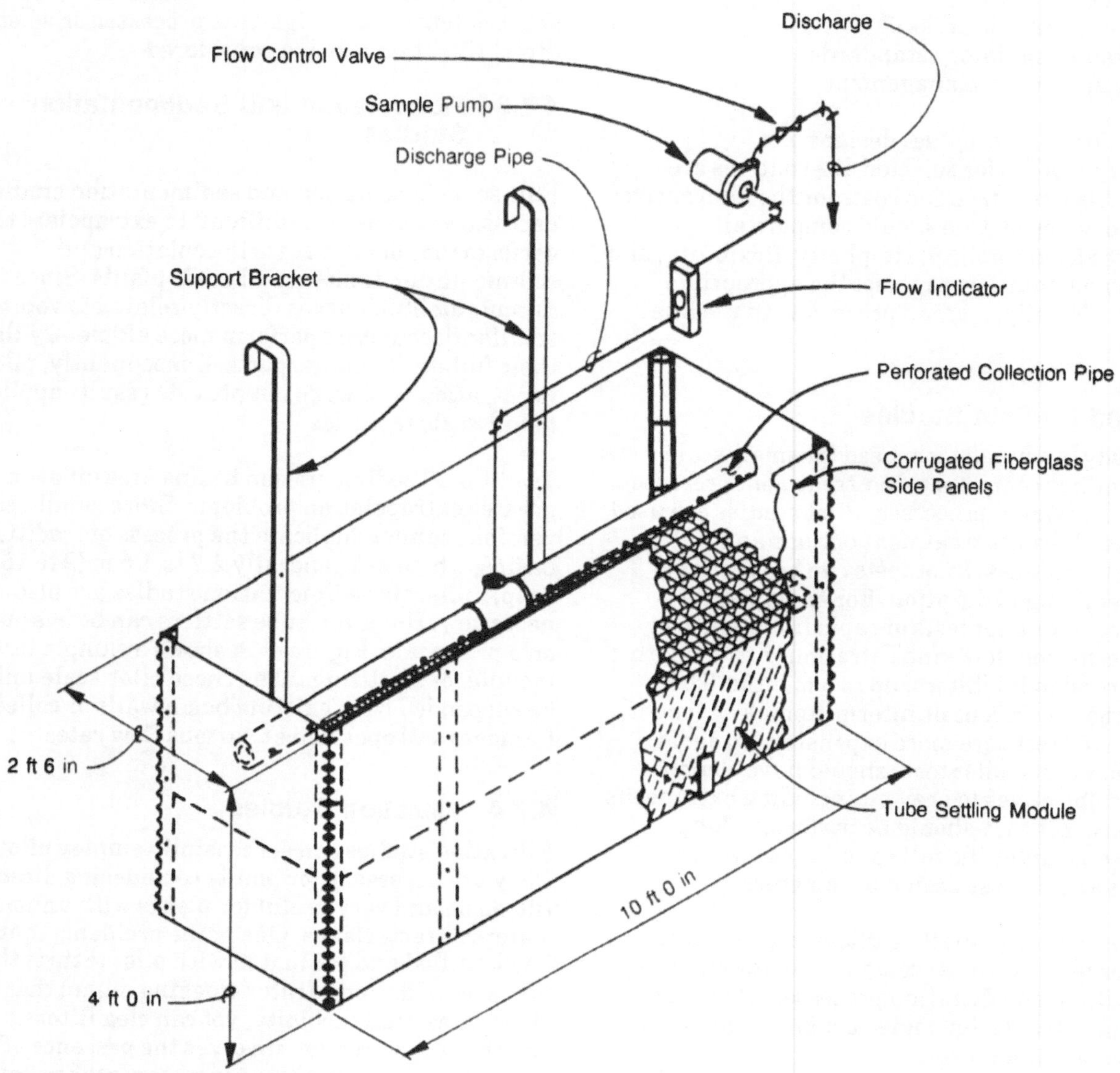

Figure 4-14. Tube settling test module.

present if a maladjustment of the system occurs. Due to the small permissible margin of error, pilot tests of actual site conditions are necessary to supplement system designs with extensive operational and pilot test histories.

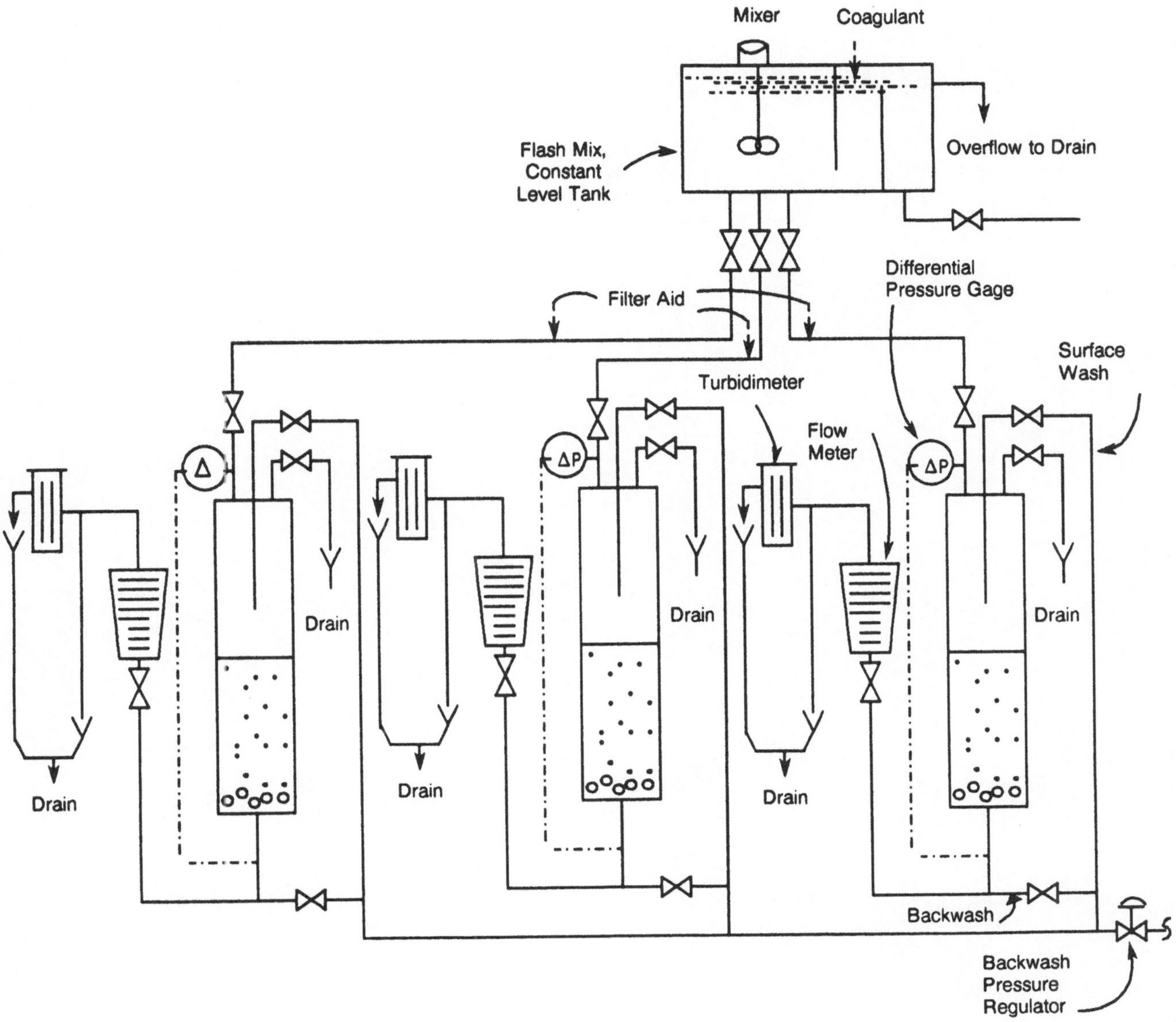

Figure 4-15. Pilot filter schematic.

Chapter 5

Disinfection and Disinfection By-Products

Disinfection is a process that kills or inactivates pathogenic microorganisms that occur in drinking water supplies. Water treatment plants employ both primary and secondary disinfection. Primary disinfection achieves the desired level of microorganism kill or inactivation. Secondary disinfection ensures a stable residual concentration of disinfectant in the finished water to prevent microbial growth in the distribution system. Chlorine, ozone, chlorine dioxide, and ultraviolet (UV) radiation are the major primary disinfectants; chlorine and monochloramine are the major secondary disinfectants. Some disinfectants can be used for both processes.

The 1986 Amendments to the Safe Drinking Water Act (SDWA) require all public water suppliers to disinfect drinking water. In addition, inorganic and organic chemicals will be regulated by means of Maximum Contaminant Levels (MCLs). Since some disinfectants can produce chemical by-products, the dual objective of disinfection is to provide the required level of organism destruction while remaining within the MCLs for by-products set by EPA.

Chlorine has been the most widely used disinfectant in the United States; however, it produces trihalomethanes (THMs) and other halogenated organic compounds in drinking water. Because of this, water suppliers are beginning to utilize other disinfectants, such as ozone, chlorine dioxide, and monochloramine, or combinations of disinfectants, such as ozone followed by chlorine. Research on the potential by-products associated with these other disinfectants and combinations of disinfectants is ongoing.

This chapter discusses the various disinfection technologies used today as well as the issues relating to disinfection by-products. Section 5.1 discusses the objectives of disinfection and Section 5.2, the problem of disinfection by-products and strategies for their control. The advantages and disadvantages of the major disinfectants are discussed in Section 5.3. Sections 5.4 and 5.5 provide an overview of primary and secondary disinfection technologies. Appendix B provides case histories of water treatment plants using various disinfection technologies.

5.1 The Objectives of Disinfection

According to the Amendments to the SDWA, all public water suppliers, including those that rely on ground water, will have to disinfect drinking water before distribution. To assure compliance with all applicable regulations (both current and anticipated), the specific objectives of disinfection are to:

- Assure 99.9 percent (3 log) and 99.99 percent (4 log) inactivation of *Giardia lamblia* cysts and enteric viruses, respectively
- Assure control of other harmful microorganisms
- Not impart toxicity to the disinfected water
- Minimize the formation of undesirable disinfection by-products
- Meet the Maximum Contaminant Levels (MCLs) for the disinfectants used and by-products that may form

Disinfection alone, or a combination of disinfection and filtration, can achieve the minimum mandatory removals and/or inactivations of 99.9 percent *Giardia* cysts and 99.99 percent enteric viruses. Primary disinfection systems that use ozone, chlorine, or chlorine dioxide can achieve greater than the above-stated inactivation of enteric viruses when 99.9 percent inactivation of *Giardia* cysts is attained. Therefore, achieving sufficient *Giardia* cyst inactivation can assure adequate inactivation of both types of organisms. This is not the case, however, when using chloramination because it is such a poor virucide. Pilot-scale tests must be conducted by the utility to assure both inactivations are achieved when relying on chloramination as a primary disinfectant.

Conventional treatment, which includes coagulation, flocculation, sedimentation, and filtration, along with disinfection, can achieve 99.9 percent inactivation of *Giardia* cysts and 99.99 inactivation of enteric viruses when properly designed and operated. Direct filtration, slow sand filtration, and diatomaceous earth filtration systems, each combined with disinfection, have also achieved these reductions.

Ground-water systems that apply disinfection to comply with regulations may have to add filtration if they contain iron and manganese. Insoluble oxides form when chlorine, chlorine dioxide, or ozone are added to these systems; thus, filters would be needed for their removal. In addition, both ozonation and chlorination may cause flocculation of dissolved organics, thus increasing turbidity and necessitating filtration. The presence of such insolubles will require the use of secondary disinfection after filtration as well.

5.1.1 CT Values

"CT values" indicate the effectiveness of disinfectants in achieving primary disinfection. They describe the attainable degree of disinfection as the product of the disinfectant residual concentration (in mg/L) and the contact time (in minutes). For chlorine, chlorine dioxide, or monochloramine, the contact time can be the time required for the water to move from the point at which the disinfectant is applied to the point it reaches the first customer (at peak flow). This is the total time the water is exposed to the chlorinous residual before being used. Ozone, however, has a short half-life in water; therefore, the contact time is considered the time water is exposed to a continuous ozone residual during the water treatment process only.

The Final Surface Water Treatment Rule (SWTR) (U.S. EPA, 1989b) states:

> Systems may measure "C" (in mg/L) at different points along the treatment train, and may use this value, with the corresponding "T" (in minutes), to calculate the total percent inactivation. In determining the total percent inactivation, the system may calculate the CT at each point where "C" was measured and compare this with the $CT_{99.9}$ value (the CT value necessary to achieve 99.9 percent inactivation of *Giardia* cysts) in the rule for specified conditions (pH, temperature, and residual disinfectant concentration). Each calculated CT value (CT_{calc}) must be divided by the $CT_{99.9}$ value found in the SWTR tables to determine the inactivation ratio. If the sum of the inactivation ratios, or
>
> $$\Sigma\, CT_{calc}/CT_{99.9}$$
>
> at each point prior to the first customer where CT was calculated is equal to or greater than 1.0, i.e., there was a total of at least 99.9 percent inactivation of *Giardia lamblia*, the system is in compliance with the performance requirement of the SWTR.

The final Guidance Manual for the SWTR is expected to recommend that systems determine contact time based on the time it takes water with 10 percent of an approximate tracer concentration (T_{10}) to appear at the sampling site at peak hourly flow. For ground water not under direct influence of surface water, CT is determined in the same manner using enteric viruses or an acceptable viral surrogate as the determinant microorganism, since *Giardia* cysts will not be present.

Table 5-1 presents the CT values required to attain 1-log reductions of *Giardia* cysts, for four disinfectants. As shown, lower temperatures require higher CT values; with chlorine, an increase in pH also increases necessary CT values. If more than one disinfectant is used, the percent inactivation achieved by each is additive and can be included in calculating the total CT value.

Table 5-1. CT Values for Achieving 90 Percent Inactivation of *Giardia Lamblia*

		Temperature					
Disinfectant	pH	≤1°C	5°C	10°C	15°C	20°C	25°C
Free Chlorine[a] (2 mg/L)	6	55	39	29	19	15	10
	7	79	55	41	28	21	14
	8	115	81	61	41	30	20
	9	167	118	88	59	44	29
Ozone	6-9	0.97	0.63	0.48	0.32	0.24	0.16
Chlorine dioxide	6-9	21	8.7	7.7	6.3	5	3.7
Chloramines[b] (preformed)	6-9	1,270	735	615	500	370	250

[a] CT values will vary depending on concentration of free chlorine. Values indicated are for 2.0 mg/L of free chlorine. CT values for different free chlorine concentrations are specified in tables in the Guidance Manual.

[b] To obtain 99.99 percent inactivation of enteric viruses with preformed chloramines requires CT values > 5,000 at temperatures of 0.5, 5, 10, and 15°C.

Source: U.S. EPA (1989a).

When direct filtration is included in the water treatment process, disinfection credit can be taken by the filtration step for a 2-log inactivation of *Giardia* cysts and a 1-log inactivation of viruses. This means that the primary disinfectant must provide an additional 1-log inactivation of *Giardia* cysts and 3-log inactivation of viruses. In the specific instance of a conventional treatment process that includes coagulation, flocculation, sedimentation, and filtration, an inactivation credit of 2.5 logs for *Giardia* cysts and 2 logs for viruses may be taken. This means that the primary disinfectant must provide an additional 0.5 log inactivation of *Giardia* cysts but a 2-log inactivation of viruses.

If a water supply system does not use filtration, the 99.9 percent inactivation of *Giardia* and 99.99 percent inactivation of enteric viruses must be achieved by the primary disinfecting agents alone. Table 5-2 presents CT values for the four disinfectants for achieving 99.9 percent reductions of *Giardia* cysts. Table 5-3 presents the CT values for virus inactivation. Although ground-water disinfection regulations have not been finalized, these values will probably apply to systems treating ground water determined by the State not to be under direct influence of surface water.

Table 5-2. CT Values for Achieving 99.9 Percent Inactivation of *Giardia Lamblia*[a]

		Temperature					
Disinfectant	pH	≤1°C	5°C	10°C	15°C	20°C	25°C
Free	6	165	116	87	58	44	29
Chlorine[b]	7	236	165	124	83	62	41
(2 mg/L)	8	346	243	182	122	91	61
	9	500	353	265	177	132	88
Ozone	6-9	2.9	1.9	1.4	0.95	0.72	0.48
Chlorine dioxide	6-9	63	26	23	19	15	11
Chloramines[c] (preformed)	6-9	3,800	2,200	1,850	1,500	1,100	750

[a] These CT values for free chlorine, chlorine dioxide, and ozone will guarantee greater than 99.99 percent inactivation of enteric viruses.
[b] CT values will vary depending on concentration of free chlorine. Values indicated are for 2.0 mg/L of free chlorine. CT values for different free chlorine concentrations are specified in tables in the Guidance Manual (U.S. EPA, 1989a).
[c] To obtain 99.99 percent inactivation of enteric viruses with preformed chloramines requires CT values > 5,000 at temperatures of 0.5, 5, 10, and 15°C.
Source: U.S. EPA (1989a).

In the final SWTR (U.S. EPA, 1989b), the CT values for ozone have been lowered to levels such that the CT values required to provide 0.5-log inactivation of *Giardia* cysts at the higher water temperatures are below those required to provide 2 or 3 logs of inactivation of enteric viruses. Consequently, the 2- or 3-log virus inactivation CT requirement becomes the pacing parameter for the amount of additional primary disinfection to be provided by ozone during conventional treatment, rather than the 0.5-log inactivation of *Giardia* cysts.

5.2 Disinfection By-Products

Disinfection by-products are formed by two basic mechanisms: (1) reduction, oxidation, or

Table 5-3. CT Values for Achieving Inactivation of Viruses at pH 6 Through 9

	Log Inac-	Temperature					
	tivation	0.5°C	5°C	10°C	15°C	20°C	25°C
Free	2	6	4	3	2	1	1
Chlorine[a]	3	9	6	4	3	2	1
	4	12	8	6	4	3	2
Ozone[b]	2	0.9	0.6	0.5	0.3	0.25	0.15
	3	1.4	0.9	0.8	0.5	0.4	0.25
	4	1.8	1.2	1.0	0.6	0.5	0.3
Chlorine	2	8.4	5.6	4.2	2.8	2.1	–
dioxide[c]	3	25.6	17.1	12.8	8.6	6.4	–
	4	50.1	33.4	25.1	16.7	12.5	–
Chlor-	2	1,243	857	643	428	321	214
amines[d]	3	2,063	1,423	1,067	712	534	356
	4	2,883	1,988	1,491	994	746	497

[a] Data adapted from Sobsey (1988a) for inactivation of Hepatitis A Virus (HAV) at pH = 6, 7, 8, 9, and 10 and at 5°C. CT values include a safety factor of 3.
[b] Data adapted from Roy et al. (1982) for inactivation of poliovirus at pH 7.2 and 5°C. CT values include a safety factor of 3.
[c] CT values for chlorine dioxide are based on laboratory studies at pH 6.0 and 5°C (Sobsey, 1988a). CT values include a safety factor of 3.
[d] Data from Sobsey (1988a) for inactivation of HAV for pH = 8.0, 5°C, and assumed to apply to pH in the range of 6.0 to 10.0. These CT values apply only for systems using combined chlorine where chlorine is added prior to ammonia in the treatment sequence. CT values given here should not be used for estimating the adequacy of disinfection in systems applying preformed chloramines, or applying ammonia ahead of chlorine.
Source: U.S. EPA (1989a).

disproportionation of the disinfecting agent and (2) reaction of oxidation by the disinfectant with materials already in the water. Reduction, oxidation, or disproportionation can occur when the disinfecting agent is added to water. Three examples of this reaction are the formation of chlorite and chlorate ions associated with chlorine dioxide, the formation of dissolved oxygen associated with ozone, and the formation of chloride ions associated with chlorine.

Oxidation of humic acids (in the water from organic materials) produces aldehydes, ketones, alcohols, and carboxylic acids upon the addition of ozone, chlorine, chlorine dioxide, or potassium permanganate. Halogenation of organic materials can occur in the presence of free chlorine to produce trihalomethanes and other halogenated organics. Chlorine can also form organic chloramines by reacting with nitrogen-containing organic compounds (amino acids and proteins). In addition, monochloramine can produce organic chloramines in the presence of organonitrogen compounds.

If bromide ion is present in the untreated water, it may be oxidized by ozone or chlorine (but apparently not by chlorine dioxide or chloramine) to form hypobromous acid, which in turn can brominate organic materials. Bromine-containing trihalomethanes, for example, are known to form in this manner.

By-products are also produced when oxidants, like ozone or chlorine, are used for oxidation purposes other than disinfection. For instance, breakpoint chlorination is sometimes used early in the water treatment process to remove ammonia. In the presence of organic compounds considered precursors, the same by-products that are formed during chlorine disinfection are also formed in this oxidation step.

As another example, ozone is used as an oxidant to improve turbidity, color, taste, odor, or microflocculation; or to oxidize organic compounds, iron, or manganese. The addition of ozone early in the treatment process as an oxidant may produce the same by-products as when added later in an ozone disinfection process. Potassium permanganate, also used as an early oxidant, can produce oxidation by-products as well. The maximum concentration of by-products is usually produced when oxidants are used at the point in the treatment process where the concentration of organics capable of being oxidized is greatest and/or when large amounts of oxidizing agents are employed for long contact times.

Even when oxidants are used in the treatment process for purposes other than disinfection, some degree of disinfection occurs. In some cases, especially in treatment processes involving ozone, chlorine dioxide, and chlorine under lower pH conditions, the primary disinfection requirement may be satisfied during the preoxidation procedure (prior to filtration).

Since oxidation is so important in determining disinfection by-products, a brief description of the chemistry of oxidation is provided in Section 5.2.1. This is followed in Section 5.2.2 by a short overview of the occurrence and nature of disinfection by-products. Lastly, Section 5.2.3 discusses the strategies for controlling disinfection by-products.

5.2.1 The Chemistry of Oxidation

The measure of an agent's ability to oxidize organic material is its oxidation potential (measured in volts of electrical energy). Oxidation potential indicates the degree of chemical transformation to be expected when using various oxidants. It gauges the ease with which a substance loses electrons and is converted to a higher state of oxidation. For example, if substance A has a higher oxidation potential than substance B, substance B theoretically can be oxidized by substance A. Conversely, a particular substance cannot oxidize another with a higher oxidation potential. For example, ozone and chlorine can oxidize bromide ions to hypobromous acid, but chlorine dioxide cannot. The oxidation potentials of common oxidants and disinfectants associated with water treatment are listed in Table 5-4.

Table 5-4. Oxidation Potentials of Water Treatment Oxidants

Species		Oxidation Potential (Volts)
Hydroxyl free radical	$(OH)^{\cdot}$	2.80
Ozone[a]	O_3	2.07
Hydrogen peroxide	H_2O_2	1.76
Permanganate ion	MnO_4	1.68
Hypochlorous acid[a]	HOCl	1.49
Chlorine[a]	Cl_2	1.36
Hypobromous acid[a]	HOBr	1.33
Bromine[a]	Br_2	1.07
Hypoidous acid	HOI	0.99
Chlorine dioxide[a]	$ClO_2(aq)$	0.95
Iodine[a]	I_2	0.54
Oxygen	O_2	0.40

[a]Excellent disinfecting agents.

An agent's effectiveness as a disinfectant is not always related to its effectiveness as an oxidant. For example, whereas ozone is both a powerful oxidant and disinfectant, hydrogen peroxide and potassium permanganate are powerful oxidants but poor disinfectants. Chlorine dioxide and iodine are weak oxidants but strong disinfectants.

Oxidation potential does not indicate the relative speed of oxidation nor how complete the oxidation reactions will be. Complete oxidation converts a specific organic compound to carbon dioxide and water. Oxidation reactions that take place during water treatment are rarely complete; therefore, partially oxidized organic compounds, such as aldehydes, ketones, acids, and alcohols, normally are produced during the relatively short reaction periods.

The behavior of a disinfectant as an oxidant will also depend on the particular organic compounds in the water supply. The level of total organic carbon (TOC) and the total organic halogen formation potential (TOXFP), when chlorine is used, indicate the likelihood that undesirable halogenated by-products will be formed. Simply monitoring the reduction in concentration of a particular organic compound, however, is insufficient to indicate how completely oxidation reactions are taking place. Unless a compound is totally oxidized to carbon dioxide and water, the TOC level may not change; therefore, the concentrations of oxidation products must also be measured. The TOXFP and the nonvolatile TOXFP, referred to as the nonpurgeable TOXFP (NPTOXFP), indicate the potential for halogenated by-products to be formed from a specific raw water source.

5.2.2 The Presence of Disinfection By-Products in Drinking Water

EPA has recently surveyed 10 operating water utilities for the presence of 22 halogenated disinfection by-products in chlorine-treated water (Stevens et al., 1987). Table 5-5 presents the frequency and range of concentrations of those by-products of greatest concern. Table 5-6 summarizes the current knowledge of health effects of selected chlorination by-products. EPA is currently studying the by-products associated with ozonation. To date, however, extensive studies of by-products of treatment with ozone, chloramination, and chlorine dioxide have not been conducted.

Table 5-5. Occurrence of Chlorinated Disinfection By-Products at 10 Water Utilities[a]

Compound	Number of Locations	Range of Values µg/L
High Confidence		
Chloroform	10 of 10	2.6 to 594
Bromodichloromethane	10 of 10	2.6 to 77
Chlorodibromomethane	10 of 10	0.1 to 31
Bromoform	6 of 10	0.1 to 2.7
Dichloroacetonitrile	10 of 10	0.2 to 9.5
Dibromoacetonitrile	3 of 7	0.4 to 1.2
Bromochloroacetonitrile	7 of 7	0.2 to 4.0
Chloropicrin	8 of 10	0.2 to 5.6
Low Confidence		
Chloroacetic acid	6 of 10	<10
Dichloroacetic acid	10 of 10	<10 to >100
Trichloroacetic acid	6 of 10	10 to 100
Trichloroacetaldehyde (as chloral hydrate)	10 of 10	10 to 100
1,1,1-Trichloropropanone	10 of 10	10 to 100
2-Chlorophenol	0 of 10	---
2,4-Dichlorophenol	0 of 10	---
2,4,6-Trichlorophenol	0 of 10	---
Qualitative Only		
1,1-Dichloropropanone	0 of 8	---
1,1-Dichloro-2-butanone	0 of 8	---
3,3-Dichloro-2-butanone	1 of 8	---
1,1,1-Trichloro-2-butanone	0 of 8	---
Cyanogen chloride	1 of 7	---
Dichloroacetaldehyde	0 of 10	---

[a] In the first two groups, contaminants are grouped according to whether current knowledge of health effects indicates a high or low confidence that adverse health effects exist; in the third group, current knowledge of health effects is only qualitative to date.
Source: Stevens et al. (1987).

Table 5-6. Summary of Health Effects Associated with Chlorination By-Products

Chemical Class	Example	Toxicological Effects
Trihalomethanes	Chloroform	C, H, RT
	Dichlorobromomethane	H, RT
	Dibromochloromethane	H, RT
	Bromoform	H, RT
Haloacetonitriles	Chloroacetonitrile	G, D
	Dichloroacetonitrile	M, G, D
	Trichloroacetonitrile	G, D
	Bromochloroacetonitrile	M, G, D
	Dibromoacetonitrile	G, D
Haloacid derivatives	Dichloroacetic acid	MD, C, N, OL, A
	Trichloroacetic acid	HPP
Chlorophenols	2-Chlorophenol	F, TP
	2,4-Dichlorophenol	F, TP
	2,4,6-Trichlorophenol	C
Chlorinated ketones	1,1-Dichloropropanone	M
	1,1,1-Trichloropropanone	M
	1,1,3,3-Tetrachloropropanone	M
Chlorinated furanones	MX	M, Cl
Chlorinated aldehydes	2-Chloroacetaldehyde	G

Key to Toxicological Effects:

C = Carcinogenic	N = Neurotoxic
H = Hepatotoxic	OL = Ocular Lesions
RT = Renal Toxic	A = Aspermatogenesis
G = Genotoxic	HPP = Hepatic Peroxisome Proliferation
D = Developmental	F = Fetotoxic
M = Mutagenic	TP = Tumor Promoter
MD = Metabolic Disturbance	Cl = Clastogenic

Source: Akin et al. (1987).

5.2.3 Strategies for Controlling Disinfection By-Products

The formation of halogenated by-products is affected by a number of factors, including the concentration and types of organic materials present when chlorine is added, the dosage of chlorine, the temperature and pH of the water, and the reaction time. EPA has identified three strategies for controlling formation of halogenated materials during chlorination:

1. Remove the by-products after they are formed.

2. Use alternate disinfectants that do not produce undesirable by-products.

3. Reduce the concentration of organics in the water before oxidation or chlorination to minimize the formation of by-products.

The first approach, removing the by-products after they are formed, can be difficult and costly. Section 6 discusses the treatment technologies available for organic contaminant removal. The second approach, using alternative disinfectants, is often the most cost-effective. The third approach, reducing the concentrations of organic precursors before adding chlorine or other oxidants, will provide the highest quality finished water.

Using Alternative Disinfectants

The second approach, using other than chlorine for disinfection, is sound if the replacements do not produce undesirable by-products of their own and if they perform equally as both primary and secondary disinfectants. Cost is also a consideration. Alternative disinfectants currently being considered by water treatment specialists are chlorine dioxide, monochloramine, UV radiation, and ozone. Both ozone and UV radiation do not provide stable residuals for the distribution system and, therefore, cannot be used as substitute disinfectants by themselves.[1]

Although extensive studies of ozone by-products have not yet been conducted, many immediate oxidation products of naturally occurring organic materials have been identified repeatedly. For the most part, these by-products are organic aldehydes, acids, and ketones. Oxidation of raw water containing bromide ion will produce hypobromous acid, which can brominate organic precursors.

Since ozone is employed only for primary disinfection, a chlorinated compound (chlorine or chloramine) must be added for secondary disinfection following ozonation, i.e., to provide a residual for the distribution system. Consequently, the "secondary" by-products, those formed by the reaction of chlorine or chloramine with the primary by-products of ozonation, become a concern to water treatment specialists. Although some studies have examined by-products produced by two-step oxidation sequences of this type, no compounds have yet been reported that are not produced by one of the two oxidation processes acting alone.

For example, preozonation may affect the yields of THMs formed by subsequent chlorination. Usually these THM yields are lowered by preozonation, but in some cases, usually with high ozone dosages or at high pH values, they can be enhanced. The yield of chloropicrin (nitrotrichloro-methane) can be enhanced if chlorination is preceded by ozonation. Chloramine is known to react with acetaldehyde to produce acetonitrile under drinking water treatment conditions. This and other nitriles might be expected to be produced upon direct chloramination of ozonated waters containing aldehydes.

Chlorine dioxide is effective as a primary and secondary disinfectant, but some chlorite ion is produced. (See Section 5.5.1. for a discussion of the chemistry of chlorine dioxide.) The use of chlorine dioxide has been associated with hematological effects in laboratory animals, which may result from the production of chlorite and chlorate ions. Neurological effects have also been identified. Due to these concerns, EPA currently recommends maximum total oxidant levels (total chlorine dioxide plus chlorite ion plus chlorate ion) in finished water of 1 mg/L. Thus, chlorine dioxide normally can be used as a primary disinfectant only in very clean waters, requiring low dosages of no more than 1.2 to 1.4 mg/L.

If a strong chemical reducing agent is added somewhere in the treatment process after chlorine dioxide primary disinfection, then chlorine dioxide and chlorite ions can be reduced to chloride ion. This would leave only traces of chlorate ion in the water. This chemical reduction technique will allow much higher chlorine dioxide dosages to be applied for oxidation and/or primary disinfection than the 1.2 to 1.4 mg/L currently recommended

At present, granular activated carbon (GAC) or sulfur dioxide are known to chemically reduce chlorine dioxide and chlorite ion (but not chlorate ion) to the innocuous chloride ion. This approach to the application of chlorine dioxide will be discussed in more detail in Section 5.4.3.2.

The remaining alternative, monochloramine, is a weak disinfectant. The very high CT values required to inactivate 99.9 percent *Giardia* and 99.99 percent enteric viruses make monochloramine impractical for use as a primary disinfectant. Therefore, monochloramine should only be considered as a secondary disinfectant.

Minimizing Precursor Concentrations

The third approach for controlling disinfection by-products is to reduce the concentration of organic materials before adding chlorine or any oxidant. This approach will minimize the formation of by-products. Another option is to use an oxidant that does not contain chlorine, such as ozone, potassium permanganate, or hydrogen peroxide, before or during rapid mix and/or filtration to partially oxidize organics. This will improve the flocculation and filtration processes that follow. However, if the water

1 Ozone has been shown to be an effective secondary disinfectant in a study of European and Canadian applications. However, four conditions must be met simultaneously, which were found to coexist only rarely. The conditions are: (1) water temperatures must be cool to slow biological regrowths, (2) water must be free of iron and ammonia, (3) total organic carbon values must be less than 1 mg/L, and (4) residence time in the distribution system must be less than 12 hours (Miller et al., 1978).

contains substantial amounts of bromide ion, brominated organics may be produced.

Because all disinfectants and oxidants produce some types of by-products, the most efficient approach to lowering organic by-product precursors is to optimize physical process treatment steps before adding oxidants. For example, if alum is used as the coagulant, it should be recognized that the optimum pH is about 6.5. In addition, coagulant dosage should be tested to maximize removal of organics.

It is important to note that extensive oxidation converts natural organic materials (and some SOCs) into simpler oxidation products (aldehydes, acids, ketones, etc.), which are much more biodegradable than the initial organic materials. Consequently, a biological treatment step following oxidation can mineralize the oxidized materials, removing them from solution, thus avoiding the incorporation of these by-products into the finished water.

Examples of effective biological treatment steps are filtration, specifically through sand; dual media filters; GAC/sand filters (GAC on top of sand); and postfiltration GAC adsorbers. The biological efficiencies of these types of filters increase in the order listed. To allow biological filtration, it is critical that no residual disinfectant be present in solution. Otherwise, microbial activity present in the filter media will be eliminated or at least adversely affected.

5.3 Comparing Disinfectants

This section summarizes the advantages and disadvantages of the various disinfectants. Since most disinfectants also function as oxidizing agents, both oxidizing and disinfecting properties must be considered when selecting a disinfecting agent. The important characteristics of chlorine, chlorine dioxide, monochloramine, ozone, and UV radiation are described in Sections 5.3.1 through 5.3.5. Section 5.3.6 describes advanced oxidation processes that use combinations of ozone with UV radiation or hydrogen peroxide. Sections 5.4 and 5.5 describe the technologies used to apply each disinfectant.

5.3.1 Chlorine

Chlorine is an excellent disinfectant and oxidant, and is also a good chlorinating agent. It provides a stable residual for the distribution system if the water is free of chlorine-demanding ammonia and organic materials. Since chlorine can produce THMs and other halogenated (TOX) and nonhalogenated organic compounds, the use of chlorine should be minimized, particularly when THM and TOX precursors are present. Some of the nonhalogenated oxidation products of chlorination of humic and fulvic acids are identical to those produced by potassium permanganate, ozone, and chlorine dioxide.

5.3.2 Chlorine Dioxide

Chlorine dioxide is an excellent disinfectant, but is not as strong an oxidant as free chlorine. Because it is unstable, it must be generated on site. In its pure state, chlorine dioxide does not produce THMs in the presence of organic materials. Some procedures for synthesizing chlorine dioxide (e.g., sodium chlorite and elemental chlorine procedures) use excess chlorine, which can produce THMs. Using less excess chlorine in these procedures will lower the resulting THM concentrations. Generating chlorine dioxide with mineral acid and sodium chlorite solution avoids the presence of excess free chlorine.

Chlorine dioxide treatment of organic pollutants generally produces nonchlorinated oxidation products and some chlorinated oxidation products, but in smaller quantities than chlorine. Chlorine dioxide does not oxidize bromide ion to hypobromous acid, as do ozone and chlorine, apparently because of its low oxidation potential.

During oxidation and disinfection reactions, as much as 90-95 percent of the chlorine dioxide reverts back to chlorite ion, which, along with chlorate ion, has been associated with undesirable health effects. EPA currently recommends that finished water supplies contain a maximum total of 1 mg/L combined chlorite and chlorate ions, and chlorine dioxide.

5.3.3 Monochloramine

Monochloramine is a weak cysticidal disinfectant and a poorer virucide. Therefore, the contact times and concentrations required for adequate primary disinfection are much longer and higher than with chlorine, chlorine dioxide, or ozone. When monochloramine is generated, dichloramine can be present. Water containing chloramines can be fatal to individuals on kidney dialysis, so local hospitals and treatment centers must be warned against using water containing chloramines for these patients. Monochloramine is also a weak oxidant, and its slow dissociation in water to free chlorine produces traces of halogenated organic materials. Monochloramine is not recommended as a primary disinfectant because its inactivation of *Giardia* cysts is slow and it is a poor viricide.

Monochloramine may be produced in three ways: (1) adding ammonia to water containing chlorine, (2) adding chlorine to water containing ammonia, and (3) using a preformed solution of monochloramine. Each technique is discussed in more detail in Section 5.5.2.

Adding ammonia to water treated with chlorine (Method 1) will form THMs and other by-products associated with chlorination during the chlorine contact time used for disinfection or oxidation purposes. In this case, the benefit from using an alternative disinfectant to chlorine is negated. Adding ammonia will arrest THM generation, however, and the THM level will remain as produced from the initial chlorination contact time.

Adding chlorine to water already treated with ammonia (Method 2), assuming proper mixing, will produce insignificant concentrations of free chlorine and, consequently, lower concentrations of halogenated materials. However, disinfection is less effective because the weak disinfectant monochloramine is performing the disinfection function, rather than free chlorine.

This problem is exacerbated if organic nitrogen materials are already in the water. They react with free chlorine and chloramines to form organic chloramines, and all organic chloramines are even weaker disinfectants than monochloramine. Available field analytical methods do not distinguish between inorganic monochloramine and organic chloramines in water. A utility with raw water containing organic nitrogen materials that adds ammonia and then chlorine to produce monochloramine for primary disinfection may seriously overestimate the degree of disinfection attained.

Using a preformed monochloramine solution (Method 3), creates the same problem of producing less effective organic chloramines, if organo-nitrogen compounds are present.

5.3.4 Ozone

Ozone is the strongest disinfectant and oxidizing agent available for water treatment; however, it is an unstable gas and must be generated on site. In addition, it is only partially soluble in water, so efficient contact with the water must be established and excess ozone from the contactor must be handled properly. Section 5.4.2 discusses the specifics of ozone generation and contacting methods. Ozone cannot be used as a secondary disinfectant because it is unable to maintain an adequate residual in water for more than a short period of time.

Although the capital costs of ozonation systems are high, their operating costs are moderate. Because of its high oxidation potential, ozone requires short contact times and dosages for disinfection and oxidative purposes. As a microflocculation aid, ozone is added during or before the rapid mix step and its usage is followed by coagulation and direct or conventional filtration. Higher dosages are used to oxidize undesirable inorganic materials, such as iron, manganese, sulfide, nitrite, and arsenic; or to treat organic materials responsible for tastes, odors, color, and THM precursors.

Ozone does not directly produce any halogenated organic materials, but if bromide ion is present in the raw water, it may do so indirectly. Ozone converts bromide ion to hypobromous acid, which can then form brominated organic materials. The primary by-products of ozonation are oxygen-containing derivatives of the original organic materials, mostly aldehydes, ketones, alcohols, and carboxylic acids. Ozone, however, produces toxic oxidation products from a few organic compounds. For example, the pesticide heptachlor forms high yields of heptachlor epoxide upon ozonation. Therefore, when selecting ozone for oxidation and/or disinfection purposes, one must know the specific compounds present in the raw water so as to provide the appropriate downstream treatment to cope with by-products. Researchers are continuing to study ozonation by-products and their potential health effects.

Even when ozone is used to oxidize rather than disinfect, primary disinfection is attained simultaneously provided contact times and dissolved ozone concentrations are appropriate. Consequently, both oxidation and primary disinfection objectives can be satisfied with ozone prior to filtration, after which only secondary disinfection is needed.

There are two cases in which this one-step oxidation/disinfection with ozone is not feasible: (1) when high concentrations of iron or manganese are in the raw water, and (2) when ozone is used for turbidity control. In both cases, measurement of the degree of disinfection (dissolved ozone concentrations) is impractical. When iron or manganese are in the water, ozonation precipitates dark insoluble oxides that interfere with the measurement of dissolved ozone. When ozone is used for turbidity control, such low dosages of ozone are used that a measurable concentration of dissolved ozone may never be attained. In these two cases, ozone oxidation and disinfection must occur separately.

After being partially oxidized by ozone, organic materials become more biodegradable and, therefore, more easily mineralized during biological filtration. Preozonation of water fed to slow sand filters increases the ease of biodegradation of organic materials and enhances biological removal of organic materials during GAC filtration. The adsorptive efficiency of the GAC is extended because it only has to adsorb the organics unchanged by ozone, while the partially oxidized organics are biologically converted to carbon dioxide and water.

Primary disinfection (or oxidation) with ozone produces a significant amount of assimilable organic carbon (AOC) comprised of readily biodegradable aldehydes, acids, ketones, and alcohols. Many of these are also precursors of THMs and TOX compounds. Consequently, if ozone disinfection is immediately followed by chlorination, higher levels of THMs and TOX compounds may be produced than without ozonation.

5.3.5 Ultraviolet Radiation

UV radiation (254 nm) disinfection of bacteria and viruses has several important advantages: (1) it is readily available; (2) it produces no toxic residuals; (3) required contact times are relatively short; and (4) the equipment is easy to operate and maintain, although maintenance must be performed on a regular basis to prevent fouling of certain components. Section 5.4.3. discusses the specifics of UV radiation equipment operation and maintenance.

UV radiation disinfection is inappropriate for:

- Inactivation of *Giardia lamblia* cysts
- Water containing high suspended solids concentrations, color, and turbidity
- Water with high concentrations of soluble organic matter that can react with or absorb the UV radiation, thus reducing the disinfectant's performance

Since UV radiation is ineffective against *Giardia* cysts, but effective against viruses and bacteria, it is a good candidate for disinfecting ground water not directly influenced by surface water. If the amount of radiation received by a target organism is not a lethal dose, however, reconstitution of the organism and reinfection of the water can occur.

Since UV radiation disinfection provides no disinfecting residual, a secondary disinfectant is needed. Very little oxidation of organic materials occurs with typical UV radiation systems used for drinking water disinfection; consequently if oxidation is required (for iron, manganese, sulfide, nitrate, etc.), a strong oxidizing agent may be necessary and can serve as a primary disinfectant as well. However, higher energy intensities and lower UV wavelengths (184.9 nm) can produce oxidation reactions.

UV bulbs that produce radiation at 184.9 nm generate some quantities of ozone which, in turn, can provide some oxidation of organic materials. The combination of UV radiation and ozone produces the hydroxyl free radical, which is a more powerful oxidizing agent than is ozone itself (see Section 5.3.6, Advanced Oxidation Processes).

The 184.9 wavelength radiation is not as effective for UV disinfection as the 254 nm wavelength, except by the amount of ozone generated, which will provide some CT value.

5.3.6 Advanced Oxidation Processes

Ozone used in combination with UV radiation or hydrogen peroxide can adequately disinfect and, at the same time, oxidize many refractory organic compounds such as halogenated organics present in raw water. Although contact times for ozone disinfection are relatively short, they are quite long for oxidizing organic compounds. This combination process accelerates the oxidation reactions.

Advanced oxidation processes involve combining ozonation with UV radiation (UV_{254} bulbs submerged in the ozone contactor) with hydrogen peroxide (added prior to ozonation) or simply by conducting the ozonation process at elevated pH levels (between 8 and 10). Under any of these conditions, ozone decomposes to produce the hydroxyl free radical, which has an oxidation potential of 2.80 V compared with 2.07 V for molecular ozone. However, hydroxyl free radicals have very short half-lives, on the order of microseconds, compared with much longer half-lives for the ozone molecule.

Many organic compounds that normally are stable under direct reaction with the ozone molecule can be oxidized rapidly by the hydroxyl free radical. Chlorinated solvents such as trichloroethylene (TCE) and tetrachloroethylene (PCE) can be destroyed rapidly and cost effectively by hydroxyl free radicals (Glaze et al.,1988; Aieta et al., 1988).

5.4 Primary Disinfection Technologies

Primary disinfection is a key step in the water treatment process. Typically, this step occurs either just before or just after filtration in a conventional surface water treatment plant. Chlorine dioxide can be used for primary disinfection. However, the maximum recommended residual for chlorine dioxide and its decomposition products (chlorite and chlorate ions) may limit the use of this chemical as a primary disinfectant, except in cases of very clean waters and/or short distribution systems that need only a small amount of disinfectant. Monochloramine cannot achieve the required inactivations of *Giardia* cysts and enteric viruses within reasonable contact times, and, therefore, is not appropriate for primary disinfection.

For surface water treatment, chlorine and ozone are the predominant candidates. For ground water (not directly influenced by surface water), chlorine, ozone, and UV radiation are potential primary disinfectants. Chlorine dioxide appears to be a good primary disinfection candidate for both types of waters, provided that a means of reducing chlorine dioxide and chlorate ions is used (e.g., GAC or some strong chemical reducing agent).

Because chlorine produces undesirable by-products regulated by EPA, its use for pretreatment or primary disinfection must be carefully scrutinized. It is anticipated that EPA will lower the current MCL for THMs (100 /g/L), making chlorine use more difficult. In addition, regulation of some of the other halogenated by-products of chlorination listed in Table 5-7 may place even greater restrictions on chlorine use.

Table 5-7. Disinfectants and Disinfectant By-Products Listed in the First Drinking Water Priority List (DWPL)

Disinfectants:	
	Chlorine
	Hypochlorite ion
	Chlorine dioxide
	Chlorite ion
	Chlorate ion
	Chloramine
	Ammonia
Haloacetonitriles:	
	Bromochloroacetonitrile
	Dichloroacetonitrile
	Dibromoacetonitrile
	Trichloroacetonitrile
Trihalomethanes:	
	Chloroform
	Bromoform
	Bromodichloromethane
	Dichlorobromomethane
Halogenated Acids, Alcohols, Aldehydes, Ketones, and Other Nitriles	
Others:	
	Chloropicrin (trichloronitromethane)
	Cyanogen chloride
	Ozone by-products

The four primary disinfection technologies, chlorine, chlorine dioxide, ozone, and UV radiation, will be discussed in Sections 5.4.1 through 5.4.4. Since most of the utilities that are affected by the Surface Water Treatment and upcoming Ground-Water Disinfection Treatment Rules serve less than 10,000 persons, this discussion will emphasize smaller utilities.

5.4.1 Chlorine

Chlorine is the most common primary and secondary disinfectant used in the United States. It is available as a gas, solid, or aqueous solution. Chlorine gas is used most frequently, especially by large utilities, because of its lower cost. Chlorine in its solid form is calcium hypochlorite ($Ca[OCl]_2$); the liquid form is available as sodium hypochlorite (NaOCl) solution.

Section 5.4.1.1 describes the chlorination process, including the physical and chemical factors affecting its efficiency and applicability to specific sites. The equipment, chemical, and operating and maintenance considerations relevant to the three physical forms of chlorine are discussed in Sections 5.4.1.2 through 5.4.1.4.

5.4.1.1 Process Description

Chlorine undergoes chemical reactions when added to water, and the resulting compounds inactivate or kill undesirable microorganisms. Chlorine gas will form hydrochloric and hypochlorous acids according to the following reaction:

Cl_2	+	H_2O	$\longleftrightarrow$	HCl	+	HOCl
chlorine		water		hydro-chloric acid		hypo-chlorous acid

The hypochlorous acid reacts further depending on the pH of the solution. The higher the pH, the more it will react, as shown below:

HOCl	$\longleftrightarrow$	(OCl)	+	H^+
hypochlorous acid		hypochlorite ion		hydrogen ion

The concentration of hydrogen ions in the water determines the pH; the more hydrogen ions present, the lower the pH. At neutral pH (pH = 7.0), almost 80 percent of the chlorine is in its most effective disinfecting form, hypochlorous acid; the remainder exists in the less effective disinfecting form, hypochlorite ion. Increasing pH reduces the total disinfecting strength of the solution because it causes an increasing amount of hypochlorous acid to form more hypochlorite ion. At pH 8.0, for example, nearly 80 percent of the chlorine is present as the hypochlorite ion.

Table 5-2, presented earlier, gives CT values for inactivation of *Giardia lamblia* cysts with free chlorine (2.0 mg/L). At any given concentration of chlorine, the CT values increase rapidly as the pH rises above 7.0. This is also true at each temperature listed. Figure 5-1 shows the relationship between pH and the concentration of hypochlorous acid. Effective pH control is essential to achieve a desired level of disinfection for systems relying upon chlorination.

When sodium hypochlorite (liquid) or calcium hypochlorite (solid) is used for chlorination, the resulting chemical reactions produce alkaline (basic) compounds as follows:

The resulting hydroxides increase the pH of the water, thereby lowering the concentration of hypochlorous acid and diminishing disinfection

$NaOCl$	+	H_2O	----->	$HOCl$	+	$NaOH$
sodium hypo-chlorite		water		hydro-chlorous acid		sodium hydroxide

$Ca(OCl)_2$	+	H_2O	----->	$2HOCl$	+	$CaOH$
calcium hypo-chlorite		water		hypo-chlorous acid		calcium hydroxide

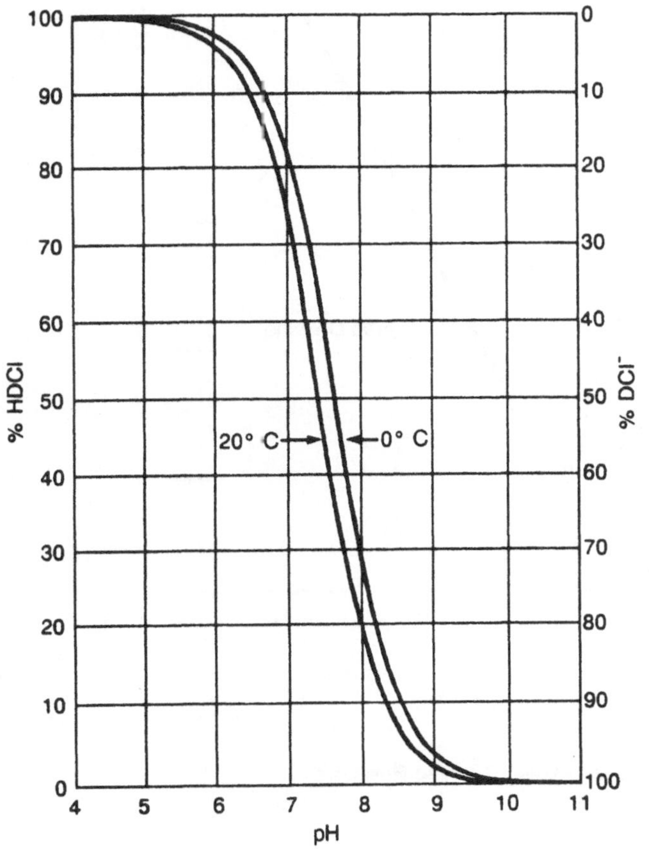

Figure 5-1. **Distribution of hypochlorous acid and hypochlorite ions in water at different pH values and temperatures of 0°C and 20°C.**

efficiencies. Therefore, the ability to adjust and control pH is critical when using the hypochlorite forms of chlorine.

Hypochlorous acid, a strong disinfecting agent, is one of the most powerful oxidizing agents, and an effective chlorinating agent. In addition to acting on target organisms, it reacts with many substances in water, as evidenced by the production of THMs and other halogenated compounds associated with chlorination. Chlorine also produces considerable quantities of nonhalogenated organic oxidation products, e.g., aldehydes, acids, and ketones.

Adequate chlorine concentration will achieve effective disinfection of currently regulated microorganisms. Since chlorine will react with many substances in the water, the "chlorine demand" of these other substances must be satisfied before an excess of free chlorine is available for disinfection. Thus, the amount of chlorine necessary to effectively disinfect must be greater than the chlorine demand of the water.

The "available chlorine residual" is the amount of chlorine that remains available for disinfection after the chlorine demand is satisfied. This is quantified by an approximate analytical testing procedure. The residual may be either a free available residual, a combined available residual, or a combination of the two. Free available chlorine refers to the total concentration of hypochlorous acid and hypochlorite ions. Combined available chlorine is the total concentration of mono- and dichloramines, plus nitrogen trichloride and organic nitrogen chlorine-containing compounds. (See discussion of chloramines in Section 5.5.2.)

Because of the many complex reactions that take place, the relationship between the amount of chlorine added and the available residual does not become linear until a certain minimum amount of chlorine has been added. In other words, increasing the amount of chlorine does not result in a proportional increase in the available residual until that "chlorine breakpoint" is reached.

A series of chemical reactions causes the breakpoint phenomenon. Water may contain small amounts of reduced substances such as sulfides and ferrous iron, as well as organic materials, organic nitrogen materials (amino acids and proteins), and some ammonia, all of which exert a chlorine demand. The initial amount of chlorine added is taken up by reactions with these contaminating substances, leaving no free available chlorine. After the chlorine demands of the reduced substances have been satisfied, the hypochlorous acid will begin to react with ammonia, organic nitrogen materials, and some of the organics present to yield chloramines, oxidized organics and chlorinated organic compounds. Next, the addition of more chlorine may induce the hypochlorous acid to oxidize or chlorinate some of the same materials it has just created. At this point, a decrease in the amount of residual (combined residual) is observed. When these oxidation reactions are complete, the breakpoint is reached, and adding more chlorine finally increases the available chlorine measured.

Figure 5-2 shows a "chlorine breakpoint curve," with the amount of chlorine shown on the horizontal scale and the amount of available chlorine shown on the vertical scale. According to the curve, the chlorine residual will not appear until 3 mg/L of chlorine is added. After this point, additional chlorine will result in an increase in residual. However, at about 6 mg/L,

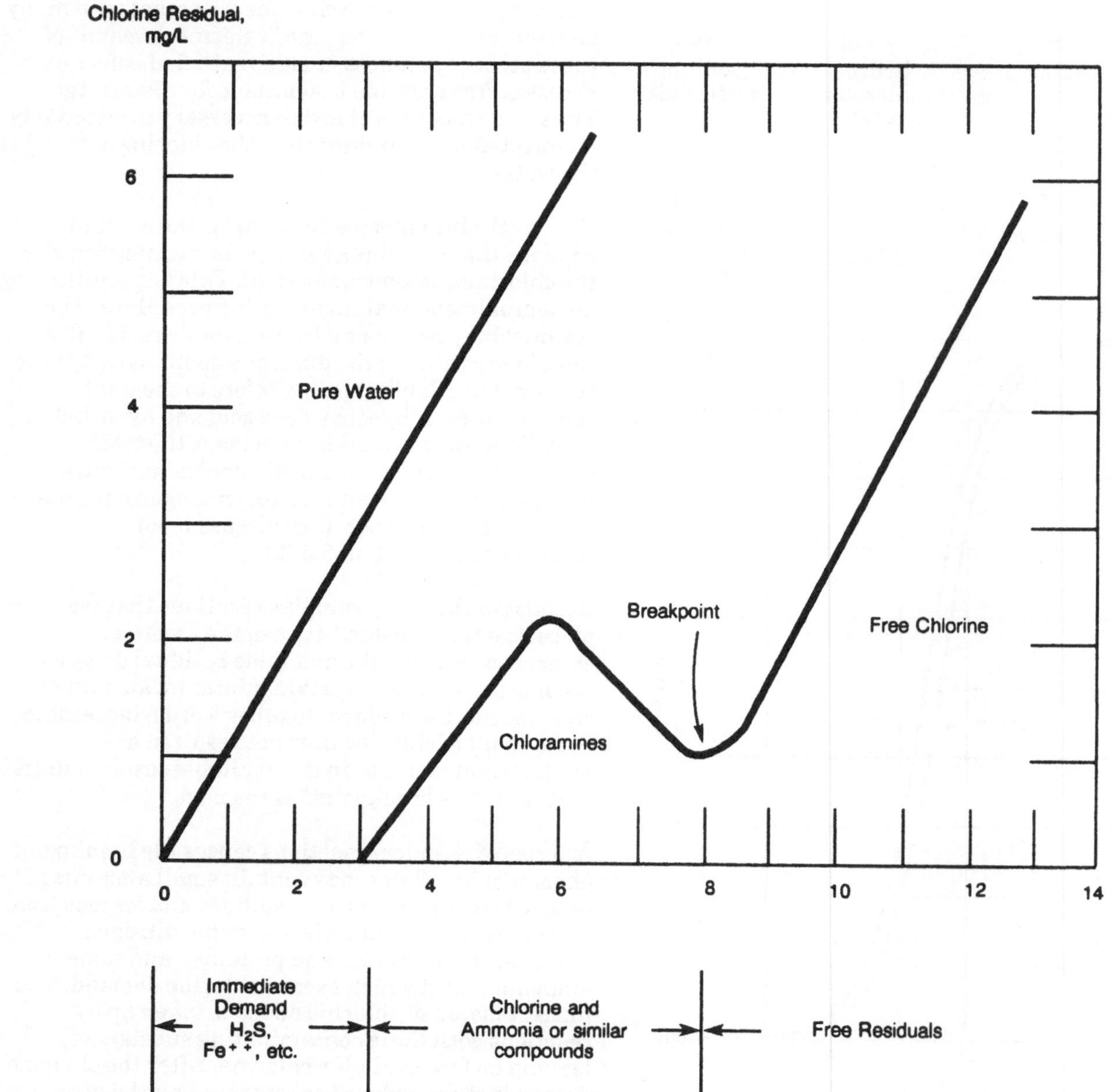

Figure 5-2. **Graphical representation of the breakpoint chlorination reaction. The straight line at the left shows that chlorine residual is proportional to dosage in pure water. When impurities are present, they exert a chlorine demand.**
Source: U.S. EPA (1983).

further additions of chlorine actually bring about a decrease in residual until the breakpoint is reached (8 mg/L in this diagram). After breakpoint is achieved, additional chlorine finally results in a proportional accumulation of residual free available chlorine.

Actual reactions are considerably more complex than described above because of the time and concentration dependencies of these processes. For these reasons, a breakpoint curve is difficult to recreate and predict; thus, individual tests must be run seasonally, and the data plotted to define the breakpoint for each water supply.

Factors Affecting the Efficiency of Chlorine Disinfection

Chlorine in its free state ($HOCl$ + $[OCl]^-$) is an effective disinfectant and inactivates most microorganisms in a matter of minutes. However, effective disinfection with chlorine requires careful attention to the following factors:

- *Free available chlorine concentration.* The concentration must be high enough to always be detectable at the farthest points in the distribution system to effect both primary and secondary disinfection.

- *pH.* The pH should be maintained as close to 7.0 as is practical or consistent with other water quality aspects. This is necessary to maintain as much of the chlorine residual as possible in the hypochlorous acid form. New EPA regulatory initiatives, however, are encouraging utilities to adjust the pH of their product water to 8.0 in order to minimize corrosion effects (U.S. EPA, 1988a). This higher pH will necessitate higher doses of chlorine to attain primary disinfection.

- *Contact time.* Contact time must be long enough to achieve the desired degree of microbial inactivation (i.e., attain the CT value that applies to the concentration of chlorine and the pH and water temperature).

- *Mixing.* The chlorine contactor should either contain sufficient baffling to eliminate the possibility of short-circuiting or an external mixing device should be added.

Temperature also affects the disinfection rate; the higher the temperature, the faster the rate of disinfection. Consequently, at higher temperatures, the CT values become lower.

The choice of chlorination system – gas, solid, or liquid – depends on a number of site-specific factors including:

- Availability of chlorine source chemical
- Capital cost of the chlorination system
- Operation and maintenance costs of the equipment
- Chemical costs
- Location of the facility
- Operator skills
- Safety considerations
- Local regulations regarding the storage of chlorine gas

Each chlorination method provides the same disinfecting power on a pound for pound basis of available chlorine at the same pH. The choice of method depends primarily on the availability of each chemical and the construction and annual operating costs for the different systems.

5.4.1.2 Disinfection with Chlorine Gas

Elemental chlorine is a toxic, yellow-green gas at standard temperatures and pressures. It is supplied as a liquid in high-strength high-pressure steel cylinders, and vaporizes rapidly when released. As the liquid evaporates, its temperature falls and slows evaporation rate, necessitating use of a container manifold or vaporizer.

Chlorine gas can be supplied in cylinders with capacities of 45.4 to 907.2 kg (100 to 2,000 lb), or in tank cars. The quantities required by small water systems can be purchased from local chemical or swimming pool chemical suppliers.

There are two basic types of gas chlorinators: (1) pressure-operated direct gas feed units and (2) vacuum operated solution-feed units. Direct gas feed units supply pressurized chlorine gas to the water and are used only when electrical power is unavailable or the water pressure differentials are insufficient to operate a solution feed unit. The solution feed units mix the gas with a side stream of water to form a solution of hypochlorous acid and hypochlorite ion, which then is mixed with the main stream. These units operate on a vacuum-controlled basis, automatically shutting off if the side stream flow is interrupted. The solution feed system is safer to operate and, therefore, is preferred by most operators. Figure 5-3 shows a solution feed system.

Equipment Costs

Table 5-8 presents a range of equipment costs for one basic and five increasingly complex solution feed gas chlorination systems. The basic system includes equipment to handle two 68-kg (150-lb) chlorine cylinders, two cylinder-mounted chlorine gas regulators, an automatic changeover valve, and a chlorine gas flow and rate valve ejector (with system backup).

The most sophisticated system includes the basic system plus two scales, a gas mask, a diffuser corporation cock (to allow connection under water line pressure), a flow-pacing chlorine addition system, a flow meter, a booster pump and piping, and a chlorine leak detector.

The costs are based on a small water treatment system sized to treat water volumes up to at least 0.044 m^3/sec (1 MGD). They include equipment, installation, safety enclosure, contractor's overhead and profit, and a 10 percent engineering fee. In May 1980 dollars, the least expensive gaseous chlorination system costs were about $9,350. The most sophisticated gaseous chlorination system with all the options added cost $16,050.

Operating and Maintenance Costs

For small solution-feed systems treating from 9.5 m^3/day to 0.044 m^3/sec (2,500 GPD to 1 MGD), operating and maintenance costs for gas chlorination systems are approximately the same. About 1,630 kWh each year is required to run the booster pump and approximately 2,560 kWh annually is required for the building housing the system, assuming a 58.1-m^2 (625 ft^2) building. Maintenance materials for miscellaneous repair of valves, electrical switches, and other equipment cost about $40/yr. Labor

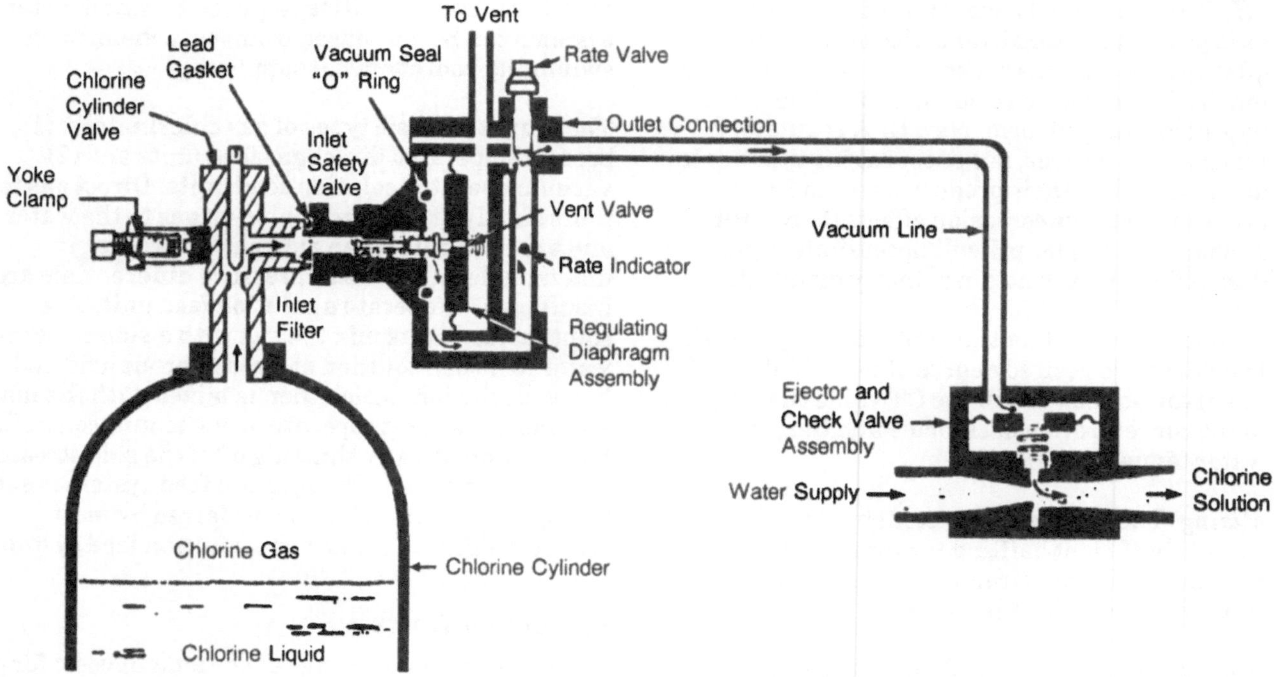

Figure 5-3. Solution feed-gas chlorination system.
Source: Capital Controls Co., Inc.

Table 5-8. Capital Costs for Gas Chlorination[a] ($1980)

Equipment Costs for a System Producing 100 Lb/Day or Less

System	
Basic System[b]	$ 1,873
Complex System	8,579
Installation	1,167
Safety Enclosure	3,500
Contractor's Overhead and Profit (20%)	1,869
Engineering Fees (10%)	934
Total Capital Cost	
Basic System	$ 9,343
Complex System	$16,049

[a] May 1980 quotes (three vendors).
[b] Basic system includes two 150-lb chlorine cylinders, two cylinder-mounted regulators, automatic changeover valve, chlorine gas flow rate valve, and ejector (system with backup).
1 lb = 0.4536 kg
Source: U.S. EPA (1983).

requirements of 1/2 hr/day (or 183 hr/yr) cover daily checks on equipment.

Table 5-9 estimates total annual operating and maintenance costs of $2,163 for a solution feed gas chlorination system. The estimate assumes energy costs of $0.07/kWh and labor costs of $10/hr, which were the prevailing rates in 1982.

Chemical Costs

The cost of chlorine for a system can be estimated using the following formula:

$$\frac{Cl_2 \text{ dosage (mg/L} \times \text{water treated (L/day)} \times Cl_2 \text{ cost (\$/lb)}}{1{,}000 \text{ (mg/g)} \times 454 \text{ (g/lb)}}$$

$$= Cl_2 \text{cost (\$/day)}$$

Table 5-9. Capital Costs for Gas Chlorination[a] ($1980)

Item	Requirements	Costs in $
Electrical Energy		
Process	1,630 kWh/yr	114.10
Building	2,560 kWh/yr	179.20
Subtotal	4,190 kWh/yr	293.30[a]
Maintenance Materials		40.00
Labor	183 h/yr	1,830[b]
Total Annual O&M Cost		2,163

[a] Assumes $0.07/kWh.
[b] Assumes $10/hour labor cost.
Source: U.S. EPA (1983).

In the New York metropolitan area (1989), a 68-kg (150-lb) cylinder of chlorine costs about $1.65/kg ($0.75/lb). Applying the above equation, assuming a dosage of 5 mg/L, gaseous chlorine would cost $30/yr for a 9.5 m³/day (2,500-GPD) plant and $11,400/yr for a 0.044 m³/sec (1-MGD) plant.

5.4.1.3 *Disinfection with Sodium Hypochlorite Solution*

Sodium hypochlorite solution is usually supplied commercially in concentrations of 5 and 15 percent chlorine. It is easier to handle than gaseous chlorine or calcium hypochlorite. Metered chlorinators deliver the solution directly into the water.

Sodium hypochlorite solutions lose their disinfecting (oxidizing) power during storage, and thus should be stored in a cool, dark, dry area. No more than a 1-month supply of the chemical should be purchased at one time to prevent loss of available chlorine. The material is supplied in glass or plastic bottles, carboys, or lined drums ranging in size from 1.89 to 208.2 L (0.5 to 55 gal). Bulk shipment by tank truck is also a common form of transport.

Sodium hypochlorite solution is more costly per pound of available chlorine and does not contain the high concentrations of chlorine available from chlorine gas. However, the handling and storage costs are lower than for chlorine gas.

An onsite generation technique for hypochlorite solutions recently has been developed. This system consists of a two-cell unit, in which a brine solution (salt in water) is electrolyzed, producing a solution of hypochlorous acid in one cell and a solution of caustic (sodium hydroxide) in the other, according to the following equation:

$$\underset{\text{sodium}}{Na^+} + \underset{\text{chloride}}{Cl^-} + \underset{\text{water}}{2H_2O} + \underset{\text{electron}}{e^-} \longrightarrow$$

$$\underset{\text{hypochlorous acid}}{HOCl} + \underset{\text{sodium hydroxide}}{NaOH} + \underset{\text{hydrogen}}{H_2}$$

Using onsite generation avoids the purchase and storage of large volumes of chlorine gas or hypochlorite solutions, but there are significant disadvantages. The generation process produces hydrogen, which poses fire and explosion hazards, and sodium hydroxide, which is a caustic solution that requires proper disposal. Also, the cost per pound of available chlorine is typically much higher for onsite generation (e.g., $0.66 to $0.77/kg [$0.30 to $0.35/lb] for onsite generation compared to $0.18 to $0.33/kg [$0.08 to $0.15/lb] for chlorine gas). However, certain site-specific considerations may make onsite generation a preferred disinfection technique.

Equipment Costs

Table 5-10 presents estimated capital costs for sodium hypochlorite chlorination systems. This table provides estimates for basic and complex systems, both electrically and hydraulically activated. The basic liquid hypochlorination systems include two metering pumps (one serving as a standby), a solution tank, diffuser, and appropriate quantities of tubing. The more complex system adds a diffuser corporation cock, antisiphon backflow preventer, a safety housing enclosure, a flow pacing system, and a flow meter and signal.

Table 5-10. Capital Costs for Liquid Chlorinators[a] ($1980)

	Electrically Activated	Hydraulically Activated
Equipment Cost (basic system[b])	$ 1,800	$ 2,266
	500	1,000
Installation		
Site Work	250	250
Contractor's Overhead & Profit (20%)	729	1,004
Engineering Fees	364	503
Add Ons:		
Alternate #1: add diffuser corporation cock and antisiphon backflow preventer	165	231
Alternate #2: add safety enclosure (housing)	6,930	6,930
Alternate #3: add flow pacing existing signal	-	1,485
Alternate #4: add flow meter signal, 8 in. or less	-	1,452
Total Capital Cost		
Basic System	$ 3,643	$ 5,023
Most Sophisticated	$10,738	$15,121

[a] May 1980 quotes (two vendors).
[b] Basic System includes two metering pumps (one standby), tubing, solution tank, and diffuser.
Source: U.S. EPA (1983).

Total capital costs for an electrically activated system range from $3,643 for the basic system to $10,738 for the most sophisticated system. The comparable range for the hydraulically activated systems is $5,023 to $15,121.

Operating and Maintenance Costs

As with solution-feed gas chlorinators, operating and maintenance costs for systems in the 9.5 m^3/day to 0.044 m^3/sec (2,500 GPD to 1 MGD) size range are roughly the same. The annual estimated energy requirements for the diaphragm metering pump and the housing structure, assumed to be 58.1 m^2 (625 ft^2), are 570 kWh and 2,560 kWh, respectively. Maintenance materials for minor component repairs are about $20 each year. Approximately 1 hour of labor is required each day to mix the sodium hypochlorite solution and check equipment.

Table 5-11 summarizes total annual operating and maintenance costs for the sodium hypochlorite solution feed system. The total of $3,889/yr is based on the same energy and labor cost assumptions of $0.07/kWh and $10/hr as were used for the gas chlorination system.

Table 5-11. Operation and Maintenance Cost Summary for Sodium Hypochlorite Solution Feed ($1982)

Item	Requirements	Costs in $
Electrical Energy		
Process	570 kWh/yr	39.90
Building	2,560 kWh/yr	179.20
Subtotal	3,130 kWh/yr	219.10[a]
Maintenance Materials		20.00
Labor	365 h/yr	3,650[b]
Total Annual O&M Cost		**3,889**

[a] Assumes $0.07/kWh.
[b] Assumes $10/hour labor cost.
Source: U.S. EPA (1983).

Chemical Costs

Typically, sodium hypochlorite is available as a 15 percent (by weight) solution. Four-tenths of a kg (0.9 lb) of sodium hypochlorite solution is equivalent in oxidation potential to 0.45 kg (1 lb) of gaseous chlorine, and its cost is about 3 times that of gaseous chlorine. These ratios for the two forms of chlorine can be used in conjunction with the formula provided in Section 5.4.1.2 to estimate the cost of the solution.

For example, a small water utility treating 9.5 m^3/day (2,500 GPD) with a dosage requirement of 5 mg/L of chlorine requires 242.2 L (64 gal)/yr of the 15 percent solution. If the cost of the solution was $0.13/L ($0.50/gal) (as it was in the Baltimore/Washington, D.C. area in 1987), the annual chemical cost would be $32.00. A 0.044 m^3/sec (1-MGD) plant using the same chlorine dosage would require 400 times the chemical volume and would spend $12,800 annually.

5.4.1.4 Disinfection with Solid Calcium Hypochlorite

Calcium hypochlorite is a white solid that can be purchased in granular, powdered, or tablet form. It contains 65 percent available chlorine and is readily soluble in water. The chemical is available in 0.9-, 2.3-, 3.6-, and 15.9-kg (2-, 5-, 8-, and 35-lb) cans and 362.9-kg (800-lb) drums, which are usually resealable. Calcium hypochlorite is a corrosive material with a strong odor and requires proper handling.

When packaged, calcium hypochlorite is very stable; therefore, an annual supply can be purchased in a single procurement. However, it is hygroscopic (readily absorbs moisture), and reacts slowly with atmospheric moisture to form chlorine gas. Therefore, shipping containers must be emptied completely or carefully resealed. Bulk handling systems cannot be used.

Typically, the entire contents of a calcium hypochlorite container are emptied into a mixing tank where they are readily and completely dissolved in water. The resulting corrosive solution is stored in and fed from a stock solution vessel constructed of corrosion-resistant materials such as plastic, ceramic, glass, or rubber lined steel. Solutions of 1 or 2 percent available chlorine can be delivered by a diaphragm-type, chemical feed/metering pump.

Equipment, Operating, and Maintenance Costs

Equipment, operating, and maintenance costs for calcium hypochlorite solution feed systems are similar to those for sodium hypochlorite feed systems. The equipment needed to mix the solution and inject it into the water being treated is the same.

Chemical Costs

A 9.5 m^3/day (2,500-GPD) treatment plant using a 5 mg/L dosage of chlorine needs 0.104 lb chlorine/day. Because solid calcium hypochlorite contains 65 percent available chlorine, 15.95 kg (0.16 lb)/day is required.

In 1987 (in the Baltimore/Washington D.C. area), 1 kg of calcium hypochlorite cost $2.09 ($0.95/lb). For a 9.5 m^3/day (2,500-GPD) facility, 26.5 kg (58.4 lb) costing $55.48 is needed for 1 year. A 0.044 m^3/sec (1-MGD) facility, using 400 times that amount of chlorine, would spend $22,192 annually.

5.4.2 Ozone

While ozone is widely used for disinfection and oxidation in other parts of the world, it is relatively new in the United States. The process of ozone disinfection, including its chemistry, is discussed in Section 5.4.2.1. Section 5.4.2.2, system design considerations, covers issues related to the essential components of an ozone treatment system. Lastly, Section 5.4.2.3 reviews the costs of ozone system equipment, installation, housing, and operation and maintenance.

5.4.2.1 Process Description

Ozone (O_3) is a powerful oxidizing agent, second only to elemental fluorine among readily available chemical supplies. Because it is such a strong oxidant, ozone is also a powerful disinfectant. Unlike chlorine, ozone does not react with water to produce a disinfecting species. Instead, when exposed to a neutral or alkaline environment (pH above 6), UV light, or hydrogen peroxide, it decomposes in water to more reactive hydroxyl free radicals as shown in the equation below:

O_3	+	H_2O	----->	O_2	+	2(OH)
ozone		water		oxygen		hydroxyl radicals

This reaction is accelerated at pH values above 8.

In water, ozone reacts as the ozone molecule, the hydroxyl free radical, or as a mixture of both. For primary disinfection, CT values for ozone (shown in Tables 5-1 and 5-2) have been developed for molecular ozone, not for hydroxyl free radicals. These free radical species are more effective oxidizing agents than molecular ozone. However, they have extremely short half-lives (microseconds) and consequently may not be good disinfectants.[2]

Since ozone is unstable at ambient temperatures and pressures, it must be generated onsite and used quickly. Ozone is generated by applying energy to oxygen (pure oxygen or dried air). A high-energy electrical field causes oxygen to dissociate according to the equation below:

O_2	+	e^-	----->	2[O]
oxygen		electrons		oxygen "fragments"

These oxygen "fragments" are highly reactive and combine rapidly with molecular oxygen to form the triatomic molecule, ozone:

2[O]	+	$2O_2$	----->	$2O_3$
oxygen "fragments"		oxygen		ozone

The overall reaction that produces ozone is the sum of the above reactions:

$3O_2$	+	e^-	<----->	$2O_3$
oxygen		energy		ozone

[2] Very recent data (Wolfe et al., 1989) show that hydroxyl free residuals produced by the combination of ozone with hydrogen peroxide have the ability to inactivate *Giardia* cysts and enteric viruses rapidly.

This reaction is reversible; once formed, ozone decomposes to oxygen. This reverse reaction occurs quite rapidly above 35°C. Because the reactions that convert oxygen to ozone also produce a considerable amount of heat, ozone generators have cooling components to minimize ozone losses by thermal decomposition.

Ozone has a characteristic odor that is detectable even at low concentrations (0.01 to 0.02 ppm by volume). Higher levels may cause olfactory and other reaction fatigue, and much higher levels are acutely toxic in some instances. The longer the exposure to ozone, the less noticeable the odor.

Ozone is only slightly soluble in water, about 2 to 10 times more soluble than oxygen, depending on the temperature and its concentration as it enters the ozone contactor. The higher the concentration of ozone generated, the more soluble it is in water. Increasing pressure in the ozone contactor system also increases its solubility. Ozone's half-life in water ranges from 8 minutes to 14 hours, depending on the level of ozone-demanding contaminants in the water.

5.4.2.2 System Design Considerations

The five major elements of an ozonation system are:

- Air preparation or oxygen feed
- Electrical power supply
- Ozone generation
- Ozone contacting
- Ozone contactor exhaust gas destruction

Air Preparation

Ambient air should be dried to a maximum dew point of -65°C before use in the ozonation system. Using air with a higher dew point will produce less ozone, cause slow fouling of the ozone production (dielectric) tubes, and cause increased corrosion in the ozone generator unit and downstream equipment. These last two factors result in increased maintenance and downtime of the equipment.

Air feed systems can dry ambient air or use pure oxygen. Using pure oxygen has certain advantages that have to be weighed against its added expense. Most suppliers of large-scale ozone equipment consider it cost effective to use ambient air for ozone systems having less than 1,587.6 kg/day (3,500 lb/day) generating capacity. Above this production rate, pure oxygen appears to be more cost effective. Systems that dry ambient air consist of desiccant dryers, commonly used in conjunction with compression and refrigerant dryers for generating large and moderate quantities of ozone. Very small systems (up to 0.044 m^3/sec [1 MGD]) can use air drying systems with just two desiccant dryers (no compression or refrigerant drying). These systems use silica gel, activated alumina, or molecular sieves to dry air to the necessary dew point (-65°C).

Ambient air feed systems used for ozone generation are classified by low, medium, or high operating pressure. The most common type is a system that operates at medium pressures ranging from 0.7 to 1.05 kg/cm^2 (10 to 15 psig). High-pressure systems operate at pressures ranging from 4.9 to 7.03 kg/cm^2 (70 to 100 psig) and reduce the pressure prior to the ozone generator. Low- and high-pressure systems are typically used in small- to medium-sized applications. Medium- and high-pressure systems may be used in conjunction with most ozone generators and with most contacting systems. Low-pressure systems operate at subatmospheric pressures, usually created by a submerged turbine or other contactors producing a partial vacuum throughout the air preparation and ozone generation system. Creation of this vacuum results in ambient air being drawn into the ozonation system.

The decision to use a high-, medium-, or low-pressure air preparation system often is based on a qualitative evaluation of potential maintenance requirements, as well as an evaluation of capital and operating costs. High-pressure air pretreatment equipment generally has higher air compressor maintenance requirements, lower desiccant dryer maintenance requirements, and lower capital costs.

At small- to medium-sized installations, the lower capital costs may offset the additional maintenance required for the air compressors and associated equipment, such as filters for the oil-type compressors. Typical low- and high-pressure feed gas pretreatment systems are shown in Figures 5-4 and 5-5. Figure 5-4 is also representative of a medium-pressure system, but may require a pressure reducing valve upstream from the ozone generator.

For many applications, pure oxygen is a more attractive ozone feed gas than air for the following reasons:

- It has a higher production density (more ozone produced per unit area of dielectric).
- It requires lower energy consumption (energy supplied per unit area of dielectric).
- Essentially double the amount of ozone can be generated per unit time from oxygen than from air (for the same power expenditure); this means that ozone generation and contacting equipment can be halved in size when using oxygen, to generate and contact the same amount of ozone.
- Smaller gas volumes are handled using oxygen, rather than air, for the same ozone output; thus, costs for ancillary equipment are lower with oxygen feed gas than with air.
- If used in a once-through system, gas recovery and pretreatment equipment are eliminated.
- Ozone transfer efficiencies are higher due to the higher concentration of ozone generated.

However, the economic implications of these advantages must be weighed against the capital expenditure required for onsite oxygen production or operating costs associated with purchase of liquid oxygen produced off site.

Oxygen can be purchased as a gas (pure or mixed with nitrogen) or as a liquid (at -183°C or below). Normally the purity of the available oxygen gas is quite adequate, and no particular pretreatment is required. Its purity always should be better than 95 percent, and its dew point consistently lower than -60°C. When liquid oxygen is the oxygen source, it is converted to the gas phase in an evaporator, and then sent directly to the ozone generator. Purchasing oxygen as a gas or liquid is only practical for small- to medium-sized systems. Liquid oxygen can be added to dried air to produce oxygen-enriched air (as at the Tailfer plant serving Brussels, Belgium). A membrane separation method that also produces oxygen-enriched air is being developed.

There are currently two methods for producing oxygen on site for ozone generation; pressure swing adsorption of oxygen from air and cryogenic production (liquefaction of air followed by fractional distillative separation of oxygen from nitrogen). Systems for the production of oxygen on site contain many of the same elements as air preparation systems discussed above, since the gas stream must be clean and dry in order to successfully generate ozone. Gaseous oxygen produced on site by pressure swing adsorption typically is 93 to 95 percent pure, while liquid oxygen produced cryogenically generally is 99.5 percent pure. In most plants utilizing on site production of ozone, a backup liquid oxygen storage system is included.

At smaller plants, oxygen can be separated from ambient air by pressure swing adsorption using molecular sieves. During the high-pressure phase, nitrogen is adsorbed by the sieves and oxygen exits the system as product gas. When the pressure is reduced, the nitrogen desorbs and is removed from the vessel by purge gas. Precautions should be taken to avoid contamination of the oxygen prepared by this procedure with hydrocarbons, which are present as oils associated with the pressurized equipment.

Pressure swing adsorption systems for producing oxygen are manufactured to produce from 90.7 to 27,216 kg/day (200 to 60,000 lb/day) of oxygen. This production range would supply ozone for water treatment systems at 6 percent concentration in oxygen, for an applied ozone dosage of 4 mg/L and

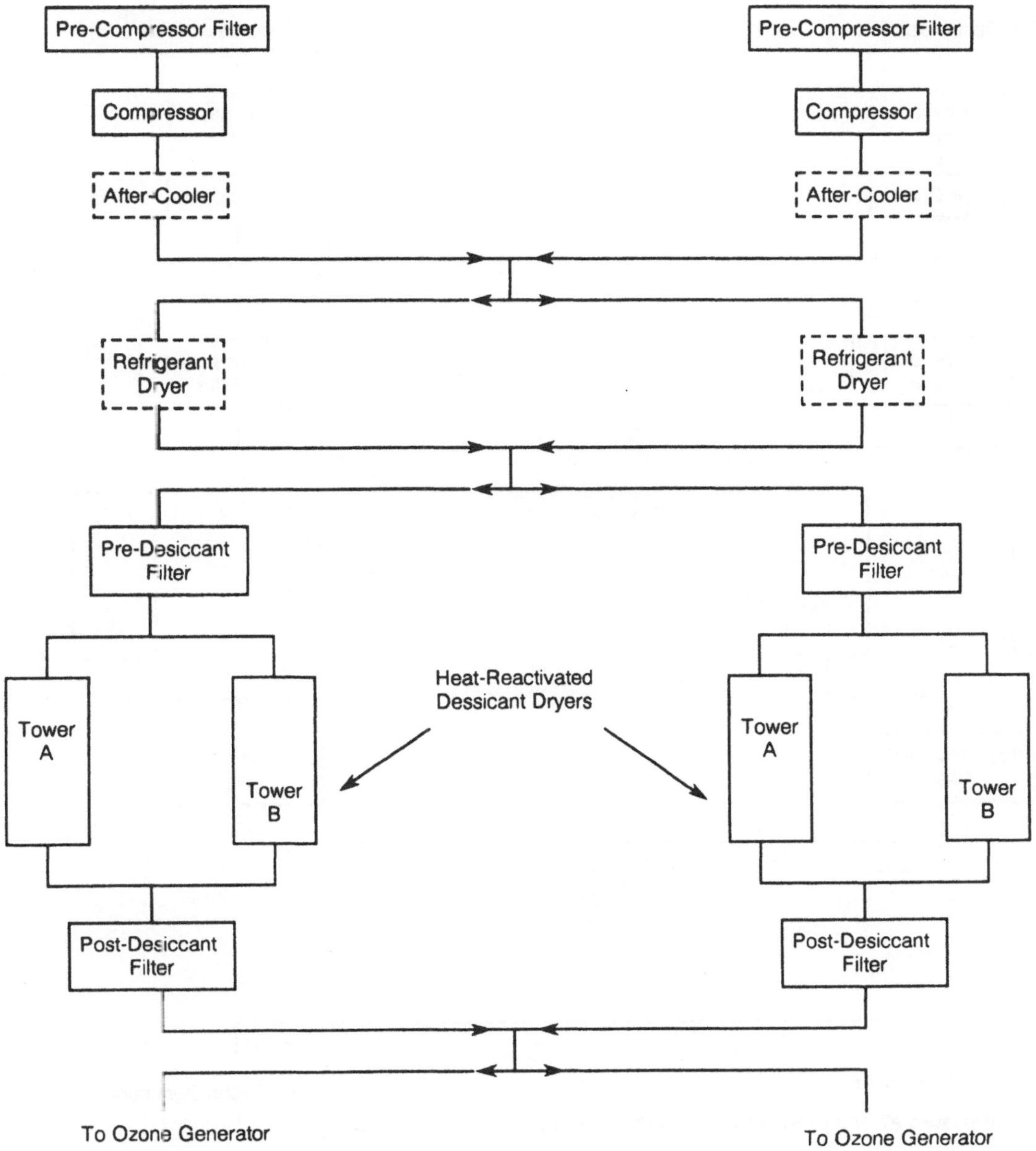

Figure 5-4. Low pressure air feed-gas treatment schematic for ozone generation.
Source: U.S. EPA (1986b).

production of water at the rates of about 0.018 m^3/sec to about 5.26 m^3/sec (0.4 MGD to about 120 MGD).

For cryogenic oxygen production, low temperature refrigeration is used to liquefy the air, followed by column distillation to separate oxygen from nitrogen. Cryogenic systems are operationally sophisticated, and operating and maintenance expertise is required. Production of oxygen by the cryogenic technique is more capital intensive than by pressure swing adsorption, but generally operation and maintenance costs are lower. For oxygen requirements of 18,144 to 18,144,000 kg/day (20 to 20,000 tons/day), cryogenic separation systems are quite practical. This would exclude their use for small water treatment plants. The Los Angeles Aqueduct Filtration Plant, which is 26.28m^3/sec (600 MGD) is using a 3,628.8-kg/day (8,000-lb/day) ozone system, with cryogenically produced oxygen as a feed gas.

Reuse or recycling of the oxygen-rich contactor offgases is possible, and requires removal of moisture and possible ammonia, carbon dioxide, and nitrogen before returning the gas to the generator.

Electrical Power Supply

The voltage or frequency supplied to the ozone generator is varied to control the amount and rate of ozone produced. Varying the power requires specialized supply equipment that should be designed for and purchased from the ozone generator manufacturer.

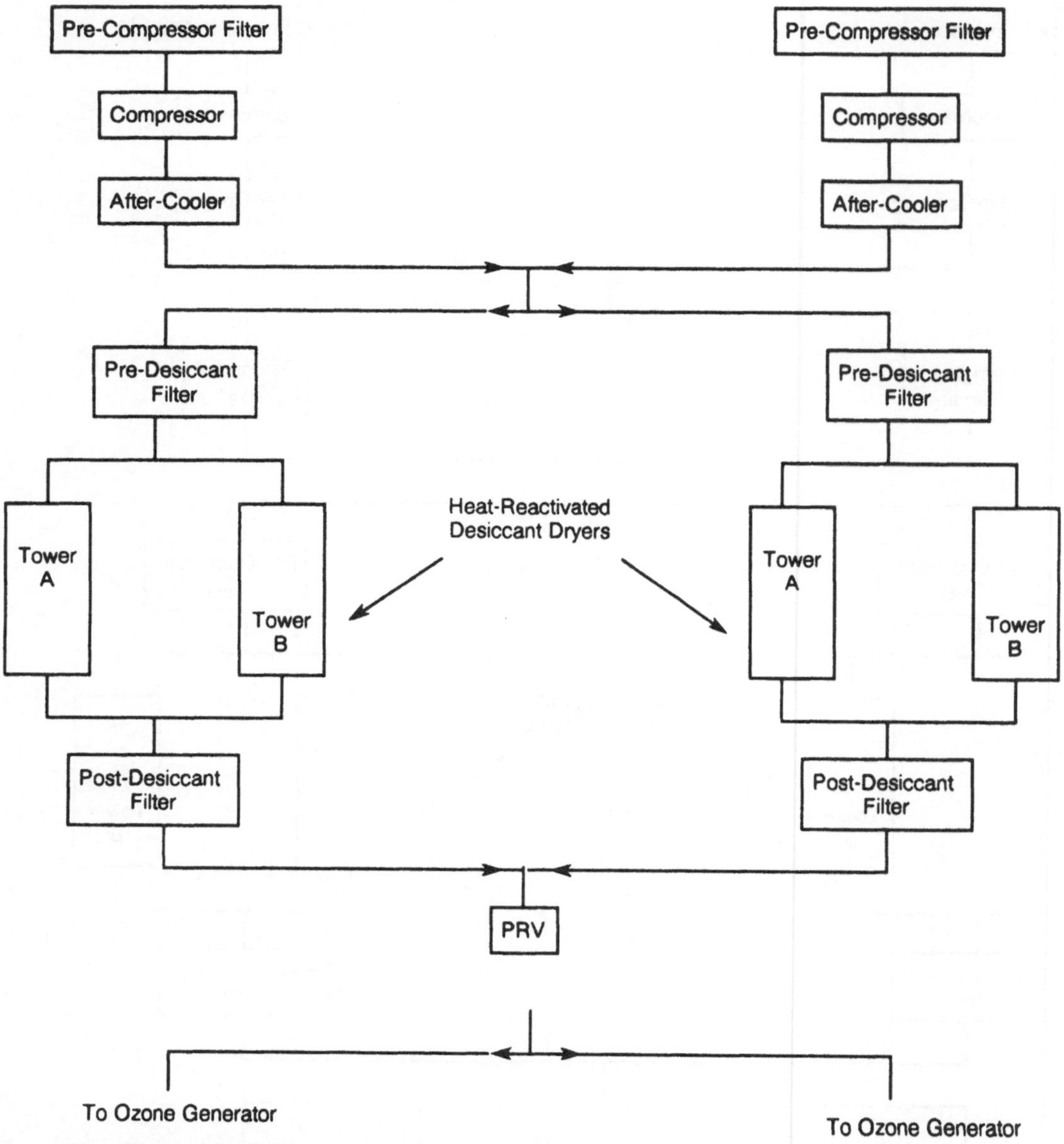

Figure 5-5. High pressure air feed-gas treatment schematic for ozone generation.
Source: U.S. EPA (1986b).

Ozone generators use high voltages (>10,000 V) or high-frequency electrical current (up to 2,000 Hz), necessitating special electrical design considerations. Electrical wires have to be properly insulated; high voltage transformers must be kept in a cool environment; and transformers should be protected from ozone contamination, which can occur from small ozone leaks. The frequency and voltage transformers must be high quality units designed specifically for ozone service. The ozone generator supplier should be responsible for designing and supplying the electrical subsystems.

Ozone Generators

Ozone used for water treatment is usually generated using a corona discharge cell.[3] The discharge cell consists of two electrodes separated by a discharge gap and a dielectric material, across which high voltage potentials are maintained. Oxygen-enriched air, pure oxygen, or air that has been dried and cooled flows between the electrodes and produces ozone. More recent designs use medium and high frequencies, rather than high voltages and low frequencies, to generate ozone.

[3] This technique produces concentrations of ozone sufficiently high (above 1 percent by weight) to solubilize enough ozone and to attain the requisite CT values necessary to guarantee disinfection of *Giardia* cysts. Ozone also can be generated by UV radiation techniques, but only at maximum concentrations of 0.25 percent by weight. At such low gas concentrations, it is not possible to solubilize sufficient ozone to guarantee disinfection of *Giarda* cysts.

Figure 5-6 depicts the essential components of a corona discharge ozone generator. Either ambient air, oxygen-enriched air, or pure oxygen is fed to the generator. If ambient air is used, the generator produces dry, cool air containing 1 to 3.5 percent ozone (by weight), which can be mixed with water. When pure oxygen is used, the concentration of ozone produced is approximately double that produced with ambient air (up to 8 to 9 percent by weight).

The most common commercially available ozone generators are:

- Horizontal tube, one electrode water cooled
- Vertical tube, one electrode water cooled
- Vertical tube, both electrodes cooled (water and oil cooled)
- Plate, water or air cooled

The operating conditions of these generators can be subdivided into low frequency (60 Hz), high voltage (> 20,000 V); medium frequency (600 Hz), medium voltage (< 20,000 V); and high frequency (> 1,000 Hz), low voltage (<10,000 V).

Currently, low frequency, high voltage units are most common, but recent improvements in electronic circuitry make higher frequency, lower voltage units more desirable.

Operating an ozone generator at 60 to 70 percent of its maximum production rate is most cost effective. Therefore, if the treatment plant normally requires 45.36 kg/day (100 lb/day) of ozone and 68.04 kg/day (150 lb/day) during peak periods, it is wise to purchase three generators, each designed for 27.22 kg/day (60 lb/day) and operate all three at about 65 percent capacity for normal production. This arrangement will satisfy peak demands and one generator will be on hand during off-peak periods for standby or maintenance.

Ozone Contacting

Ozone can be generated under positive or negative air pressure. If generated under positive pressure, the contactor most commonly used is a two- chamber porous plate diffuser, with a 4.8-m (16-ft) water column. With this method, the ozone-containing air exits the ozone generator at approximately 1.05 kg/cm^2 (15 psig) and passes through porous diffusers at the base of the column. Fine bubbles containing ozone and air (or oxygen) rise slowly through the column, the mass of ozone is transferred (dissolves), and oxidation and/or disinfection takes place. The 4.8-m (16-ft) height maximizes the amount of ozone transferred from the bubbles as they rise in this type of porous diffuser contactor.

Other types of positive pressure ozone contactors include packed columns, static mixers, and high speed agitators. An atomizer that sprays water through small orifices into an ozone-containing atmosphere also can be used.

When ozone is generated under negative pressure, vacuum action draws the ozone mixture from the generators, providing contact as the gas mixes with the flowing water as with a submerged turbine. Other common methods of creating negative pressure use injectors or Venturi-type nozzles. These systems pump water past a small orifice (injector) or through a Venturi nozzle, thus creating negative pressure.

The diffuser and packed tower contactors require no energy above that required to generate ozone at 1.05

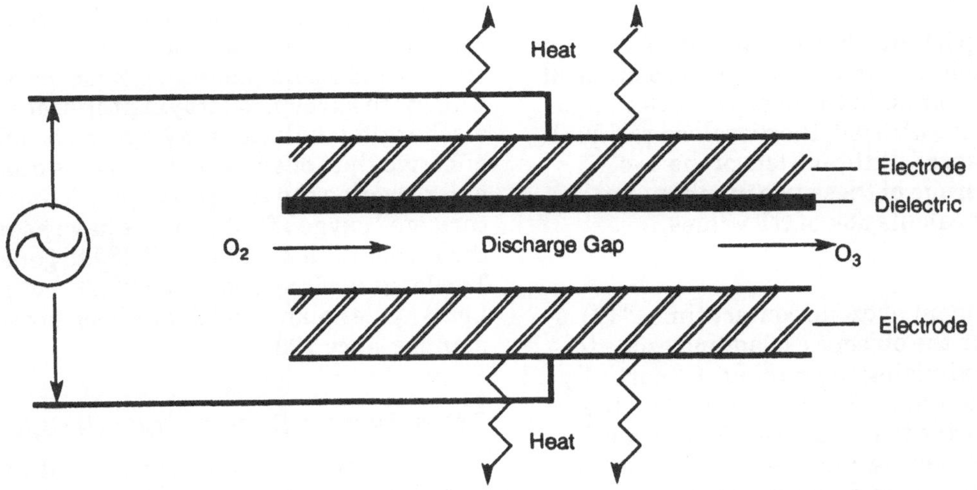

Figure 5-6. Typical ozone generating configuration for a Corona discharge cell.

kg/cm^2 (15 psig). The high speed agitators, static mixers, and all the negative pressure contactors require additional energy.

Ozone reactions are very fast to destroy or inactivate microorganisms; oxidize iron, manganese, sulfide and nitrite ions, and some organics; and lower turbidity levels. However, ozone oxidizes organic compounds such as humic and fulvic materials, as well as many pesticides and volatile organic compounds, quite slowly compared to these other solutes.

For disinfection, the initial dose of ozone is used to satisfy the ozone demand of the water. Once this demand is satisfied, a specific ozone concentration must be maintained for a specific period of time for disinfection. These two stages of ozonation are usually conducted in two different contacting chambers (see Figure 5-7). Approximately two-thirds of the total ozone required is added to the first chamber where the ozone demand of the water is met and the dissolved ozone reaches a residual level (typically 0.4 mg/L). The remaining ozone is applied in the second chamber, where it maintains the residual ozone concentration for the necessary contact time with water to attain the required CT value.

When ozone is added to water, its dissolved residual is not stable. Not only will ozone react with many contaminants in water supplies, but its half-life in water is fairly short, on the order of minutes, due to decomposition back to oxygen. At the higher pH ranges (above 8), decomposition of molecular ozone into reactive intermediates (including the hydroxyl free radical) is accelerated. Consequently, it is not possible to monitor the residual ozone concentration at any single point in the treatment train and expect a single concentration level to hold steady from the point of gas/liquid mixing throughout the ozone treatment subsystem.

Therefore, it is important when using ozone for primary disinfection to monitor for dissolved ozone at a minimum of two points. In the event that two ozone contact chambers are utilized, the dissolved residual ozone can be monitored at the outlets of the two chambers. The average of these two numbers can be used as the "C" for calculation of CT values.

Absolute measurement of ozone contact time ("T") is not simple, because the objective of adding ozone to water involves maximizing the mixing of a partially soluble gas with the liquid. The more complete the contacting, the shorter the actual residence time for the water in the contacting chamber. As a result, the more completely the gas and liquid media are mixed, the less the hydraulic residence time can be used as the value for "T". Only when water flowing through an ozone contacting system approaches plug flow does the actual ozone contact time approach the hydraulic residence time.

The greater the number of ozone contact chambers that can be connected in series, the closer the water flow will approach plug flow. In such cases, the T_{10} (time for 10 percent of an added tracer to pass through the ozone contacting system) will approach 50 percent of the hydraulic detention time for water passed through the ozone contacting system. Recently published studies of the 26.28 m^3/sec (600-MGD) Los Angeles Aqueduct Filtration Plant and of the design of the 5.26 m^3/sec (120-MGD) Tucson water treatment plant have confirmed that the actual hydraulic residence time (T_{10}) is approximately 50 percent of the theoretical hydraulic residence time.

Destruction of Excess Ozone from Ozone Contactor Exhaust Gases

The ozone in exhaust gases from the contacting unit must be destroyed or removed by recycling prior to venting. The current Occupational Safety and Health Administration standard for exposure of workers during an 8-hour work day is a maximum ozone concentration of 0.0002 g/m^3 (0.1 ppm by volume time-weighted average). Typical concentrations in contactor exhaust gases are greater than 1 g/m^3 (500 ppm by volume).

The four primary methods of destroying excess ozone are thermal destruction (heating the gases to 300° to 350°C for 3 seconds), thermal/catalytic destruction, and catalytic destruction (with metal catalysts or metal oxides). Moist granular activated carbon is used extensively at European plants treating less than 0.088 m^3/sec (2 MGD) and for swimming pools using ozone.

When ozone is generated from air, destroying ozone in exhaust gases is more cost effective than recirculating the gases through the air preparation system and ozone generator. When ozone is generated from pure oxygen, destroying the ozone and discharging the excess oxygen can be more cost effective than destroying the excess ozone, and drying and recycling the excess oxygen. A number of "once-through" oxygen feed systems have been installed throughout the world since 1980 to generate ozone. The largest of these is at the 26.28 m^3/sec (600-MGD) Los Angeles plant, which has been operating in this manner since 1987.

Other System Design Considerations

Ozonation system components that directly contact ozone-containing gas should be constructed of corrosion-resistant material, such as reinforced concrete for the contactors, stainless steel for the

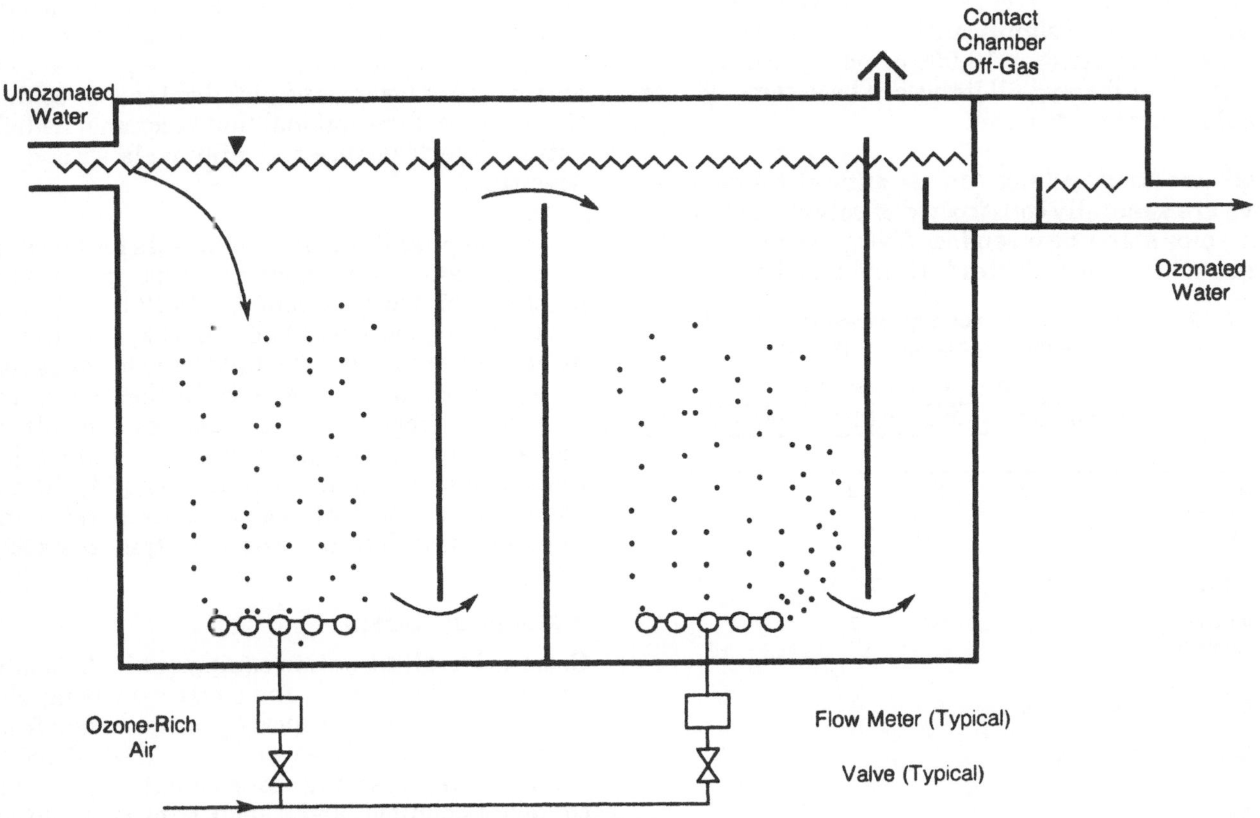

Figure 5-7. Two-compartment ozone contactor with porous diffusers.

piping, and Teflon for gaskets. Piping for dry service should be 304-L stainless steel, while piping for wet service should be 316-L stainless steel.

Proper monitors and controllers should be supplied with the ozone system, including:

- Gas pressure and temperature monitors at key points in the air preparation system. Simple pressure gauges and mercury thermometers are adequate.
- Continuous monitors/controllers for the dew point to determine the moisture content of the dried feed gas to the ozone generator. The monitors should sound an alarm and shut down the generator when high dew points are indicated. Equipment to calibrate the dew point monitor should be provided as well.
- Inlet/discharge temperature monitors for the ozone generator coolant media (water and/or oil, or air), and a means of determining whether coolant is actually flowing through the generator. An automatic system shutdown should be provided if coolant flow is interrupted or if its discharge pressure exceeds specified limits.
- Flow rate, temperature, and pressure monitors, and an ozone concentration monitor for the gas discharged from the ozone generator to determine the ozone production rate.
- Power input monitor for the ozone generator.

5.4.2.3 Costs of Ozonation Systems

The discussion of ozone system costs is divided into four sections covering equipment, installation, housing, and operating and maintenance costs.

Equipment Costs

Ozonation equipment to be purchased includes: air preparation equipment (drying and cooling), an ozone generator, an ozone contactor, an ozone destruction unit, and instrumentation and controls. Because of the many differences in air pretreatment methods, ozone contacting, contactor exhaust gas handling, monitoring, and other operational parameters, equipment costs presented in this section are general guidelines only.

For generating large quantities of ozone, 45.36 kg/day (100 lb/day) and higher, air preparation, ozone generation, and contacting equipment costs run approximately $2,866/kg ($1,300/lb) of ozone generation capacity per day. This figure does not include ozone destruction, instrumentation, control, building, and installation costs.

For smaller quantities of ozone, costs are higher per kilogram, but vary significantly from site to site. For plants serving less than 10,000 persons per day, 1.4 to 9.5 kg (3 to 21 lb/day), all items can be assembled into a single skid mounted unit.

Small ozonation systems can use diffuser contactors, which are generally constructed of polyvinyl chloride (PVC) pipe standing on end, or fiberglass reinforced plastic (FRP) tanks. Tables 5-12 and 5-13 list equipment cost estimates obtained from two ozonation system suppliers in 1982 for small water supply systems. Equipment costs are higher at higher dosages for a given flow rate.

Table 5-12. Costs of Ozonation Equipment for Small Water Supply Systems: Supplier A ($1982)

	Size of Treatment Plant					
	500,000 GPD		350,000 GPD		180,000 GPD	
Maximum dosage of ozone (mg/L) at peak flow	5	3	5	3	5	3
Daily ozone requirement (lb)	21	14	14	7	7	5
Contact chamber diameter[a] (ft)	6	6	5	5	4	4
Equipment Costs						
Air preparation and ozone generation unit	$31,500	$25,000	$25,000	$22,000	$22,000	$19,500
Contact chamber with diffusers	11,500	11,500	10,200	10,200	9,900	9,900
Monitoring Instrumentation[b]	15,000	15,000	15,000	15,000	15,000	15,000
Ozone Destruction Unit	6,700 (10 cfm)	5,000 (7 cfm)	5,000 (7 cfm)	4,200 (3 cfm)	4,200 (3 cfm)	4,200 (3 cfm)
Total Equipment Costs	$64,700	$56,500	$55,200	$51,400	$51,100	$48,600
Power Requirement (kWh)	13.3	10.1	10.1	5.0	5.0	3.65

[a] 14 ft. high, four compartments, four diffusers, Derakane fiberglass reinforced plastic.
[b] Includes monitors for ozone in generator product, ozone in ambient plant air, ozone dissolved in water, and dew point monitor in air preparation unit.
1 lb = 0.4536 kg;
1 ft = 0.3048 meters;
1 GPD = 0.003785 m^3/day.
Source: U.S. EPA (1983).

Ozone Supplier A provides four monitors with the system: dew point in the air preparation unit, ozone output of the generator, ozone in the plant ambient air (in case of leaks), and dissolved ozone residual in the water. All are optional (but recommended) for optimum performance and minimal labor and downtime.

Ozone Supplier B provided estimates for two types of air preparation equipment. One type operates at high pressures (5.6 to 8.4 kg/cm^2 [80 to 120 psig]), the other at low pressures (0.56 to 0.84 kg/cm^2 [8 to 12 psig]). The high-pressure units have lower capital costs, but require more energy for their operation. Supplier B offers two types of devices to monitor ozone output from the generator. The automatic, in-line continuous reading monitor costs $4,000; the nonautomatic monitor, which requires wet chemistry calculations to determine ozone output, costs $2,000.

Installation Costs

Costs to install ozonation equipment include labor and material for piping and electrical wiring. Piping can be extensive - transporting water to and from the ozone generators (if they are water-cooled) and the contactor, transporting ozone-containing air to the contactor chamber, and transporting contactor off-gases to and from the ozone destruction unit.

The ozonation equipment suppliers estimate that for units producing up to 13.6 kg/day (30 lb/day) of ozone, installation costs average from $9,705 to $16,175 for Supplier A and $12,750 to $21,250 for Supplier B.

Housing Costs

Installation of the power supply, air preparation, ozone generation, and turbine contacting operations require an area of approximately 3 by 5.1 m (10 by 17 ft). Diffuser contacting units are quite large and high (5.4 m [18 ft]), and are typically installed outside existing buildings or in the basement of buildings constructed for the other ozonation equipment. A Butler building with the above dimensions can be constructed for about $6,000.

Operating and Maintenance Costs

Operating costs for ozonation systems vary depending on:

- Oxygen use or air preparation method - high or low pressure, or subatmospheric pressure desiccant systems with or without an air chiller

- Generator cooling method - air or water cooled. In northern climates, water at the plant is generally cold enough to be used as a coolant all year. Southern climates must refrigerate cooling water

Table 5-13. Costs of Ozonation Equipment for Small Water Supply Systems: Supplier B ($1982)

	Size of Treatment Plant									
	0.1 MGD		0.2 MGD		0.3 MGD		0.4 MGD		0.5 MGD	
Maximum ozone dosage (mg/L), at peak flow	3		3		3		3		3	
Daily ozone requirement (lb/day)	3		6		7		12		14	
Equipment Costs	Pressure		Pressure		Pressure		Pressure		Pressure	
	Low[b]	High[c]	Low[b]	High[c]	Low[b]	High[c]	Low[b]	High[c]	Low[b]	High[c]
Air preparation and ozone generator[a]		$17,500	$33,200	$30,200	$38,500	$35,000	$43,000	$40,000	$49,800	$46,800
Power requirements (kWh/lb of O_3 generated)	10.5	20	10.5	13.5	10.5	13.5	10.5	13.5	10.5	13.5
Ozone contractor with diffusers	$8,500		$12,000		$16,000		$21,000		$29,000	
Ozone generation monitor	4,000		4,000		4,000		4,000		4,000	
Chamber exhaust monitor	2,200		2,200		2,200		2,200		2,200	
Dew point monitor	3,500		3,500		3,500		3,500		3,500	
Total Equipment Costs		$35,700	$52,900	$49,900	$62,200	$59,200	$71,700	$68,700	$86,500	$83,500

[a] Includes air preparation, ozone generation, ozone destruction, and system controls.
[b] Air preparation unit includes air filters or separators, compressor delivering air at 8 to 12 psig to a refrigerative cooler and a dual tower desiccant dryer.
[c] Same as low-pressure air preparation system, except compressor delivers air at 80 to 120 psig. High-pressure system takes less space and requires less maintenance, but requires more energy.
[d] $4,000 instrument is an automatic, continuous reading in-line monitor. A substitute is a $2,000 instrument that is not automatic and utilizes wet chemistry.
1 pound = 0.4536 kilograms; 1 MGD = 0.044 m^3/sec.
Source: U.S. EPA, 1983.

most of the year. Medium frequency generators require increased cooling.

- Contacting method - diffuser contactors do not require electrical energy as do the more compact turbine diffusers.
- Ozone dosage required
- Contactor exhaust gas handling - thermal, catalytic, or GAC destruction systems.

Maintenance costs include periodic cleaning, repair, and replacement of equipment parts. For example, air preparation systems contain air prefilters that must be replaced frequently, and tube-type ozone generators normally are shut down for annual tube cleaning and other general maintenance. Tube cleaning can require several days of labor, depending on the number and size of ozone generators in the system. Tubes, which can be broken during cleaning or deteriorate after years of operation at high voltages (or more rapidly if the air is improperly treated), must be replaced periodically.

Labor requirements, other than for periodic generator cleaning, include annual maintenance of the contacting basins and day-to-day operation of the generating equipment (average 0.5 hour/day).

Table 5-14 summarizes operating and maintenance costs for equipment from ozone Suppliers A and B. This table also includes building heating costs (assumed to be the same up to a 0.22 m^3/sec [0.5-MGD] plant). There are no chemical costs related to ozone generation, except for periodic changing of desiccant in air preparation systems (normally after 5 years of use).

5.4.3 Chlorine Dioxide

Chlorine dioxide is an unstable gas, explosive in air in concentrations above 10 percent by volume (corresponding to solution concentrations of about 12 g/L). Because of its instability, chlorine dioxide is always generated on site, in aqueous solution, and used shortly after its preparation. Solutions up to 5 g/L can be stored for up to 7 days. As long as care is taken to keep chlorine dioxide in solution and long-time storage of solution is avoided, there is no

Table 5-14. Operating and Maintenance Costs for Small Ozone Systems[a] ($1982)

Water Flow Rate (mgd)	Electrical Energy (kWh/yr) Building	Process	Total	Energy Costs[b]	Maintenance Material	Labor (hr/yr)	Labor Cost[c]	Total Cost ($/yr)
Supplier A								
0.18	6,570	6,661	13,231	$926	$120	250	$2,500	$3,546
0.35	6,570	12,775	19,345	1,354	200	250	2,500	4,054
0.50	6,570	51,611	58,181	4,073	300	250	2,500	6,873
Supplier B (High-Pressure Air Preparation)								
0.10	6,570	21,900	28,470	$1,993	$120	250	$2,500	$4,613
0.20	6,570	29,565	36,135	2,529	120	250	2,500	5,149
0.30	6,570	34,493	41,063	2,874	200	250	2,500	5,574
0.40	6,570	59,130	65,700	4,599	250	250	2,500	7,349
0.50	6,570	68,985	75,555	5,289	300	250	2,500	8,089

[a] Assumes 3 mg/L ozone dosage.
[b] Assumes 0.07/kWh.
[c] Assumes 10/hour.
1 mgd = 0.044 m^3/sec
Source: U.S. EPA (1983).

explosion hazard. Chlorine dioxide is readily soluble in water, but decomposes in sunlight.

Chlorine dioxide is a more powerful biocide than chlorine but has a lower oxidation potential. When prepared in the absence of excess free chlorine, it does not produce THMs or other chlorinated organic by-products of concern. Additionally, oxidation of bromide ion to hypobromous acid does not occur except when free chlorine is present. However, some chlorine dioxide generation methods do create conditions of excess free chlorine in which by-products are produced if their precursors are present. Excess free chlorine can also slowly produce chlorite and chlorate ions by disproportionation under pH conditions below 2 or above 11. (These conditions are not usually found in treating drinking water.)

Health effects studies have shown hematological effects in laboratory animals as a result of exposures to chlorate and chlorite ions (Abdel-Rahman et al., 1980). For this reason, EPA currently advises that the total concentration of chlorine dioxide and its decomposition products (chlorite and chlorate ions) be maintained at or below 1 mg/L, which is equivalent to a maximum applied dosage of 1.2 to 1.4 mg/L.

Gaseous chlorine dioxide has a strong, disagreeable odor, similar to that of chlorine gas, and is toxic to humans when inhaled. Its odor is detectable above concentrations of 0.1 ppm.

Chlorine dioxide can be used to preoxidize phenolic compounds and separate iron and manganese from some organic complexes that are stable to chlorination. Distribution system residuals of dissolved chlorine dioxide can be longer-lasting than those of chlorine because chlorine dioxide does not react with ammonia or form chlorinated organic materials. Depending on the types and quantities of organic materials present, periodic tastes and odors can be produced by chlorine dioxide (Routt, 1989).

Section 5.4.3.1 describes the generation of chlorine dioxide. Section 5.4.3.2 explains how excess chlorine dioxide and chlorite ion can be removed from solution,and Section 5.4.3.3 discusses system design considerations. Lastly, Section 5.4.3.4 discusses system costs.

5.4.3.1 Process Description

For drinking water treatment, chlorine dioxide is generated from solutions of sodium chlorite ($NaClO_2$), which is usually purchased as a 25 percent aqueous solution or as a solid (80 percent sodium chlorite). Historically, chlorine dioxide has been produced by treating sodium chlorite with either chlorine gas, sodium hypochlorite solution and mineral acid, or mineral acid alone. In all three cases, the appropriate aqueous solutions of reactants are metered into and mixed in a chlorine dioxide reactor, which is a cylinder containing flow distributing packings, such as Raschig rings, glass beads, or hollow glass cylinders. Residence time of the solutions in a properly sized reactor is only a few seconds. The resulting yellow solution is pumped directly into the water to be treated.

In this manner, chlorine dioxide solutions are generated as the material is required and used immediately. Reactor operations are automated with appropriate metering and instrumentation that controls the addition of chlorine dioxide according to the flow rate of the water being treated.

The three historical techniques for generating chlorine dioxide are discussed in detail by U.S. EPA (1983). However, some of these procedures can result in excess free chlorine being present. Free chlorine can oxidize chlorine dioxide to form chlorate ions, which are difficult to remove from solution. Consequently, the current recommended approach to chlorine dioxide generation is to maximize its yield while minimizing the presence of free chlorine (thus minimizing the formation of chlorate ion). Slootmaekers et al. (1989) discusses generation techniques to meet these objectives.

For water disinfection, chlorine dioxide can be generated using several reaction schemes, such as the reaction of aqueous hypochlorous acid with dissolved chlorite ion:

$$2NaClO_2 + HOCl \text{ -----> } NaCl + NaOH + 2ClO_2 \quad [1]$$

Chlorine dioxide also can be generated by the reaction of solid sodium chlorite in solution with mineral acid, with chlorine or with hypochlorous acid. The reaction for chlorine and/or hypochlorous acid with chlorite ion is:

$$2ClO_2^- + Cl_2(g) \text{ -----> } 2ClO_2(g) + 2Cl^- \quad [2a]$$

$$2ClO_2^- + HOCl \text{ -----> } 2ClO_2(g) + Cl^- + (OH)^- \quad [2b]$$

These reactions involve the formation of the unsymmetrical intermediate, Cl_2O_2:

$$Cl_2 + ClO_2^- \text{ -----> } [Cl_2O_2] + Cl^- \quad [3]$$

At high concentrations of both reactants, the intermediate is formed very rapidly. Elemental chlorine formed by Equation [4] is recycled by means of Equation [3]. Thus, primarily chlorine dioxide is produced as a result:

$$2[Cl_2O_2] \text{ ----> } 2ClO_2 + Cl_2 \quad [4]$$

or

$$[Cl_2O_2] + ClO_2^- \text{ -----> } 2ClO_2 + Cl^- \quad [5]$$

On the other hand, at low initial reactant concentrations, or in the presence of excess hypochlorous acid, primarily chlorate ion is formed, due to the following reactions:

$$[Cl_2O_2] + H_2O \text{ -----> } ClO_3^- + Cl^- + 2H^+ \quad [6]$$

and

$$[Cl_2O_2] + HOCl \text{ -----> } ClO_3^- + Cl^- + H^+ \quad [7]$$

Thus, high concentrations of excess chlorite ion favor the second order reactions (Equations 4 and 5), and chlorine dioxide is formed. At low concentrations, the second order disproportionation process becomes unimportant, and reactions 6 and 7 produce chlorate ion rather than chlorine dioxide. The reasons for the production of chlorate ion are related to the presence of high concentrations of free chlorine and the rapid formation of the Cl_2O_2 intermediate, which, in turn, reacts with the excess hypochlorous acid to form the unwanted chlorate ion.

The stoichiometry of the undesirable reactions which forms chlorate ion is:

$$Cl_2O_2^- + HOCl \text{ -----> } ClO_3^- + Cl^- + H^+ \quad [8]$$

$$ClO_2^- + Cl_2 + H_2O \text{ -----> } ClO_3^- + 2Cl^- + 2H^+ \quad [9]$$

Therefore, the most effective way to minimize chlorate ion formation is to avoid conditions that result in low reaction rates (e.g., high pH values and/or low initial reactant concentrations, and the presence of free hypochlorous acid). Clearly, the reaction forming chlorate ion (Equation 6) will be more troublesome in dilute solutions. On the other hand, whenever treatment by chlorine dioxide (which forms chlorite ion in the process) is followed by the addition of free chlorine (HOCl with a pH of 5 to 8), the unwanted chlorate ion will also be formed.

5.4.3.2 Establishing a Chlorine Dioxide Residual

Laboratory studies have shown that about 70 percent of the chlorine dioxide added to drinking water is converted to chlorite ion (Singer, 1986). Therefore, 1.2 to 1.4 mg/L chlorine dioxide is the maximum practical dosage to meet the currently recommended maximum total oxidant residual of 1 mg/L. However, Slootmaekers et al. (1989) reports that nearly all of the chlorine dioxide ion added as a primary oxidant/disinfectant is converted to chlorite ion. Because of differences in the nature of water constituents that exert demand for chlorine dioxide, this ratio should be individually determined for each water supply.

Recent studies (Pinsky and Coletti, 1988) have shown that when water containing 1 to 5 mg/L of chlorine dioxide is passed through a GAC column (with 6-minute EBCT), the residual chlorine dioxide and chlorite ions are chemically reduced to innocuous chloride ions. This implies that as long as GAC with sufficient EBCT is used after primary disinfection or oxidation with chlorine dioxide, the currently recommended maximum levels for chlorine dioxide, chlorite, and chlorate are achievable regardless of the dosages of chlorine dioxide used.

More recently, Howe et al. (1989) have confirmed the reduction of chlorine dioxide and chlorite ion on passage through GAC. GAC does not, however, remove chlorate ion. These researchers showed that breakthrough of chlorite ion occurred after approximately 94 and 100 days of passage through GAC columns having 7.5 and 15 minute EBCT, respectively. Thus, the capacity for GAC removing chlorite ions in these studies was between 14 and 15 mg of chlorite ion per gram of GAC.

These data indicate that if GAC is to be installed for postfiltration adsorption of organics, chlorine dioxide can be employed as a primary disinfectant somewhere ahead of the GAC. However, if postfiltration GAC adsorption is installed only to remove chlorine dioxide and chlorite ion, the costs will be very high (because of the postfiltration GAC installation and 90-110 days reactivation needed for breakthrough of chlorite ion).

Slootmaekers et al. (1989) have shown that it is technologically feasible to reduce chlorine dioxide and chlorite ions to chloride ion with sulfur dioxide. Once oxidant demand has been satisfied with chlorine dioxide, excess sulfur dioxide/sulfite ion can be added to the treated water to remove residual chlorine dioxide and chlorite ion. This would be followed by the addition of free chlorine to remove the excess residual sulfur dioxide/sulfite ion.

The stoichiometry of the reaction is consistent with the equation:

$$2SO_3^{2-} + ClO_2^- \text{----->} 2SO_4^{2-} + Cl^-$$

This corresponds to a stoichiometry of two moles of SO_3^{2-} consumed for every mole of $C1O_2^-$ reduced. Over a range of conditions studied by Slootmaekers et al. (1989), at room temperature, the stoichiometry deviated by less than 5% from the value shown in the equation. Furthermore, the stoichiometry appears to be independent of temperature in the 5° to 30°C range.

However, at pH 8 and above and in the presence of air, the overall stoichiometry appears to deviate from that shown by the above equation, probably due to an increase in the rate of the competing sulfite ion/oxygen reaction. Thus, less acid solutions (e.g., in the pH 5.5 to 6 range) favor the rapid reduction of chlorite ion and minimize the loss of sulfite ion from the competing sulfite ion/oxygen reaction.

Thus, Slootmaekers et al. (1989) have established that the sulfur dioxide/sulfite ion - chlorite ion reaction is greater than 95 percent effective. Furthermore, with a 10-fold excess of sulfur dioxide/sulfite ion and with chlorite ion at the 0.5 to 7 mg/L level, the removal of chlorite ion is complete in less than one minute at pH 5 and below, and completed in 15 minutes or less at pH 6.5. This means that sulfur dioxide/sulfite ion can be used to reduce the level of chlorite ion by-product in drinking water to below the 0.1 mg/L without great difficulty.

However, the rate of reductive conversion of chlorine dioxide and chlorite ion to chloride ion decreases with increasing pH as illustrated by the following data:

pH	Beginning Chloride Ion Conc. (mg/L)	Beginning Sulfite Ion Conc. (mg/L)	Time required for 99% removal of chlorite ion
5.0	0.5	5.0	0.34 minutes
5.5	0.5	5.0	4.4 minutes
5.5	1.0	10.0	1.1 minutes
6.5	1.0	10.0	15.2 minutes
6.5	1.0	20.0	3.8 minutes
7.5	1.0	10.0	15.6 hours
7.5	1.0	100.0	9.4 minutes
8.5	1.0	100.0	3.2 days

In practice, if all of the chlorite ion is not removed in the time allotted, three choices are available to the treatment plant operator:

- Increase the amount of sulfur dioxide/sulfite ion used in the removal process: doubling the amount will remove the chlorite ion four times faster since the rate is second order in the concentration of sulfur dioxide/sulfite ion.
- Decrease the pH: every decrease of pH by one unit increases the rate of chlorite removal by at least a factor of ten.
- Increase the length of time for removal: the rate of removal is logarithmic in time, since the removal of chlorite ion is first order in the concentration of chlorite ion.

The ability to remove chlorite ion by-product allows considerably higher chlorine dioxide concentrations to be used (easily up to 6 to 8 mg/L), eliminating the need for a combination of chlorine dioxide and free chlorine, the most common sources of chlorate ion.

It is important to note that chlorate ion is not chemically reduced by sulfur dioxide/sulfite ion under the conditions studied by Slootmaekers et al. (1989). The calculated half-life for the removal of mg/L levels of chlorate ion with excess sulfur dioxide/sulfite ion is in excess of many months.

Chlorate ion in the finished water can arise from one of two possible sources:

- Improperly tuned older chlorine dioxide generators.
- As a by-product during the chlorine dioxide oxidation/disinfection step, especially when free chlorine is used concomitantly.

5.4.3.3 Chlorine Dioxide System Design Considerations

Several types of generation equipment are available. Each supplier provides recipes for preparing and metering the solutions to the reactor to produce a known and constant chlorine dioxide concentration. For smaller systems treating less than 0.022 m^3/sec (0.5 MGD), the chlorine dioxide dosages required are very low. In these cases, the generating units may be operated intermittently, collecting chlorine dioxide solution in an enclosed holding tank from which a metered flow can be delivered later. Intermittent operation is recommended over continuous operation because the mixing of reactant solutions is less efficient in the reactor at consistently low flow rates; thus, the conversion of chlorite ion to chlorine dioxide will be less efficient. Selected generation systems are briefly discussed below.

Gaseous Chlorine Systems

The most common chlorine dioxide generation process uses chlorine gas and sodium chlorite (U.S. EPA, 1983). This process is convenient when chlorine gas was previously used for primary disinfection at the plant. Figure 5-8 is a schematic diagram of such a system.

Sodium Hypochlorite and Mineral Acid Systems

A manual feed system that produces chlorine dioxide using sodium hypochlorite solution with sodium chlorite and strong mineral acid is illustrated in Figure 5-9.This process is well suited to most small treatment systems. The operator can adjust solution strength for each reactant so that feed pumps of equal capacities can be used. The chlorine dioxide production and addition rates are paced according to the flow rate of the water and/or the secondary disinfectant demand.

Sodium chlorite is available either as a solid (80 percent active sodium chlorite) in 90.72-kilogram (200-pound) drums or as a solution (25 percent active sodium chlorite, 31.25 percent solids in 208.2-liter [55 gallon] drums). If not used directly from the drum, sodium chlorite solution is stored in polyethylene or fiberglass tanks and transferred by means of PVC, rubber, or Tygon tubing systems. Diaphragm pumps with PVC components are used to transfer the sodium chlorite solution. Provision must be made for immediate washdown of any chemical spills; this precaution is generic to all chlorine dioxide generating systems.

The CIFEC System for Generating Chlorine Dioxide

The CIFEC System was developed in France and is currently used by several treatment plants in the United States. Figure 5-10 is a schematic diagram of this system, which produces chlorine dioxide from chlorine gas. This system produces high yields of chlorine dioxide under conditions of minimal excess free chlorine.

In the CIFEC system, chlorine gas is injected into continuously circulating water, referred to as an "enrichment loop." This method produces much higher concentrations of dissolved chlorine (hypochlorous acid) than those achieved in a system using a single injection point. The hypochlorous acid solution (pH below 4) is pumped into the reactor along with a sodium chlorite solution. Since the pH of the hypochlorous acid solution is below 4, more chlorite ions are converted to chlorine dioxide than in other chlorine gas systems that operate in higher pH ranges. In addition, chlorine dioxide is produced with little free chlorine.

Rio Linda Chlorine Dioxide Generator

Rio Linda Chemical Co., Inc. manufactures a unique chlorine dioxide generator. In this system, shown schematically in Figure 5-11, chlorine dioxide is generated by the addition of chlorine gas to a sodium chlorite solution, as in other systems. The novel aspect of this generator is that chlorine gas is mixed with concentrated sodium chlorite solution just before the reactor. The vacuum action from the water flowing through an eductor into the reactor brings the two solutions together. Since there is no need for a pump, the system is more compact.

This system can be operated manually or automatically. Additionally, the system can be designed to mix in solutions of hydrochloric acid, hypochlorite and sodium chlorite just before the chlorine dioxide reactor. Generator efficiencies with excess chlorine are claimed to approach greater than 98 percent conversion of the sodium chlorite into chlorine dioxide, with minimal production of chlorate ion at an effluent pH between 6.5 and 7. Production rates of these generators can range from a few pounds up to 2,721.6 kilograms (6,000 pounds) per day.

Summary of Commercial Chlorine Dioxide Generators

In many older chlorine dioxide generators, hydrochloric acid is fed into the chlorine solution before reaction with the sodium chlorite. The acid shifts the chlorine solution equilibria in favor of molecular chlorine (Slootmaekers et al., 1989). (The hypochlorous acid dissociation and the chlorine hydrolysis equilibrium are shown in the equations below.)

$$HOCl \dashrightarrow (OCl)^- + H^+$$

$$Cl_2 + H_2O \dashrightarrow HOCl + H^+ + HCl^-$$

The acid must be carefully controlled so that the pH of the chlorine dioxide solution is maintained between 2 and 3. Higher or lower pH values result in decreased yields. Yields of more than 90 percent have been reported from a properly pH-adjusted system (Aieta & Berg, 1986), with approximately 7 percent excess chlorine remaining in solution (where excess chlorine is defined as the amount of unreacted chlorine remaining in the chlorine dioxide generator effluent).

High yields of chlorine dioxide, with low levels of free chlorine in solution, can be produced using a chlorine solution with a concentration greater than 4 g/L. This chlorine concentration is near the upper operating limit of commercial chlorine ejectors. Since these ejectors operate at constant water flow rates, the yield of this method of generation is dependent upon the production rate, with lower production rates resulting in lower yields. This type of generator normally operates on an intermittent basis to maintain a high yield when less than maximum production capacity is required.

The most recent development in chlorine dioxide generator technology is a system that uses the reaction of chlorine with a concentrated sodium chlorite solution under vacuum (Aieta & Berg, 1986; Aieta & Roberts, 1986). The chlorine dioxide is removed from the reaction chamber by a gas ejector, which is very similar to the common chlorine gas vacuum feed system. This technique of generation produces chlorine dioxide solutions with yields in excess of 95 percent. The chlorine dioxide concentration is 200 to 1,000 mg/L and contains less than 5 percent excess chlorine with minimum formation of chlorate ion. However, in this context, it should be noted that almost all older chlorine dioxide generators now in use in water treatment plants throughout the United States continue to produce some chlorate ion, unless they are very carefully tuned and properly monitored. A well-tuned generator may produce as little as 1 to 2 percent chlorate ion (Slootmaekers et al., 1989).

5.4.3.4 Costs of Chlorine Dioxide Disinfection

The costs of chlorine dioxide systems are presented in three categories: equipment, operating and maintenance, and chemical costs.

Equipment Costs

Quotes from two suppliers of chlorine dioxide generation equipment were obtained in 1982 (see Table 5-15). Supplier A's recirculating loop system (CIFEC) is the highest priced unit at $34,000. It operates with a special recirculating pump designed to handle hypochlorous acid below pH 4, plus a sodium chlorite solution pump and all necessary instrumentation to allow automatic operation. There are shutdown provisions in the event of interruptions in water flow.

The next lowest in price is the system from Supplier B that generates chlorine dioxide from 33 percent hydrochloric acid, 12 percent sodium hypochlorite, and 25 percent sodium chlorite solution. This wall-mounted unit costs $25,000 (installed), and includes three solution pumps, a water flow rate detector, and switches to shut down the unit if the water flow stops. For chlorine dioxide volumes sufficient to treat flows in communities with populations of 5,000 and 2,500, this unit is capable of continuous operation, with no loss in conversion efficiency of chlorite ion to chlorine dioxide. However, to supply the needs of systems serving as few as 25 persons, the unit must be operated intermittently, with chlorine dioxide solution being stored in a holding tank for later metering into the water.

Supplier C provides two types of chlorine dioxide generators for small water supply systems. One uses acid/sodium chlorite; the other uses chlorine gas and sodium chlorite. These units cost $3,600 wall-mounted, and $4,320 for a floor-mounted cabinet. The single size unit offered by this supplier is designed to generate up to 63.5 kg/day (140 lb/day). In order to produce 3.63 kg/day (8 lb/day) or less, a small water utility must install a holding tank and operate the generator intermittently.

The chlorine gas/sodium chlorite generator of Supplier C requires a gas chlorinator to feed chlorine gas. Therefore, in new plants considering this type of equipment, the cost of a chlorinator must be added to the cost of the chlorine dioxide generator. In existing plants currently using gas chlorination, the

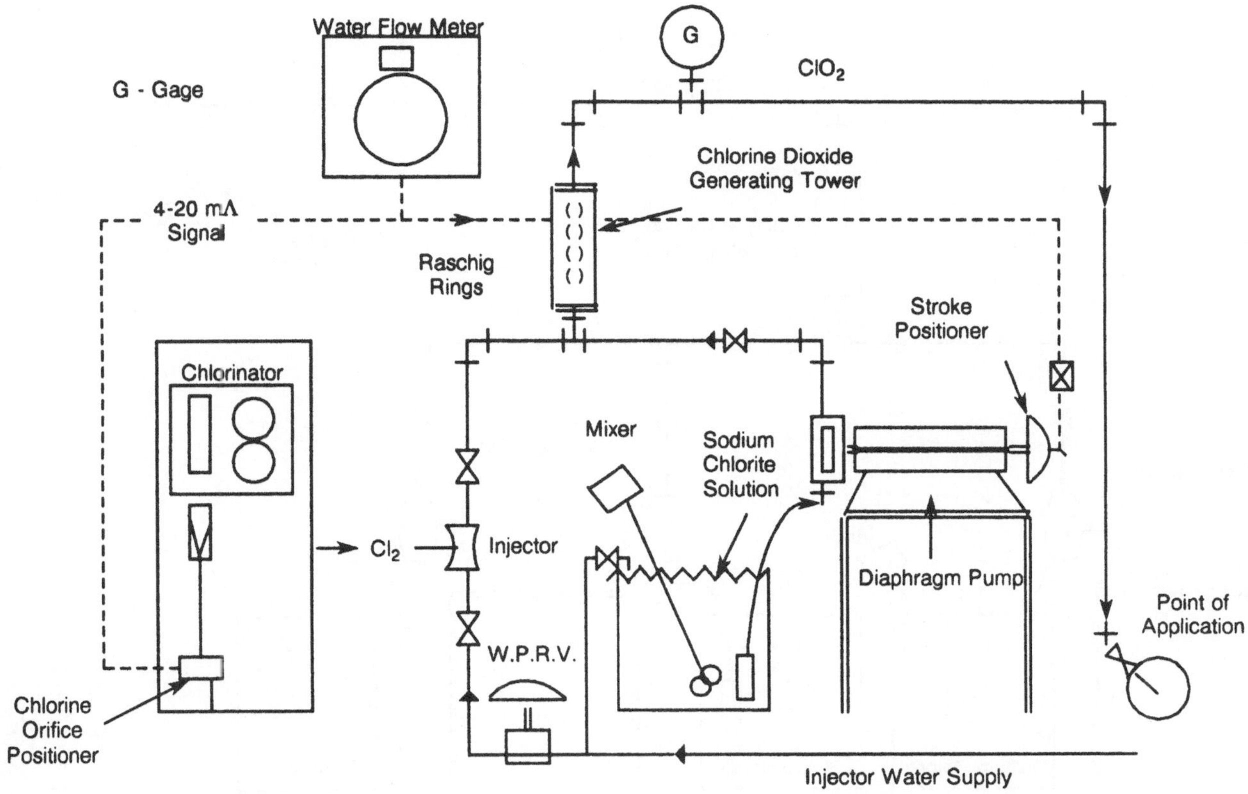

Figure 5-8. Schematic diagram of an automatic feed, automatic flow-proportional chlorine dioxide system: Generation from chlorine and sodium chlorite.
Source: Capital Controls Co., Inc.

chlorinator already is in place, and therefore would not represent additional equipment cost.

The purchaser of chlorine dioxide generating systems should consider the inclusion of additional equipment to allow automatic operation, such as microprocessor-controlled electronic valve systems, ratio-proportioning and flow-paced systems, and remote start and feed capabilities. In addition, alarm facilities for operating parameters such as low feed/production ratios, low yields, shut-offs for chlorine dioxide in air, low reactant feed for chlorine gas or sodium chlorite solution, high or low vacuum, and low water flows or pressures are advisable options.

Operating and Maintenance Costs

In general, operating and maintenance costs for generating chlorine dioxide are independent of the quantities generated. Table 5-16 summarizes these costs on an annual basis. Maintenance material costs are for minor equipment repair only. Labor required for preparation of solutions and periodic maintenance of the equipment is estimated to be 1 hour/day, or 365 hours/year. Total annual operating and maintenance costs of $4,124/year are estimated.

Chemical Costs

At a production rate of only 3.6 kg/day (8 lb/day) (maximum for a 0.044 m^3/sec [1-MGD] water treatment plant at a 1 mg/L applied chlorine dioxide dose), chemical costs are not as significant as pumping costs. Chemical costs in 1982 were as follows:

Gaseous chlorine	$1.04/kg ($0.47/lb)
Sodium chlorite	$3.42 to $3.64 ($1.55 to $1.65/lb) (100 percent solids)
Hydrochloric acid	$0.22/kg ($0.10/lb)
Sodium hypochlorite	$0.25/L ($0.93/gal) (15 percent solution)

5.4.4 Ultraviolet Radiation

UV radiation is an effective bactericide and virucide, but an ineffective cysticide. Consequently, it is recommended as a primary disinfectant only for ground waters not directly influenced by surface waters (where there is no risk of *Giardia* cyst contamination). UV radiation (254 nm wavelength) penetrates the cell wall and is absorbed by the cellular nucleic acids. Radiation absorption prevents

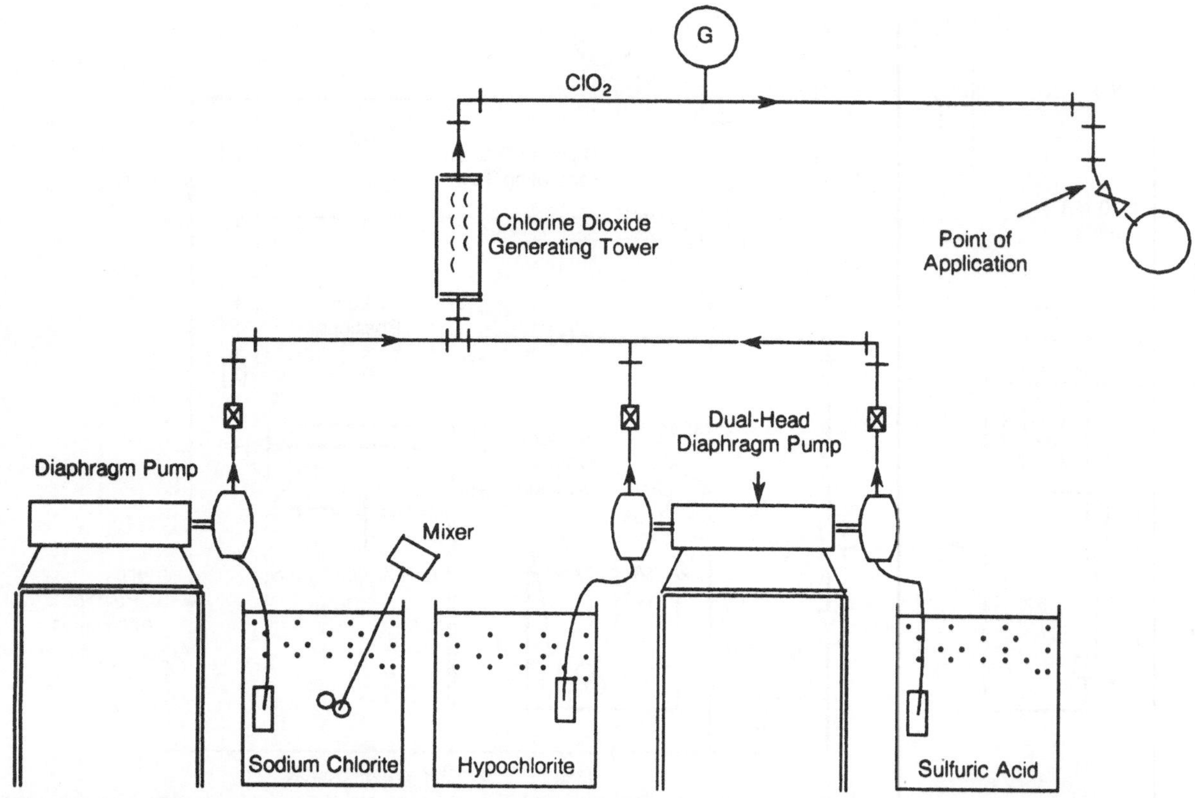

Figure 5-9. Manual feed equipment arrangement for generating chlorine dioxide from sodium hypochlorite solution and mineral acid.
Source: U.S. EPA (1983).

replication, thus killing the cell. Since UV radiation is not a chemical agent, it produces no toxic residual.

Major advantages of UV radiation are its simplicity, lack of impact on the environment and aquatic life, and minimal space requirements. In addition, required contact times are seconds rather than minutes. The equipment is simple to operate and maintain if the apparatus is cleaned properly on a regular basis.

Section 5.4.4.1 describes the process of disinfection with UV radiation. System design considerations, including lamp designs, are provided in Section 5.4.4.2. Operating and maintenance considerations are discussed in Section 5.4.4.3 and costs of UV systems are described in Section 5.4.4.4.

5.4.4.1 Process Description

UV radiation disinfection uses a special lamp to transfer electromagnetic energy to the target organism cells. The most efficient and widely used device is the mercury arc lamp. It is popular because approximately 85 percent of its energy output is of the 253.7 nm wavelength, within the optimum germicidal range of 250 to 270 nm. The lamps are long thin tubes. When an electric arc is struck through mercury vapor, the energy discharge generated by the mercury excitation results in the emission of UV radiation. This radiation then destroys the cell's genetic material and the cell dies.

The effectiveness of radiation is a direct function of the energy dose absorbed by the organism, measured as the product of the lamp's intensity and the time of exposure. Intensity is the rate at which photons are delivered to the target. The intensity in a reactor is governed not only by the power of the lamp, but also by the placement of the lamps relative to the water, and by the presence of energy sinks that consume UV radiation. Water with suspended solids, color, turbidity, and soluble organic matter can react with or absorb the UV radiation, reducing the disinfection performance. Therefore, water with high concentrations of these substances may receive inadequate disinfection.

The radiation dose absorbed by the water is the water's UV demand, which is analogous to chlorine demand and is quantified as the absorption of UV energy (wavelength of 253.7 nm) in a given depth of water. The measurement is most commonly expressed by the UV absorbance coefficient alpha:

$$\text{alpha} = 2.3 \text{ absorbance units(a.u.)/cm}$$

In addition to intensity and UV demand of the water, the exposure time also affects the energy dosage that the target organisms absorb. Exposure time is

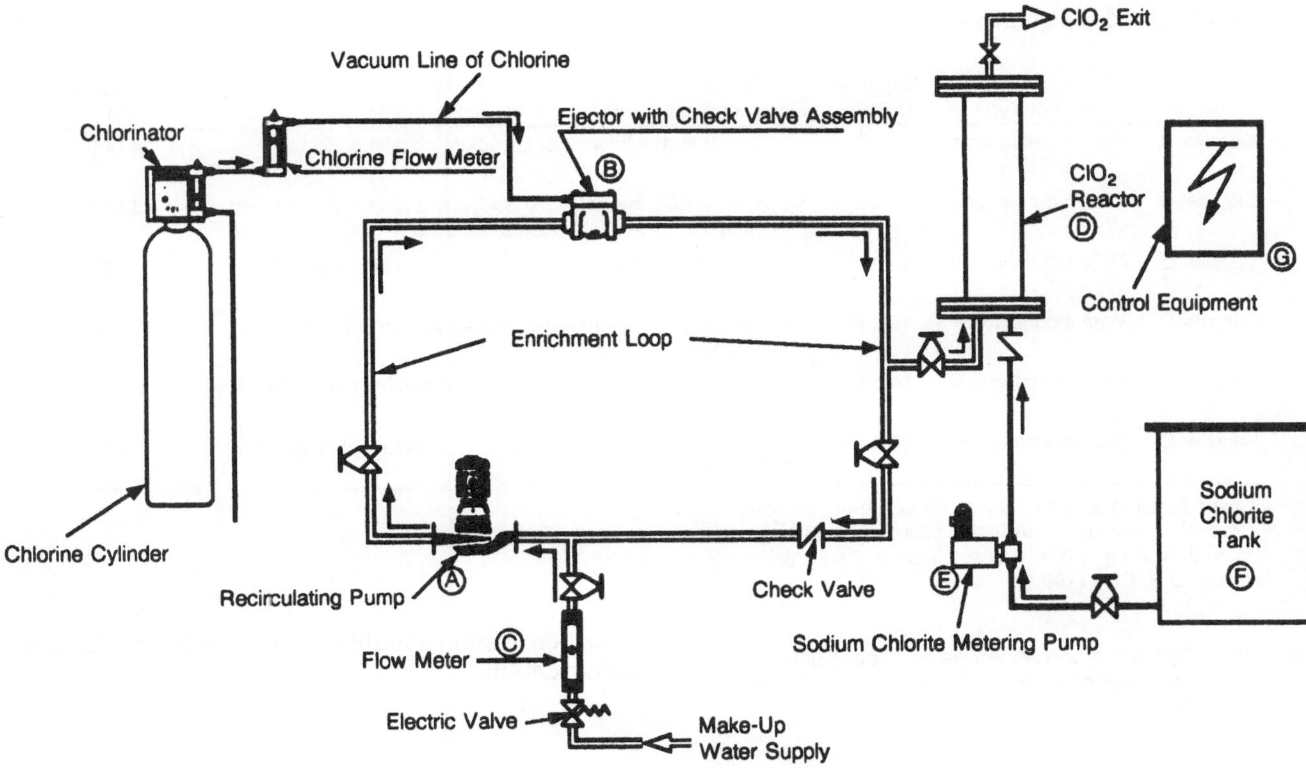

Figure 5-10. Schematic of CIFEC chlorine dioxide generating system.
Source: CIFEC, Inc., Paris, France.

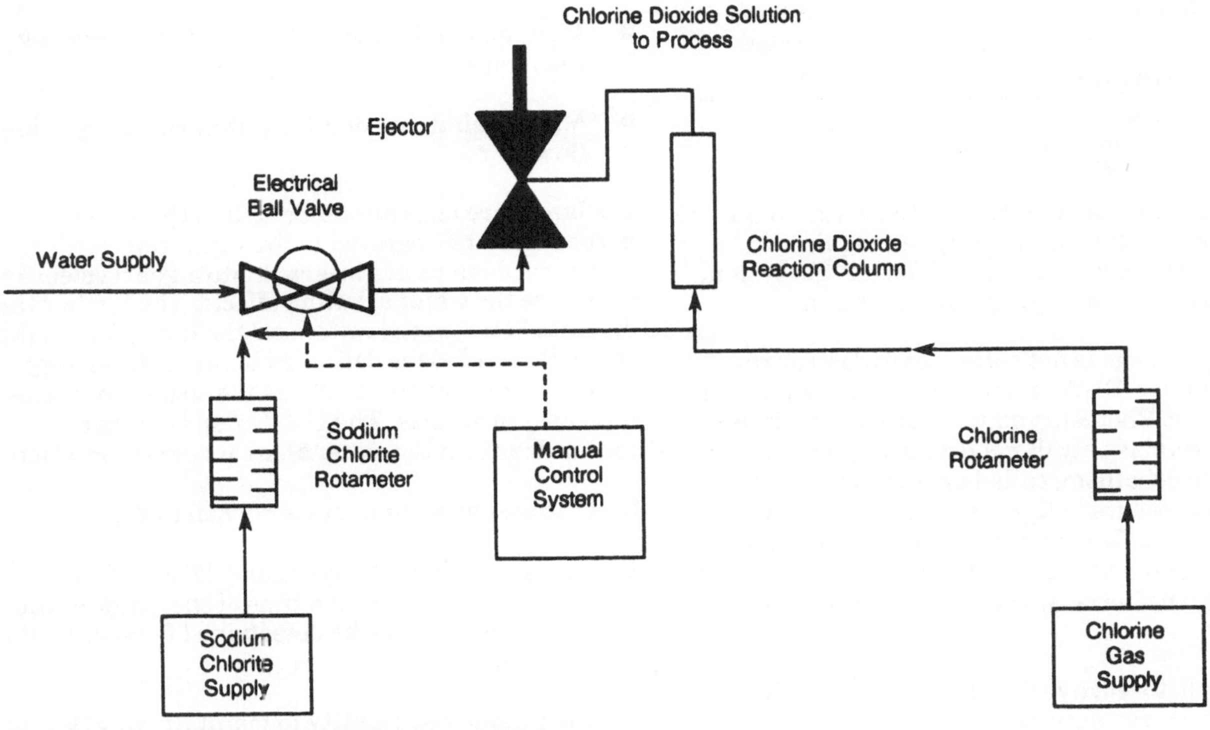

Figure 5-11. Schematic of manual chlorine dioxide system.
Source: Rio Linda Chemical Co., Sacramento, California.

Table 5-15. **Costs of Chlorine Dioxide Generating Equipment $1982**

Vendor	System Type	Production Capacity (lb/day)	Space Requirements[a]	Reactants	Unit Cost ($)
Supplier A	Recirculating loop (CIFEC)	1-10	2 × 3 × 6 feet	Chlorine gas, sodium chlorite solution	34,000
Supplier B	Wall-mounted unit	4	3.5 × 4 × 1.5 feet	Hydrochloric acid, sodium hypochlorite, and sodium chlorite solutions	25,000
Supplier C	Floor-mounted unit	14-140	4 × 3 × 1.5 6.5 inches	Chlorine gas and sodium chlorite solution	4,300[b]
Supplier C	Wall-mounted unit	14-140	37.5 × 27 × 6.5 inches	Chlorine gas and sodium chlorite solution	3,600[b]
Supplier C	Floor-mounted unit	14-140	4 × 3 × 1.5 feet	Hydrochloric acid and sodium chlorite solution	4,300
Supplier C	Wall-mounted unit	14-140	37.5 × 27 × 6.5 inches	Hydrochloric acid and sodium chlorite solution	3,600

[a] Space requirements for solution tanks are not included.
[b] Costs for a chlorinator are not included in these estimates.
1 lb = 0.4536 kg; 1 ft = 0.3 m; 1 inch = 2.54 cm.
Source: U.S. EPA (1983).

Table 5-16. **Operation and Maintenance Cost Summary for Chlorine Dioxide Generating and Feed Systems**

Item	Requirements	Annual Costs ($)
Electrical Energy		
Metering pumps and mixers	1,240 kWh/yr	86.80
Building	4,100 kWh/yr	287.00
Subtotal	5,340 kWh/yr	373.80[a]
Maintenance Materials		100.00
Labor	365 h/yr	3,650[b]
Total Annual O&M Cost		**4,124**

[a] Assumes $0.07/kWh.
[b] Assumes $10/hour labor cost.
Source: U.S. EPA (1983).

controlled by the residence time of the water in the reactor. Continually maintaining the required residence time is not always possible, but the system design should maximize plug-flow operation.

If the energy dosage is not sufficient to destroy the target organisms' DNA macromolecules, disinfection is not effective. Photoenzymatic repair occurs if the genetic material is only damaged during irradiation. This repair mechanism, called photoreactivation, occurs with exposure to light from the sun or most incandescent and fluorescent lights (wavelengths between 300 and 500 nm). Photoreactivation does not occur with all bacterial species and is difficult to predict.

To prevent photoreactivation, the rule of thumb is to increase the dosage necessary to meet required reductions in organism levels (U.S. EPA, 1986b). For example, if the disinfection criteria require a 3-log reduction of microorganism concentrations, the UV radiation system should be designed to provide a 4-log reduction.

5.4.4.2 UV Disinfection System Design Considerations

The basic design considerations for a UV system are:

- Satisfying the UV demand of the water
- Maximizing the use of UV energy delivered by the lamps
- Maintaining the conditions that encourage plug flow

UV lamps are usually submerged in the water, perpendicular or parallel to the water flow. Submerged lamps are inserted into a quartz sleeve to minimize the water's fouling effects. The further the distance between the water and the lamp, the weaker the radiation dosage delivered because the energy dissipates or becomes dilute as the space it occupies increases in volume. The UV demand of other contaminants in the water also consumes radiation.

Specific design parameters to consider are:

1. *Residence Time Distribution (RTD)* – This describes the detention time of the water in the reactor and should be determined for several flow conditions.

2. *Plug Flow* – The ability to maintain plug flow in the reactor is influenced by the inlet and exit designs. Disturbances at the inlet and exit planes of the lamp battery should be minimized and

necessary changes in the flow direction should be made outside the lamp battery.

3. *Dispersion Number* – A key goal is to minimize the dispersion number, d (cm^2/s). As a design goal, d should be between 0.02 and 0.05. This number represents a plug-flow reactor with low to moderate dispersion. This value is attained by increasing the product of the velocity (cm/s) of and distance traveled (cm) by the water as it flows through the reactor while under direct exposure to UV radiation. However, extended lengths and higher velocities cause higher head losses; therefore, adjusting the dispersion number may be necessary to meet specific criteria for both full-scale modules or pilot units. Head loss is determined over a wide velocity range and excludes entrance and exit losses.

4. *Effective Volume* – The inlet and outlet designs should achieve equivalent water velocities at all points entering and exiting the lamp battery. This maximizes the lamp battery use and improves cost effectiveness. Stilling walls (perforated baffles) and weirs in the reactor design assist in controlling water velocities.

UV Lamp Designs

Lamps used in UV disinfection systems typically have arc lengths of approximately 0.8 and 1.5 m (2.5 and 4.9 ft) and full lengths of 0.9 and 1.6 m (3 and 5.3 ft), respectively. The arc length describes the active, light-emitting portion of the lamp. Lamp diameters typically are 1.52 and 2.0 cm (0.6 and 0.8 in). A sleeve made of fused quartz or another material that is highly transparent to UV light, such as Vycor, protects lamps that are submerged. Nonsubmerged lamps are placed near the wall of the water conduit, which is made of a UV light-translucent material.

Factors Affecting the Design of the UV Disinfection System

Initial microorganism density, suspended solids (or turbidity), UV demand of the water at the disinfection point, and water flow rate all affect the size and performance of the UV disinfection system.

The performance of a UV disinfection unit relates directly to the initial density of the indicator organisms. The higher the initial density, the greater the dosage of radiation required. For this reason, microorganism density should be continually monitored. Turbidity directly affects the performance of the UV disinfection system as well. Particulates suspended in water block the UV radiation, thereby protecting bacteria and hindering disinfection. The UV demand of the water affects the radiation intensity in the reactor and, thus, affects the system size and the lamp placement that achieves the desired performance.

Water flow rate is another key factor in determining system size. Both the hydraulic load to the plant and the design of the processes preceding disinfection affect flow. The size of the UV system, however, should be based on peak flow rates and projected flows for the plant's design year rather than on average flows, which are used to predict operating and maintenance requirements.

5.4.4.3 UV System Operating and Maintenance Considerations

This chapter discusses operation and maintenance of UV lamps and reactor care. The intensity of radiation in the reactor depends on the lamp output and the reactor cleanliness. Therefore, monitoring lamp intensity and properly maintaining the reactor are essential to reliable system performance.

Operation and Maintenance of UV Lamps

Lamp output is influenced by lamp temperature, voltage potential across the lamp, and age of the lamp. Lamp temperature cannot be controlled in submerged lamp systems. In other systems, however, lamp temperatures are controlled by regulating ambient air temperatures using cooling fans or recirculating the heat generated by the lamp ballasts for warming.

Adjusting the voltage will vary the lamp output. Decreasing the voltage reduces current to the lamp and, therefore, lamp output. Voltage regulators improve system efficiency by reducing voltage and "dimming" lamps to conserve power during periods of low UV demand. Lamp intensity cannot be reduced to levels below 50 percent, however, without causing the lamps to flicker and eventually turn off.

Factors affecting deterioration of performance and aging of UV lamps include electrode failure; mercury plating or blackening in the lamp's glass tube; and tube solarization, which results in reduced energy transmission through the glass. These factors can reduce lamp output by 40 to 60 percent.

UV lamps used for disinfection are typically hot cathode lamps, which deteriorate progressively with each startup. Life expectancy is determined by the number of times the lamp is started. The lamp life cited by most manufacturers is 7,500 hours, based on a burning cycle of 8 hours; that is, the lamp will last 7,500 hours assuming it is restarted every 8 hours. The average UV output after 7,500 hours is estimated to be 70 percent of the lamp output at 100 hours.

Monitoring Lamp Intensity

Lamp intensity should be measured to monitor lamp condition and determine the need for maintenance. The monitoring procedure compares current lamp intensity to the intensity of new lamps. The operator first measures the intensity, at a fixed distance, of three to five new lamps that have burned for about 100 hours (the first 100 hours is considered a "burn-in" period). The average of these values is the benchmark from which to measure the relative output of the lamps. Each lamp is tagged so that it can be monitored individually.

A similar procedure is used to monitor the transmittance of a quartz sleeve. First, the intensity of a single lamp is measured with and without a new, clean quartz sleeve. Similar measurements are taken of the quartz unit in use and compared to the transmittance of the new quartz. Before being tested, the quartz is cleaned to assure that maximum transmittance is restored.

Operation and Maintenance of the Reactor

The most important operating factor for the UV reactor is the cleanliness of the surfaces through which radiation must pass. Surface fouling can result in inadequate performance, so a strict maintenance schedule is recommended.

An operator determines the need for reactor cleansing by draining and visually inspecting the surfaces. Open reactor systems are easily inspected. Systems with sealed vessels are inspected through portholes or manways in the reactor shell. Surfaces of submerged quartz systems become coated with an inorganic scale, very much like boiler scale. This is a particular problem in areas with hard water. Additionally, the inside surface of the quartz and the outer surfaces of the Teflon tubes eventually develop a grimy dust layer, primarily from airborne dirt and water vapor.

Fouling of the reactor's internal surfaces also is indicated by reduced performance and intensity measured by in-line probes. While these provide some indication of fouling, operators must still visually inspect the surfaces.

The fouled surfaces of lamps and quartz sleeves are cleaned manually with a mild soap solution and then swabbed with a rag soaked in isopropyl alcohol. The transmittance of the lamps and sleeves is measured after cleaning and those that have inadequate measurements are replaced. An inventory allows the plant operator to trace operation of individual components. Quartz sleeves should last between 4 and 7 years, but this varies by site.

In Teflon systems, the lamps are on removable racks and should be cleaned and monitored in the same manner as the quartz systems. The Teflon tubes should also be cleaned with mild soap and swabbed with alcohol. Each tube should be monitored for transmittance, just as with the quartz sleeves. Monitoring may not be as straightforward because of the limited accessibility to the tubes and problems in obtaining direct measurements with a UV radiometer/detector.

5.4.4.4 Costs of UV Radiation Systems

Table 5-17 summarizes construction costs developed for single and multiple UV lamp disinfecting units ranging in water throughput capacity from 54.5 to 4,251.3 m^3/day (14,400 to 1,123,200 gal/day). UV generating units are quite compact; for example, the 4,251.3 m^3/day (1,123,400 gal/day) unit occupies an area less than 2.2 m^2 (24 ft^2). Costs listed in Table 5-17 include UV unit equipment costs, and the related piping, electrical, installation, and equipment housing costs.

Operating and maintenance costs are shown in Table 5-18 for the same size plants as in Table 5-17. Costs assume continuous 24 hour/day operation, with only occasional shutdown to clean cells and replace worn out UV lamps. These costs include process energy for the mercury lamps and building energy for heating, lighting, and ventilating.

Maintenance materials costs are related to replacement of the UV lamps, which may be required about every 8 months if operated continuously. Labor requirements include occasional cleaning of the quartz sleeves surrounding the mercury vapor lamps, and periodic replacement of the weak UV bulbs.

5.5 Secondary Disinfectants

Secondary disinfectants provide an essential residual that prevents regrowth in the distribution system. Although chlorine is the most widely used secondary disinfectant, chlorine dioxide and monochloramine are appropriate as well. As secondary disinfectants, chlorine and chlorine dioxide are handled in the same manner as for primary disinfectants. (See Sections 5.4.1. and 5.4.3 for information on the process and equipment). Section 5.5.1. discusses the use of monochloramine as a secondary disinfectant.

5.5.1 Chloramination

Chloramine is recommended as a secondary disinfectant because it is ineffective as a virucide, and is only marginally effective against *Giardia* cysts. It is formed from the combination of ammonia and chlorine (hypochlorite or hypochlorous acid). The chemical is generated on site, usually by injecting ammonia gas or adding an ammonium sulfate solution to chlorinated water, or by adding chlorine to water containing ammonia. Ammonia gas can be purchased as an anhydrous liquid in cylinders for small water treatment systems; ammonium sulfate can be purchased as a powder in bags.

Section 5.5.1.1 describes the chloramination process. Sections 5.5.1.2 through 5.5.1.4 discuss establishing a chloramine residual, system design considerations, and chloramination systems costs, respectively.

5.5.1.1 Process Description

When water, chlorine, and ammonia are combined, three different species of chloramine compounds can be generated:

$$\underset{\text{ammonia}}{NH_3} + \underset{\text{hypochlorous acid}}{HOCl} \dashrightarrow \underset{\text{water}}{H_2O} + \underset{\text{monochloramine}}{NH_2Cl}$$

$$NH_2Cl + HOCl \dashrightarrow H_2O + \underset{\text{dichloramine}}{NHCl_2}$$

$$NHCl_2 + HOCl \dashrightarrow H_2O + \underset{\text{nitrogen trichloride}}{NCl_3}$$

The mix of species produced depends on the ratio of chlorine to ammonia and the pH of the water. In the pH range of 7 to 8 with a chlorine-to-ammonia ratio (by weight) of 3 to 1, monochloramine is the principal product. At higher chlorine-to-ammonia ratios or at lower pH values (5 to 7), some dichloramine will be formed. If the pH drops below 5, some nitrogen trichloride (often erroneously called "trichloramine") may be formed. Nitrogen trichloride formation should be avoided because it imparts undesirable taste and odor to the water.

Figure 5-12 shows the relative percentages of monochloramine and dichloramine produced as the pH changes, for different weight ratios of chlorine to ammonia. At a pH value of about 5.7, approximately equal amounts of mono- and dichloramines are present in solution.

Care should be taken not to exceed chlorine-to-ammonia ratios of 5 to 1. This is the "breakpoint" curve above which all ammonia is removed, chloramines are absent, and free residual chlorine is present.

5.5.1.2 Establishing a Chloramine Residual

Establishing a chloramine residual involves a period of mixing the chlorine and ammonia with the water, followed by a short holding time to allow the reactions to take place. Usually, chloramine-forming reactions are at least 99 percent complete within a few minutes.

The National Academy of Sciences (NAS) recommends adding ammonia to chlorinated water rather than adding chlorine to water containing ammonia. The recommended process produces a residual of free chlorine above that required to oxidize nitrogen (particularly the organic nitrogen compounds). Organic nitrogen compounds will compete successfully with ammonia-nitrogen for chlorine, forming organic chloramines, which are weaker disinfectants than monochloramine. Normal field analytical techniques cannot distinguish between inorganic and organic chloramines. Thus, formation of inorganic chloramines in the presence of organic nitrogen compounds can seriously overstate the actual capability of the chloramine system to provide secondary disinfection.

5.5.1.3 Chloramination System Design Considerations

Ammonia is available as an anhydrous gas (NH_3), a 29 percent aqueous solution (aqua ammonia), or as ammonium sulfate powder ($[NH_4]_2SO_4$). Gaseous ammonia is supplied in 68-kg (150-lb) cylinders, aqua ammonia in 208.2-L (55-gal) drums, and ammonium sulfate in 45.4-kg (100-lb) bags (98 percent pure, 25 percent available ammonia).

Ammonia gas is injected into treated water using systems and equipment similar to those used for chlorine gas. Aqua ammonia is handled using systems similar to those used for sodium hypochlorite. This form of ammonia is basic and has a strong odor, but is not corrosive. For ammonium sulfate powder, a 25 to 30 percent solution is prepared in a plastic or fiberglass container and added to the water by means of a chemical metering pump. Equipment similar to that used for handling calcium hypochlorite can be used for this process. Solutions of ammonium sulfate are stable, but acidic, and, therefore, corrosive to some metals. Materials that withstand dilute sulfuric acid will also withstand the corrosion effects of dilute ammonium sulfate solutions.

5.5.1.4 Costs for Chloramination

Generation of chloramines requires the same equipment as chlorination (gaseous or aqueous hypochlorination), plus equipment for adding ammonia (gaseous or aqueous). Costs for chlorination equipment and for its operation and maintenance were presented earlier (Section 5.4.1). This section presents the chemical costs for generating chloramine from ammonia and chlorine gas.

In the Baltimore/Washington, D.C., area during 1983, costs for cylinders of liquid ammonia were

Table 5-17. Construction Costs for Ultraviolet Light Disinfection ($1978)

Cost Category	Plant Capacity (GPD) 14,400	28,800	187,200	374,400	748,800	1,123,200
Excavation and sitework	$60	$60	$60	$60	$80	$110
Manufactured equipment	800	1,125	4,485	8,685	17,365	26,050
Concrete	250	250	250	250	280	300
Labor	110	170	250	300	400	500
Pipe and valves	60	150	350	450	750	1,000
Electrical and instrumentation	430	430	430	430	480	480
Housing	1,500	1,500	1,500	1,500	1,800	2,000
Subtotal	3,210	3,885	7,225	11,675	21,155	30,440
Miscellaneous and contingency	470	560	1,010	1,580	2,830	4,060
Total	$3,680	$4,445	$8,335	$13,255	$24,085	$34,500

1 GPD = 0.003785 m^3/day
Source: U.S. EPA (1979).

$0.88/kg ($0.40/lb), $1.54/kg ($0.70/lb) for drums of contained ammonia in 28 percent solution, and $1.12/kg ($0.51/lb) for bags of solid ammonium sulfate.

Chemical costs were derived for a 9.46 m^3/day (2,500-gal/day) water treatment plant using a maximum dose of 2.5 mg/L monochloramine (the maximum level currently recommended by EPA). These costs are presented in Table 5-19. Chemical costs for both chlorine gas and the ammonia-based chemical range from $14.75 to $24.55 annually, depending on which ammonia source chemical is used. Anhydrous ammonia is the least expensive, while ammonia sulfate is the most expensive.

The direction of EPA's regulatory initiative under the SDWA Amendments of 1986 is to reduce the levels of secondary disinfectants in treated water. For example, the newly promulgated Surface Water Treatment Rule (EPA, 1989b) requires that a minimum of 0.2 mg/L of secondary disinfectant be present as the treated water enters the distribution system. This level does not have to be maintained in the distribution system, as long as heterotrophic plate counts remain below 500/mL. Consequently, the costs for chloramination can be lower than the costs presented here.

Table 5-18. Operating and Maintenance Costs for Ultraviolet Light Disinfection ($1978)

Water Flow Rate (GPD)	Electrical Energy (kWh/yr) Building	Process	Total	Energy Costs[a]	Maintenance Material ($/yr)	Labor (hr/yr)	Labor Cost[b]	Total Cost ($/yr)
14,400	10,260	440	10,700	$749	$100	24	$240	$1,089
28,800	10,260	800	11,140	780	140	24	240	1,160
187,200	10,260	5,260	15,520	1,086	600	24	240	1,926
374,400	10,260	10,510	20,770	1,454	1,120	30	300	2,874
748,800	12,310	21,020	33,330	2,333	2,250	36	360	4,943
1,123,200	13,340	31,540	44,880	3,142	3,300	42	420	6,862

[a] Assumes 0.07/kWh.
[b] Assumes 10/hour.
1 GPD = 0.003785 m^3/day
Source: U.S. EPA (1979).

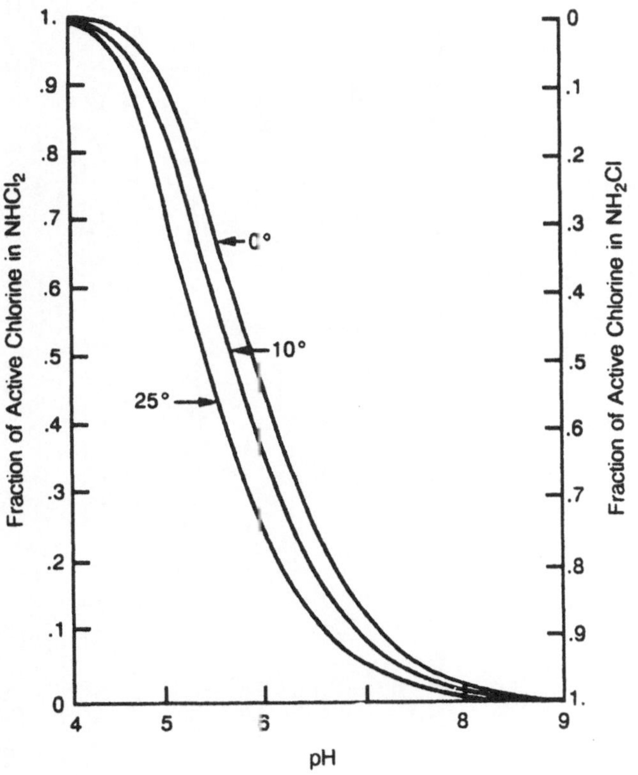

Figure 5-12. Proportions of mono- and dichloramines in water with equimolar concentrations of chlorine and ammonia.
Source: National Academy of Sciences (1980).

Table 5-19. Chemical Costs for Generating Monochloramine for 2,500 GPD Water Treatment Plant ($1983)

Source Chemicals	Amount Needed (lb/yr)	Unit Cost ($/lb)	Annual Cost ($)
Anhydrous ammonia	6.2	0.40	2.48
and chlorine gas	26.1	0.47	12.27
		Total	14.75
Aqua ammonia	6.2	0.70	4.35
and chlorine gas	26.1	0.47	12.27
		Total	16.62
Ammonia sulfate	24.1	0.51	12.29
and chlorine gas	26.1	0.47	12.27
		Total	24.55

1 lb = 0.4536 kg; 1 GPD = 0.003785 m^3/day.

Chapter 6

Treatment of Organic Contaminants

Treatment of organic contaminants in drinking water depends on the nature of the contaminants targeted for removal. The two largest categories of organic contaminants are synthetic organic compounds (SOCs) and natural organic materials (NOMs). Of special interest are SOCs and NOMs having significant potential health effects, including volatile organic compounds (VOCs), aromatic hydrocarbons, halogenated aromatic hydrocarbons (including polychlorinated biphenols [PCBs]), polynuclear aromatics, halogenated nonaromatics, hydrocarbons, and organic pesticides.

Natural organic material found in raw water is in a dissolved or particulate phase; most SOCs are in this dissolved phase. The particulate phase NOM can be removed using sedimentation and filtration processes. To enhance NOM removal, pretreatment processes preceding sedimentation and filtration can be optimized to convert as much of the dissolved NOM as possible to particulate NOM. Section 6.1 contains a discussion of these enhancements.

Twelve treatment technologies are discussed in three general categories of applicability. The first category, discussed in Sections 6.2 and 6.3, covers the most frequently used technologies, which include granular activated carbon (GAC) and packed tower aeration (PTA), also referred to as packed column aeration (PCA). GAC and PTA are classified as best available treatment (BAT) technologies for removal of VOCs under the U.S. EPA regulations promulgated in July 1987.

GAC and PTA have different removal efficiencies for different organic compounds. PTA is only effective for VOC removal. Treatment technologies frequently are chosen based on potential removal efficiencies for a specific organic contaminant. However, if a decision is based only on cost, then PTA is often chosen for VOCs because of its lower cost. Sometimes, both technologies are necessary to remove a particular combination of organic compounds. In these cases, the resultant total costs are not always additive; installing both systems may result in some cost savings.

The second category discussed includes treatment processes that have some record of performance, but are not as widely applicable as GAC or PTA to be classified as BAT. These treatments include powdered activated carbon (PAC) in a conventional treatment process train, covered in Section 6.4; diffused aeration, covered in Section 6.5; and multiple tray aeration, covered in Section 6.6.

The final category, discussed in Section 6.7, consists of technologies that are emerging from the laboratory and pilot-level stages as promising alternatives for the near future. This category embraces a number of candidates including oxidation (ozone, ozone/ultraviolet radiation, and ozone/hydrogen peroxide), reverse osmosis, mechanical aeration, catenary grid aeration, Higee aeration, resins, and oxidation followed by a biological filtration step. Appendix C describes case experiences with GAC, PTA, and PAC.

There are so many organic compounds that performance data concerning each technology's removal ability for every organic compound are not available. Table 6-1 contains a partial matrix of five of the above technologies and their removal efficiencies for 33 compounds.

6.1 Pretreatment for Natural Organic Contaminant Removal

Water treatment processes used prior to sedimentation and filtration may be designed to enhance NOM removal. The most important of these is the coagulation process, which is used primarily for controlling turbidity and is routinely applied to surface water. While the portion of organic

Table 6-1. Performance Summary for Five Organic Technologies

Organic Compounds	Granular Activated Carbon Adsorption Filtrasorb 400[b]	Packed Column Aeration	Reverse Osmosis Thin Film Composite	Ozone Oxidation (2-6 mg/l)	Conventional Treatment
	Removal Efficiency[a]				
Volatile Organic Contaminants					
Alkanes					
Carbon tetrachloride	+ +	+ +	+ +	0	0
1,2-Dichloroethane	+ +	+ +	+	0	0
1,1,1-Trichloroethane	+ +	+ +	+ +	0	0
1,2-Dichloropropane	+ +	+ +	+ +	0	0
Ethylene dibromide	+ +	+ +	+ +	0	0
Dibromochloropropane	+ +	+	NA	0	0
Alkenes					
Vinyl chloride	+ +	+ +	NA	+ +	0
Styrene	NA	NA	NA	+ +	0
1,1-Dichloroethylene	+ +	+ +	NA	+ +	0
cis-1,2-Dichloroethylene	+ +	+ +	0	+ +	0
trans-1,2-Dichloroethylene	+ +	+ +	NA	+ +	0
Trichloroethylene	+ +	+ +	+ +	+	0
Aromatics					
Benzene	+ +	+ +	0	+ +	0
Toluene	+ +	+ +	NA	+ +	0
Xylenes	+ +	+ +	NA	+ +	0
Ethylbenzene	+ +	+ +	0	+ +	0
Chlorobenzene	+ +	+ +	+ +	+	0
o-Dichlorobenzene	+ +	+ +	+	+	0
p-Dichlorobenzene	+ +	+ +	NA	+	0
Pesticides					
Pentachlorophenol	+ +	0	NA	+ +	NA
2,4-D	+ +		NA	+	0
Alachlor	+ +	+ +	+ +	+ +	0
Aldicarb	NA	0	NA	NA	NA
Carbofuran	+ +	0	+ +	+ +	0
Lindane	+ +	0	NA	0	0
Toxaphene	+ +	+ +	NA	NA	0
Heptachlor	+ +	+ +	NA	+ +[c]	NA
Chlordane	+ +	0	NA	NA	NA
2,4,5-TP	+ +	NA	NA	+	NA
Methoxychlor	+ +	NA	NA	NA	NA
Other					
Acrylamide	NA	0	NA	NA	NA
Epichlorohydrin	NA	0	NA	0	NA
PCBs	+ +	+ +	NA	NA	NA

[a] + + = Excellent 70-100%
\+ = Average 30-69%
0 = Poor 0-29%
NA = Data not available or compound has not been tested by EPA Drinking Water Research Division.
[b] Excellent removal category for carbon indicates compound has been demonstrated to be adsorbable onto GAC, in full- or pilot-scale applications, or in the laboratory with characteristics suggesting GAC can be a cost-effective technology.
[c] Ozone oxidation of heptachlor produces a high yield of heptachlor epoxide, which is not suitable for further oxidation.

contamination removed by effective coagulation is generally small, in conjunction with other treatments, coagulation can significantly add to a plant's removal effectiveness.

In general, coagulation is most effective in removing organic contaminants that are hydrophobic, acidic, and high in molecular weight. While coagulation commonly removes less than half of the organic content of the water, it tends to remove contaminants that are not removed by GAC. These contaminants include color, some trihalomethane (THM) precursors, hydrophobic SOCs, and toxic metals.

Oxidation is another pretreatment step that can have a pronounced effect upon the removal of organic contaminants. Ozone or other advanced oxidation processes employed before filtration drastically change the chemical structure of many organic contaminants.

Section 6.1.1 focuses on how adjusting the coagulant type, coagulant dosage, coagulant aid application procedures, pH, and point of coagulant addition can improve organic contaminant removal. Section 6.1.2 describes the use of oxidation pretreatments to enhance organics removal.

6.1.1 Coagulant Pretreatment

All coagulants will remove some organic contamination, especially contaminants associated with macromolecules. However, different coagulants with equivalent turbidity removal efficiencies may have different organic contaminant affinities. Coagulants should be chosen by matching their particular affinities to the specific organic contaminants of concern.

Metallic salt coagulants, most commonly alum and ferric chloride, are more effective than cationic polymers in removing lower molecular weight organic compounds with acidic functional groups (such as carbonyl or carboxyl groups). Newer metallic salt coagulants, such as polyaluminum chloride and polymeric iron chloride, are also effective at removing organics. Polymer coagulants are effective in enhancing the removal of NOMs, such as humic and color compounds.

Determining the optimal dosage of coagulant for treatment of organic contaminants parallels the derivation of the optimal conditions for turbidity removal. The first step is to fully characterize the plant influent using laboratory jar tests and pH assessments. Care should be taken to avoid errors in performing the laboratory tests. One common laboratory mistake is to overlook the adsorption of organic contaminants onto glass and filter paper. In addition, laboratory procedures should take into account contaminant volatilization. Under optimum circumstances, laboratory tests should be followed by full-scale plant tests. See Section 4.1 for a more complete discussion on determining optimum coagulant dosages.

Coagulant Dosage

The most effective method of applying coagulants depends on the characteristics of the raw water. For raw water with low pH (about 5 to 6 for alum), containing high and homogeneous concentrations of NOM, organic contaminant removal is characterized by a sharp increase in removal effectiveness around a specific coagulant dosage. For raw water with high pH, containing low and heterogenous concentrations of NOM, the removal pattern is characterized by a gradual increase in removal effectiveness with increasing coagulant dosage that asymptotically approaches maximum removal.

Overdosage is possible with the first pattern of removal, but not with the second. The second pattern is typical for surface waters with moderate to high turbidity and alkalinity. SOC removal tends to follow the second pattern of removal.

Coagulant Aid Application Procedures

Coagulant aids include substances that improve the nature of the flocculated particles, and reduce turbidity. Among these substances are acids, bases, anionic and cationic organic polymers, activated silica, and bentonite clay. Except for acids and bases, the mere presence of coagulant aids has not been found to dramatically affect organic contaminant removal. However, the order of application of coagulant aids has been found to have a significant impact.

Chemical application procedures are subject to many confounding site specific conditions. In addition, few published data exist regarding optimum chemical application strategies. However, the following four recommendations were derived from assumptions concerning removal mechanisms for anionic contaminants:

- Add acids before metal-salt coagulants to lower the pH toward an optimal level and promote the production of positively charged ions.
- Add bases after metal-salt coagulants to maintain the lower optimal pH levels and avoid encrustation of the mixer or injection nozzle.
- Add clay and PAC before metal-salt coagulants to allow adsorption of NOMs and SOCs before clay and PAC coagulation.
- Add polymers, especially anionic polymers, after metal-salt coagulants.

pH

The optimal pH for the removal of NOM with alum is frequently between 5 and 6, but can be higher with iron salts. While SOC removal is clearly associated with pH, the optimal pH level varies with each SOC. Experience thus far indicates that negatively charged SOCs generally are removed most easily in water with a pH of 5 to 7. Furthermore, positively charged SOCs are most easily removed in water with a pH between 7 and 8.5, but more data are required to confirm this finding.

Point of Coagulation Application

The point at which coagulants are added can significantly affect the final level of organic contaminants. Oxidation typically is used for disinfection purposes, yet it also reduces organic contaminant levels. Oxidation that precedes the coagulation improves its effectiveness by removing turbidity and changing the nature of the organic contaminants. However, when chlorine disinfection/pretreatment is used, it is better to add the coagulant first to reduce the level of organic contamination that can form halogenated by-products during chlorination. This is not as much of a concern with preoxidation using ozone or chlorine dioxide, where halogenated by-products are not as much of a problem.

6.1.2 Oxidation Pretreatment

Several oxidants are available for treating organics in drinking water, including ozone, chlorine, chlorine dioxide, permanganate, hydrogen peroxide and ultraviolet radiation. Organic compounds can be completely oxidized into carbon dioxide and water, or partially oxidized into intermediate reaction products. Complete oxidation is not always possible because the intermediate products formed may be more resistant to further oxidation than the original organic chemical. Pretreatment with an oxidant may convert dissolved organic material to particulate matter, thereby enhancing removal by sedimentation and filtration. In addition, oxidation of many SOCs may be accomplished during pretreatment.

Because of the concern for halogenated disinfection by-products, the use of nonchlorinous oxidants may be preferred in the early stages of water treatment. In such cases, the nonchlorinous oxidants do not produce halogenated by-products, except in those waters containing substantial amounts of bromide ion. Some waters, when treated with a preoxidant, have been shown to produce more chlorinated by-products because oxidation of the organics in the raw water produces precursors to by-product materials.

Recently, studies have been conducted to investigate the use of advanced oxidation processes (AOPs) where several oxidants are used in combination (e.g., ozone and hydrogen peroxide, ozone and ultraviolet radiation, ultraviolet radiation and hydrogen peroxide). These processes involve the generation of the hydroxyl radical in sufficient quantities to impact treatment. AOPs may achieve treatment at a lower cost than conventional oxidation. The use of AOPs has been applied to the removal of SOCs and also NOMs.

6.2 Granular Activated Carbon

Activated carbon is used for water treatment either as a granular adsorption medium or as a powder added to the water like a coagulant, which later settles out in sedimentation basins or clarifiers. Granular activated carbon (GAC), the more common method for removing organic contaminants, is discussed in this section. Powdered activated carbon (PAC) is covered in Section 6.4.

Granular activated carbon unit; Smyrna, DE.

Activated carbon works on the principle of adsorption. Dissolved contaminants (adsorbate) are transferred from the water solution to the microporous surface of the carbon particles (adsorbent). Activated carbon's large internal surface area and porosity are the primary reasons for its excellent adsorption capabilities. One gram of activated carbon has a surface area equivalent to that of a football field.

The adsorption process is primarily a physical process that can be reversed relatively easily. The ease of reversing adsorption is another key factor in activated carbon's usefulness because it facilitates the recycling or reuse of the carbon.

Contaminant characteristics greatly affect GAC's adsorption ability. GAC has an affinity for contaminants that are:

- Branch-chained, rather than straight-chained
- High in molecular weight
- Low solubility

- Nonpolar
- Present in higher concentrations

GAC's affinity for larger molecular weight compounds is illustrated by its reduced effectiveness when preceded by ozonation. Ozonation breaks down contaminants and thus can reduce GAC's ability to adsorb them. Table 6-2 lists readily and poorly adsorbed contaminants.

Table 6-2. Readily and Poorly Adsorbed Organics

Readily Adsorbed Organics

- Aromatic solvents (benzene, toluene, nitrobenzenes)
- Chlorinated aromatics (PCBs, chlorobenzenes, chloronaphthalene)
- Phenol and chlorophenols
- Polynuclear aromatics (acenapthene, benzopyrenes)
- Pesticides and herbicides (DDT, aldrin, chlordane, heptachlor)
- Chlorinated nonaromatics (carbon tetrachloride, chloroalkyl ethers)
- High molecular weight hydrocarbons (dyes, gasoline, amines, humics)

Poorly Adsorbed Organics

- Alcohols
- Low molecular weight ketones, acids, and aldehydes
- Sugars and starches
- Very high molecular weight or colloidal organics
- Low molecular weight aliphatics

The outer surfaces of macropores on GAC particles are large enough to house colonies of bacteria that feed on biodegradable organics as they pass in and out of the macropores. Consequently, considerable mineralization of organic material occurs after a few weeks of operating unused GAC. Ozonation and other advanced oxidation (ozone/hydrogen peroxide or ozone/ultraviolet radiation) convert organic contaminants into more readily biodegradable materials.

Raw water characteristics also affect GAC's adsorption ability. The most important characteristic is the presence of competing contaminants or dissolved solids which can adversely affect adsorption.

Process, least cost, and facility design considerations are covered in Sections 6.2.1, 6.2.3, and 6.2.4, respectively. Section 6.2.2 contains a discussion of tests for deriving the optimum activated carbon usage rates for different applications. The last four sections cover operation and maintenance (6.2.5), system performance (6.2.6), and system costs (6.2.7).

6.2.1 Process Design Considerations

Key process design considerations include:

- GAC type
- Surface loading rate of the GAC filter
- Empty bed contact time (EBCT)
- Contaminant type and concentration
- Contaminant competition
- Carbon depth and usage

GAC Type

Various types of GAC are available for removing organics from drinking water. The most frequently used carbon in U.S. treatment plants is coal-based carbon because of its hardness, adsorption capacity, and availability. Some peat and lignite carbons have been used. Several sizes of carbon are available, and the size selected for a particular application is based on backwash and headloss characteristics, rate of adsorption, and cost. Headloss is lessened with the larger carbons, while the rate of adsorption is increased with the smaller carbons. The relationship of carbon cost and total production cost for a wide range of reactivation frequencies is shown on Figure 6-1.

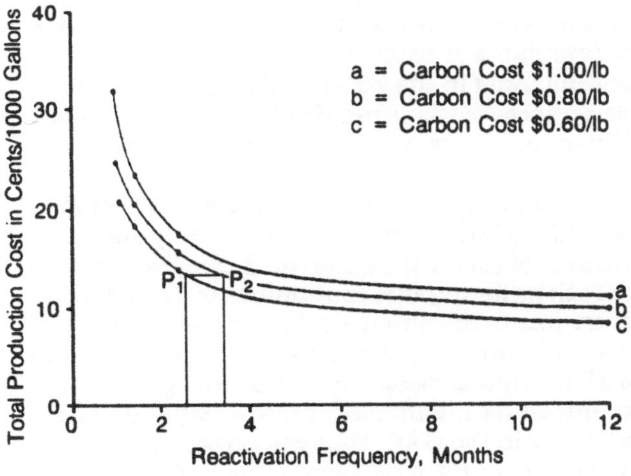

Figure 6-1. Effect of carbon cost on facility cost.

Surface Loading Rate

The surface loading rate of the GAC filter is related to an individual plant's design capacity. Surface loading rate is the amount of water passing through a one square foot area of the activated carbon filter bed per unit of time, and it typically ranges from 2 to 10 GPM/sq ft.

Empty Bed Contact Time

Contact between the influent and GAC is the primary factor in determining the size and capital cost of a GAC treatment system. The empty bed contact time (EBCT), the time required for water to pass through the empty column or bed (absent of GAC), is determined by the following equation:

EBCT (min) = GAC (m^3)/flow rate (m^3/min)

While most EBCTs range from 5 to 30 minutes, EBCTs of less than 7.5 minutes have been ineffective. Typical effective EBCTs are around 10 to 15 minutes and 10-minute EBCTs are typical for removal of most organic compounds. GAC is effective in removing radon with 180- to 200-minute EBCTs (Lowry and Brandown). However, these long EBCTs are only practical for point-of-use applications, not for many community systems.

Contaminant Type and Concentration

GAC's removal efficiency varies for different organic compounds. To date, removal tests for only a fraction of the myriad of organic compounds have been conducted using GAC technology. Table 6-1 (presented earlier) shows examples of GAC's removal ability.

Contaminants in the water can occupy GAC adsorption sites, whether they are targeted for removal or not. Therefore, the presence of other contaminants may interfere with the removal of the contaminants of concern. A profile of all contaminants in the water will help predict the design specifications required to achieve mandatory effluent levels for regulated contaminants.

Radon requires an extremely long EBCT, ranging from 100 to 200 minutes; therefore, only small amounts of radon are adsorbed during typical GAC operations for organic contaminants. In addition, radon decays continuously during the removal process (Lowry and Brandown). Consequently, the small amount of radon adsorbed during a typical organic contaminant EBCT also decays during its residency in the GAC, thus the carbon is renewed on a continual basis, as shown in Figure 6-2. Section 7.3.1 contains a more detailed discussion of radon removal.

Carbon Depth and Usage

Carbon depth is related to the amount of carbon necessary to achieve a desired EBCT and filter life with a specific level of contamination. Typical carbon depth reaches 3 to 9 m (10 to 30 ft).

Carbon usage, expressed in grams of carbon per cubic meter (pounds of carbon per 1,000 gallons) of treated water, largely determines the system's operating expenses. The carbon is considered to be exhausted when the effluent organic concentration approaches the influent organic concentration, at which time regeneration becomes necessary. Determining when regeneration is necessary, however, is a site-specific decision. It may be necessary either when contamination is detected in the finished water or when the level of contamination exceeds the regulated level.

Typical carbon usage ranges from 6 to 120 g/m^3 (0.05 to 1.0 lb/1,000 gal) of treated water, although SOCs are removed with carbon usage rates as low as 1.2 g/m^3 (0.01 lb/1,000 gal) of treated water (O'Brien et al., 1981). Table 6-3 provides carbon usage rates for the removal of eight organic compounds.

Carbon usage varies with the type of contamination. VOCs utilize the most carbon, and quickly shorten the carbon's useful life as an adsorption medium. Pesticides generally use less carbon than VOCs, however, their demands on carbon usage vary. Of the organics listed in Table 6-3, chlorinated aromatic organic compounds use the least amount of carbon. VOC carbon usage rates typically translate into carbon replacement intervals of 3 to 6 months, while other organic compounds require carbon replacement rates of from 1 to 2 years depending on contaminant concentrations. Pretreatment can significantly extend carbon's longevity. For example, PTA preceding GAC removes a large portion of VOCs. In addition, oxidation can lower the amount of adsorbable organics by converting some organics into materials that are biochemically mineralized during passage through the GAC.

In addition to the specific type of organic contamination present, an influent's dissolved organic carbon (DOC) level can increase GAC usage. DOC is especially a problem for surface water supplies. Pretreatment of surface water with coagulation and filtration reduces the burden of turbidity and DOC and thus extends GAC's longevity.

The relationship between carbon longevity, influent concentrations, and effluent contamination levels for the contaminant trichloroethylene is illustrated in Figure 6-3. For this example, the EBCT is 10 minutes. To achieve an effluent concentration of 1 μg/L from an influent with 100 μg/L, the carbon life is estimated at 130 days. To allow a greater effluent concentration of 10 μg/L from an influent of 100 μg/L, the life of the carbon is extended to about 180 days.

Figure 6-4 compares carbon longevity in the removal of three different organic compounds. The curves, for each compound, illustrate the correlation between influent concentrations and the useful life of the carbon filter. The curves assume an EBCT of 10 minutes and an effluent concentration of 10 μg/L. At influent concentrations of 200 μg/L, the carbon life ranges from about 1 month for 1,1,1-trichloroethane

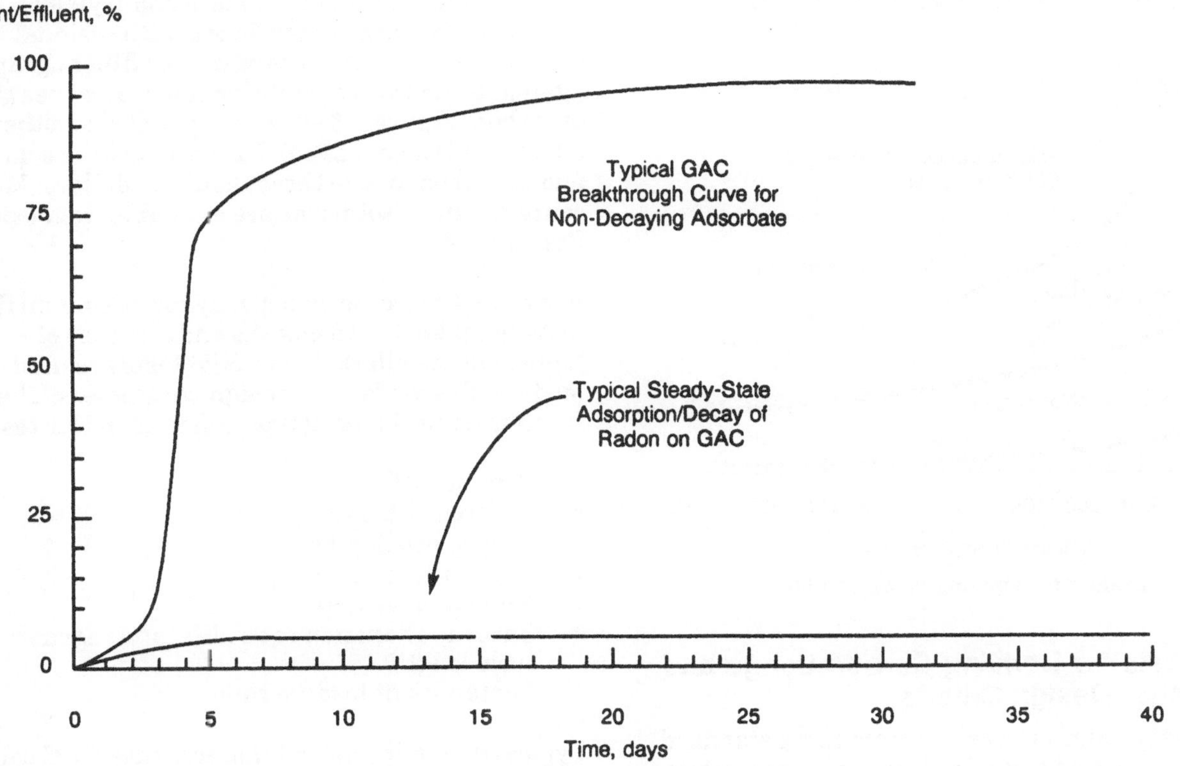

Figure 6-2. Steady-state adsorption/decay curve for radon.

Table 6-3. Summary of Carbon Usage Rates

	Contaminant Concentration (µg/L)		
	Influent	Effluent	Carbon Usage (lb/1,000 gal)[a]
Volatile Organic Compounds (VOCs)			
Tetrachloroethylene (PCE)	100	2	0.08
Trichloroethylene (TCE)	100	2	0.16
Trichloroethane (TCA)	100	2	0.96
Pesticides			
Chlordane	100	1	0.012
Dibromochloropropane (DBCP)	100	1	0.016
Aldicarb	100	1	0.02
Chlorinated Aromatics			
Dichlorobenzene	100	2	0.01
PCB (Aroclor 1016)	100	2	0.015

[a]1 lb/1,000 gal = 120 g/m^3

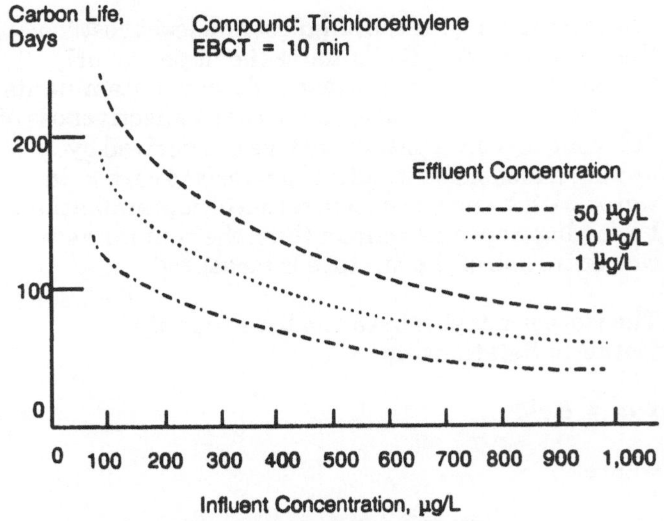

Figure 6-3. Effect of contaminant on carbon life.

removal to over 1 year for tetrachloroethylene removal.

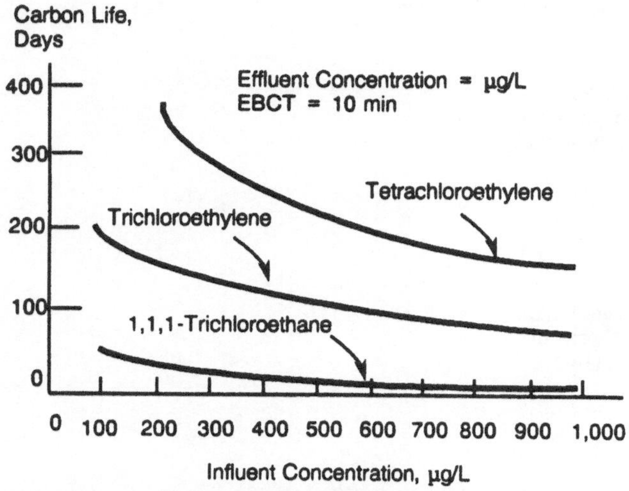

Figure 6-4. Effect of compound on carbon life.

6.2.2 Tests for Deriving Carbon Usage and Other Design Criteria

Carbon usage rates are derived from three standard laboratory and field tests: isotherm, dynamic column, and mini-column tests. The test results are critical in deriving and evaluating GAC process design parameters. Isotherm and mini-column tests are laboratory tests, and the dynamic column test is a field method that determines carbon usage.

The isotherm test is the simplest and least costly of the three methods. It evaluates the impact of pH, temperature, and the presence of other contaminants on adsorption. It can also compare the effectiveness of different carbon types. Isotherms are derived by mixing a measured weight of pulverized carbon in water with a known concentration of contamination. Following a specific contact time, the contaminant concentration of the mixture is measured.

The isotherm test is based on the Freundlich Isotherm Relationship:

$$x/m = Kc^{1/n}$$

where:

x/m	=	equilibrium capacity (mg of contaminant per gram of carbon)
k	=	capacity at 1 mg/L of contaminant concentration
c	=	contaminant effluent concentration in mg/L
$1/n$	=	exponent

Figure 6-5 shows an application of this equation in making carbon usage estimates.

The Freundlich Isotherm Relationship yields isotherms for each contaminant, with residual contaminant concentrations in the effluent graphed against the ratio of contaminant adsorbed per gram of carbon. Figure 6-6 shows graphs of six isotherms (Dobbs and Cohen, 1980). Isotherm tests are quick and cost about one to three thousand dollars. Many contaminants' isotherms are available from existing literature.[1]

In contrast to isotherm tests, dynamic column field tests can take 6 to 10 months and cost tens of thousands of dollars. These pilot tests are used to model full-scale facility design parameters. The seven most useful design parameters in this test are:

- Type of GAC
- Different EBCTs
- Carbon bed depths
- Hydraulic loadings
- Number of vessels
- Carbon exhaustion rates, life, and regeneration cycle length
- Contaminant loading rates

Typical dynamic pilot column tests use GAC columns that are about 5 feet deep with 4-inch internal diameters (see Figure 6-7).

The mini-column laboratory test is between the isotherm and dynamic column tests in accuracy and complexity. The cost is about the same as the cost for an isotherm test. This test is used to determine the feasibility of using GAC, establish preliminary design criteria, and approximate cost. Test procedures involve pumping raw water through a short column of GAC about 70 mm deep. A schematic diagram of a mini-column test is shown in Figure 6-8.

6.2.3 Least Cost Design Criteria

To derive the optimal design criteria for an effective GAC system, the design must consider EBCT, carbon usage, and column configuration. EBCT has greatest impact on capital costs, while carbon usage has the greatest impact on operating costs. The first step in determining the most cost-effective EBCT and carbon usage rates is to examine the breakthrough curves for the organic compounds contaminating the influent. The breakthrough curves indicate how long a GAC filter can produce a desired effluent concentration for an organic contaminant.

The next step is to calculate the carbon usage (CU) rate with the following formula:

1 Data for several compounds are contained in: *Carbon Adsorption Isotherms for Toxic Organics*, EPA-600/8-80-023, April 1980.

$$CU = \frac{\text{Mass of Carbon}}{\text{Volume of Water Treated through Breakthrough}}$$

Carbon usage is then plotted against EBCT to select the optimal combination.

Column configuration is the next important factor to consider in determining the length of the carbon regeneration period. Parallel column configurations require carbon replacement or regeneration when a filter breakthrough occurs, while a series column configuration allows replacement to take place at the point of carbon exhaustion. Carbon exhaustion occurs when the saturated carbon can no longer hold contamination. Series configuration extends the operating period between regeneration cycles.

Selection of the appropriate system involves a tradeoff between capital and operating costs. Parallel systems require fewer contactors, and their associated capital expense. Series configuration has a longer carbon regeneration cycle with lower operating costs.

6.2.4 Facility Design Criteria

The three major components of a GAC system are the carbon contactor, carbon transfer system (which moves the carbon in and out of the contactors), and carbon regeneration system (see Figure 6-9).

The two important operational characteristics of contactors are flow direction (upflow or downflow) and water feed mode (gravity or pressure). Upflow and downflow carbon contactors are used in either series or parallel column configuration. Pressure contactors usually cause a minimum pressure drop of 10 psi.

Upflow contactors are used for contact periods of up to 2 hours (U.S. EPA, 1973). The upflow design permits suspended solids to pass through without excessive drops in pressure. These extended contact periods are used to remove suspended solids and organic compounds. The disadvantage of upflow contactors is that they sometimes carry fine carbon particles into their effluent. Upflow designs are more typical of wastewater treatment plants than drinking water treatment facilities. Downflow contactors typically are used for contact periods of 30 minutes or less, and are able to remove only a small amount of suspended solids during backwashing.

Pressure-fed contactors are used for systems with capacities between 0.04 and 0.44 m^3/sec (1 and 10 MGD); packaged plants using pressure-fed contactors are rarely over 0.04 m^3/sec (1 MGD). The pressure-fed contactor operates with higher head loss levels than the gravity-fed contactor. In the pressure-fed contactor, pumps move the water from the well to the contactors and into the distribution system. Existing pump systems usually are adequate to maintain pressures in the distribution systems.

Gravity-fed contactors typically are used for systems with greater than 0.44 m^3/sec (10 MGD) capacity. By using common wall construction, they are able to reduce costs. Gravity-fed GAC systems are constructed either by modifying existing gravity-fed sand filters or by using new concrete contactors.

The foremost consideration in carbon transfer system design is minimizing carbon loss due to abrasion. The hydraulics of the slurry system, system velocities, and construction materials all affect carbon abrasion.

Carbon regeneration can take place either on site or offsite, or the carbon may be used on a disposable basis, depending on the system size. Systems with carbon exhaustion rates between 226.8 and 907.2 kg (500 and 2,000 lb)/day will generally use offsite regeneration, which is routinely performed on a contract basis. Systems exhausting over 907.2 kg (2,000 lb) carbon/day may consider onsite regeneration, which involves heating the carbon to destroy the organics. To avoid unnecessary energy costs, dewatering of the carbon is required. Systems using less than 226.8 kg (500 lb)/day generally dispose of their carbon, rather than regenerate it.

Both carbon replacement and regeneration produce wastes. Disposing of carbon with contaminants classified as hazardous waste will dramatically increase disposal costs. Carbon regeneration operations must meet all applicable air quality regulations. Gas emissions from these operations are sometimes categorized as significant point sources.

6.2.5 Operation and Maintenance

Several operational and maintenance factors affect the performance of GAC units, including those involving the nature of the influent, dynamics of the GAC process, and management of the resultant wastes. These factors require consistent and careful monitoring.

A significant drop in influent contaminant concentration will cause a GAC filter to desorb, or slough off, the contaminant, because GAC is an equilibrium process (McKinnon and Dyksen, 1982). Another operational problem related to influent characteristics is the competition among contaminants for adsorption sites. Adsorbed contaminants are displaced by other contaminants with which GAC has greater affinities. As a result, water with frequently changing quality characteristics will produce effluent with unpredictable levels of contamination.

Bacterial growth on the carbon may be one potential problem. However, the nature and implications of

Isotherm equation: $x/m = Kc^{1/n}$

where:

$$K = 28 \left(\frac{(mg)}{(gm)}\,\frac{(L)}{(mg)}\right)^{1/n} \quad \text{(From isotherm data)}$$

$1/n$ = 0.62 (From isotherm data)
(C_i) = TCE influent concentration = 100 µg/L
(C_o) = TCE effluent concentration = 5 µg/L
SF = Safety Factor = 0.75

Rearranging:

$$\text{Carbon Usage (lb/1,000 gal)} = \frac{Ci - Co}{K\,(C_i)^{1/n}\,SF} \times 8.34\ \frac{(lb)}{(1,000\ gal)}\ \frac{(L)}{(mg)}$$

$$= \frac{(100 - 5)\ \frac{\mu g}{L} \times \frac{1}{1,000}\ \frac{mg}{\mu g}}{28\left(\frac{100}{1,000}\right)^{0.62} \times 0.75} \times \frac{8.34\ lb}{(1,000\ gal)} \times \frac{L}{mg}$$

$$= 0.16\ lb/1000\ gal$$

Note: 1 lb/1,000 gal = 120 g/m^3

Figure 6-5. Application of Freundlich Isotherm Relationship.

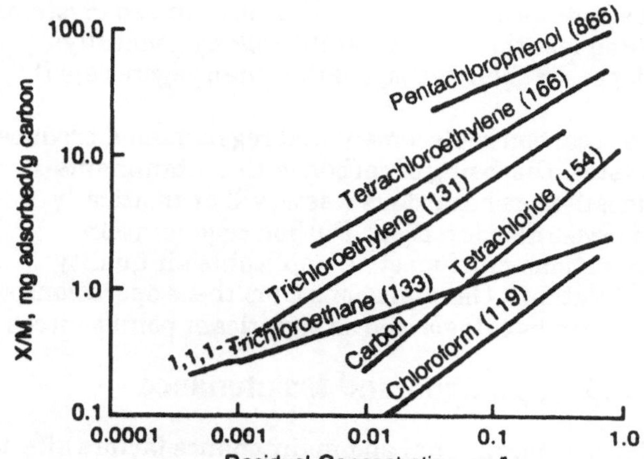

Numbers in parentheses indicate the molecular weight of the compound.

Figure 6-6. Adsorption isotherms for several organic compounds found in ground-water supplies.

this growth are not clearly defined and require further study. Excessive bacterial growth could cause clogging and higher bacterial counts in the effluent. However, it is also possible that such growth improves GAC's removal efficiencies through biodegradation.

Another potential problem is the disruption of the adsorption zone procession through a GAC contactor. This zone of saturated GAC typically moves in a wave formation through the carbon depth, as shown in Figure 6-10. If the wave of used GAC is interrupted, then the contactor may experience premature breakthrough. Changes in influent composition and concentration,as well as the method and frequency of filter backwashing all can affect the movement of the wavefront.

The final operational issue concerns the proper management of GAC wastes. Wastes from carbon regeneration and backwashing sometimes require treatment prior to disposal.

6.2.6 System Performance

GAC effectively removes most organic compounds from drinking water. GAC's precise removal effectiveness for a large number of organic compounds is unknown because of the multitude of organic compounds in existence. Table 6-1 presents removal efficiencies for 33 compounds.

GAC is not effective in removing vinyl chloride from water. In addition, the long EBCT required for radon removal makes it infeasible at the treatment plant scale. However, at the residential scale GAC systems are cost-effective for radon removal.

6.2.7 System Costs

The cost of removing many organic compounds with GAC is well within the range of general costs for delivering treated drinking water. Within the United States, that cost generally ranges from \$0.26 to \$0.40/m^3 (\$1.00 to \$1.50/1,000 gal) (1989 dollars) of treated water. Most regulated organic contaminants are managed with GAC systems having carbon bed lives of at least 3 months and EBCTs of 15 minutes.

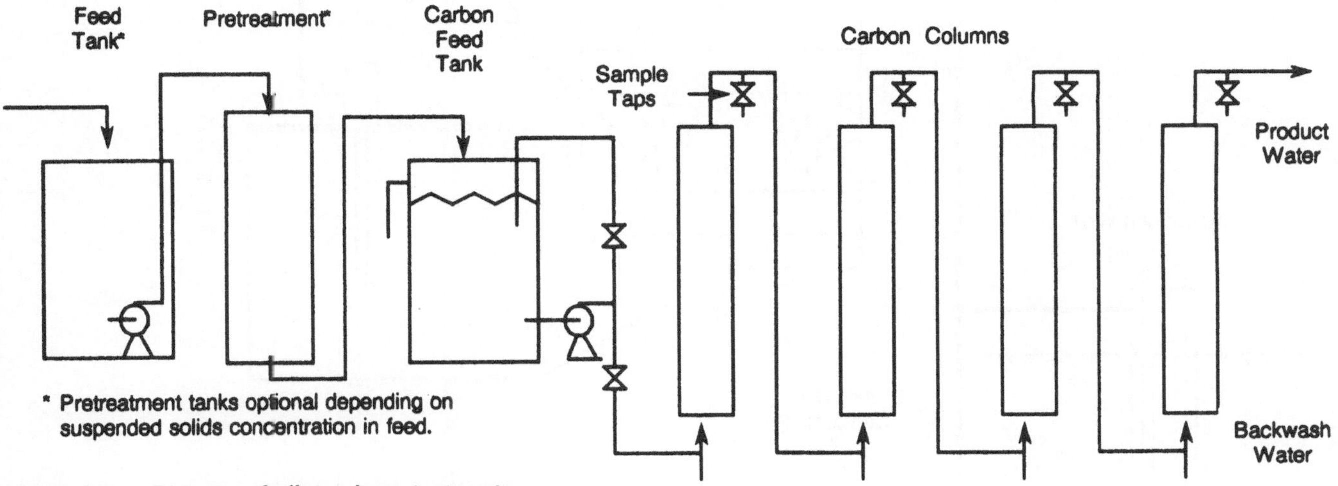

Figure 6-7. Diagram of pilot column test system.

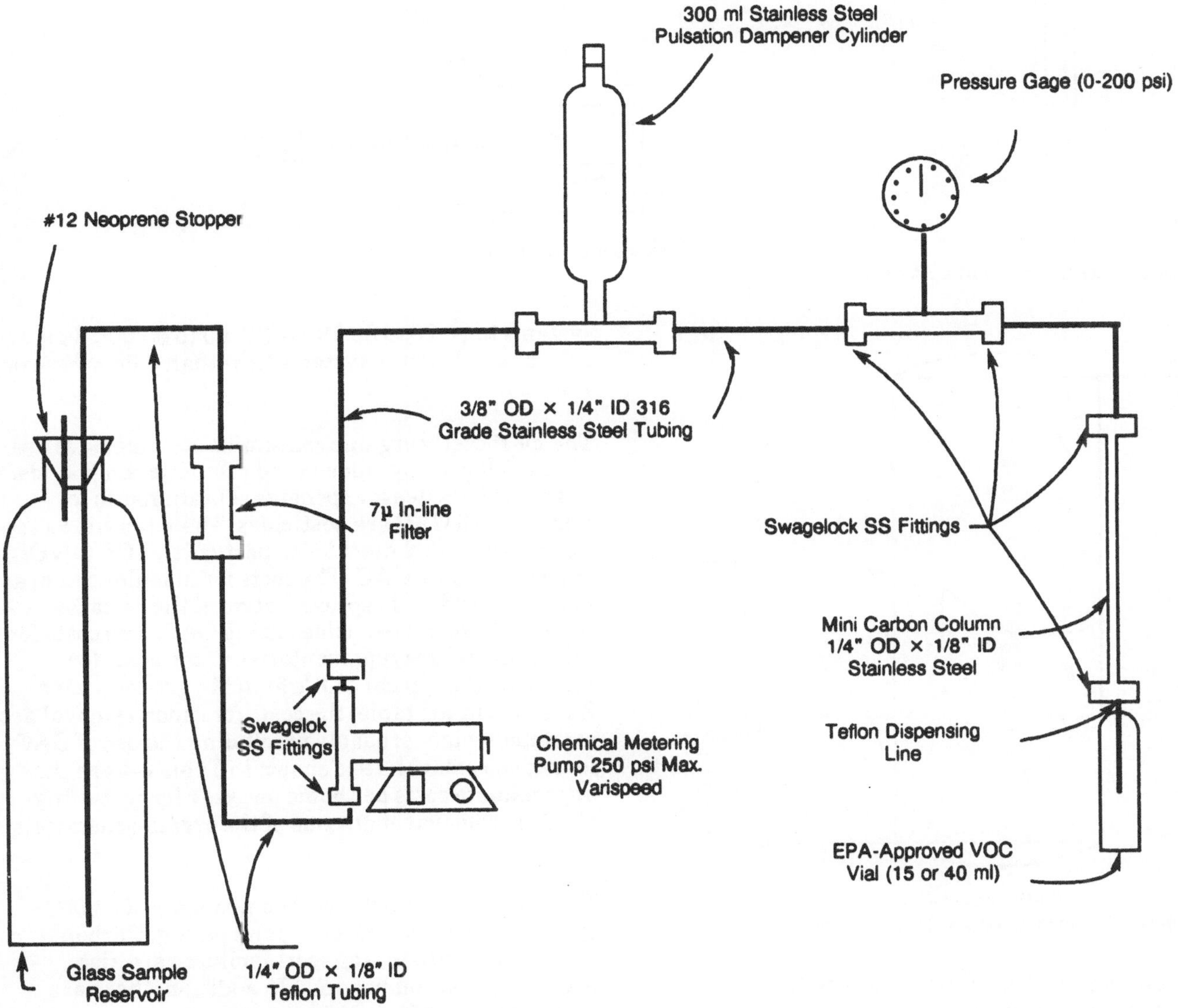

Figure 6-8. Diagram of dynamic mini-column adsorption system.

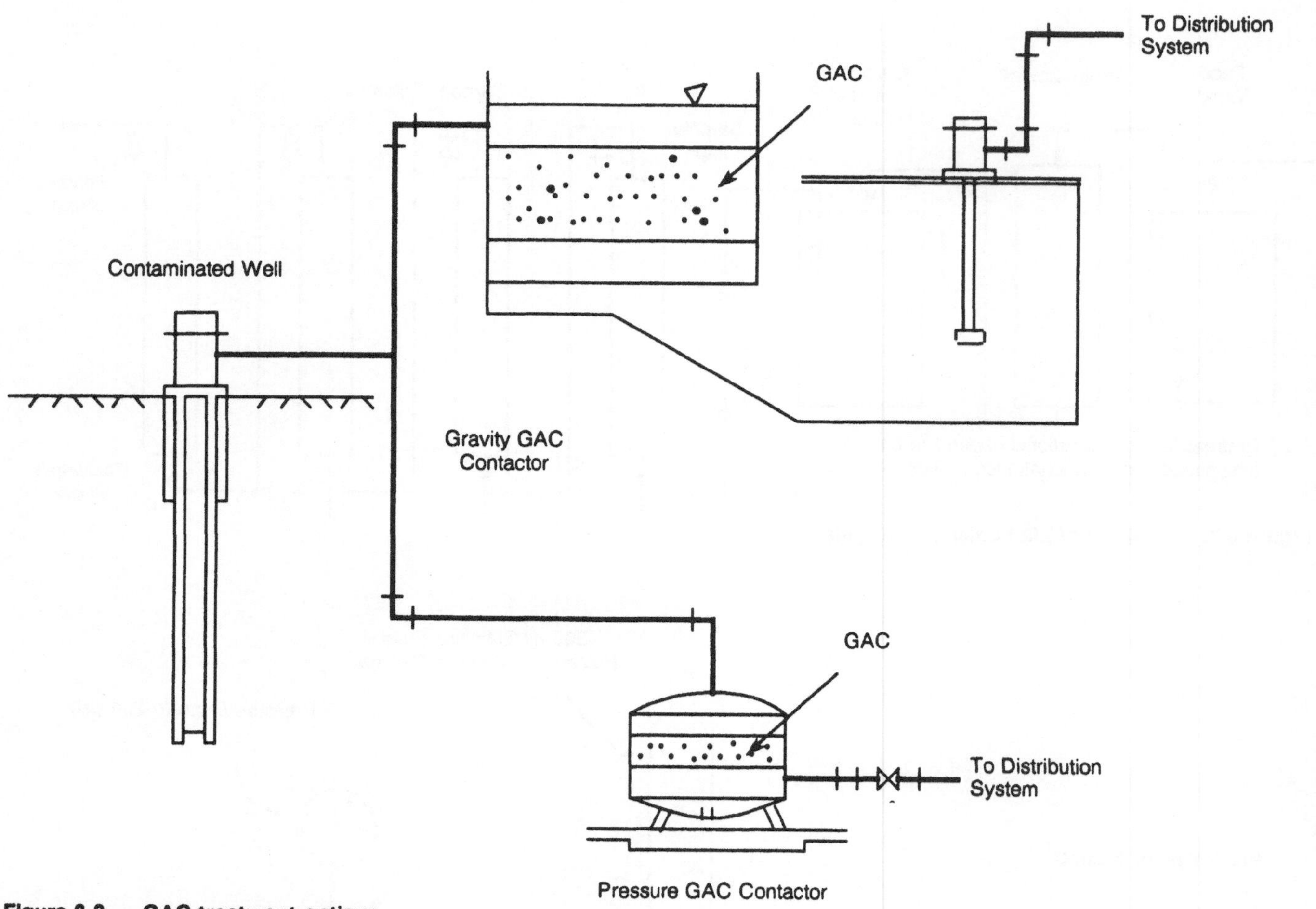

Figure 6-9. GAC treatment options.

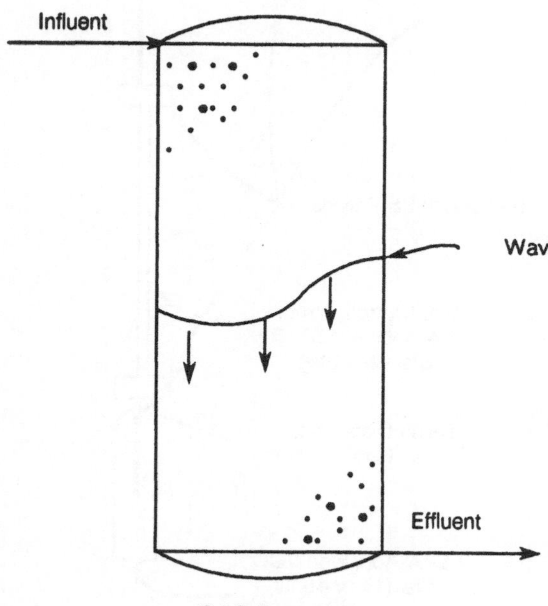

Figure 6-10. Wavefront within GAC contactor.

Water systems with GAC units having these operational characteristics cost from \$0.03 to \$0.04/m^3 (\$0.10 to \$0.15/1,000 gal) for very large systems to \$0.26 to \$0.79/m^3 (\$1.00 to \$3.00/1,000 gal) for small water systems (less than 0.04 m^3/sec or 1 MGD).

The most costly organic contaminants to remove are VOCs, followed by chlorinated aromatic compounds. In general, the least expensive contaminants to remove with GAC are pesticides. Table 6-4 shows the costs of removing alachlor (a pesticide), TCE (a VOC), and radon using GAC. The costs for alachlor removal may be considered representative of the costs for removing typical pesticides using GAC. The costs for TCE removal are representative of the costs for removing the typical VOCs found in ground water. As shown in the table, the costs for radon removal are extremely high, probably precluding the use of GAC for radon removal. Also shown in Table 6-4 are the approximate costs per home per year for installing GAC for removal of any one of the organic chemicals.

The three fundamental cost components of a GAC system are contactors, GAC, and piping. Carbon storage and carbon transport facilities are also necessary cost components. In addition, there are many site-specific cost elements to consider including:

Table 6-4. Cost for 95 Percent Removal of Several Organics and Radon Using GAC Adsorption ($1989)

	Costs by System Size				
System Capacity (MGD)	0.1	0.5	1.0	10.0	50.0
Average Daily Flow (MGD)	0.032	0.22	0.4	4.32	22.65
Population Served	500	1,500	3,000	22,000	100,000
Alachlor					
Capital Cost	69,000	138,000	190,000	1,500,000	5,600,000
Annual O&M Cost	8,500	19,100	26,500	117,000	329,000
Total Cost ($/1,000 gal)[a,b]	1.41	0.44	0.33	0.19	0.12
Cost/Home/Year[c]	100	70	50	40	30
Trichloroethylene					
Capital Cost	106,000	212,000	318,000	2,200,000	10,100,000
Annual O&M Cost	16,000	38,200	54,100	265,000	930,000
Total Cost ($/1,000 gal)	2.44	0.80	0.50	0.30	0.25
Cost/Home/Year	170	125	90	70	60
Radon					
Capital Cost	445,000	1,300,000	2,100,000	d	d
Annual O&M Cost	106,000	318,000	530,000	d	d
Total Cost ($/1,000 gal)	13.55	6.00	5.30	d	d
Cost/Home/Year	950	900	800	d	d

[a] Total cost is calculated based on amortizing the capital cost over 20 years at 10 percent interest rate, adding the annual O&M cost, and dividing by the total annual flow.
[b] 1,000 gallons = 3.78 m^3.
[c] Cost per home per year is calculated based on total cost per year divided by the number of homes served using 3.0 people per home.
[d] These costs were not calculated because they are too high for practical municipal application.

- Raw water holding tank (for ground-water systems)
- Restaged well pump to address excessive pressure drops caused by the contactors
- Contactor housing, required in cold climates
- Chemical feed equipment
- Clear well storage tank and pumps
- Backwash storage tank

The capital costs increase directly with system size. Consequently, the capital costs can be derived from the estimated EBCT, a direct measure of system size because of its relationship with the carbon contactor. The predominant elements of capital costs, as well as operating and maintenance costs, are shown in Figure 6-11.

The two primary capital cost considerations are (1) construction type and (2) carbon and reactivation strategy. Steel pressurized GAC units are most economical for small systems, generally less than 0.44 m^3/sec(10 MGD), while concrete gravity-fed systems are more appropriate for larger systems.

The carbon and reactivation system, including either onsite or offsite carbon reactivation, compose nearly half of the GAC system costs. Infrared carbon reactivation systems are cost effective for smaller systems, and fluid bed reactivation systems are suited for plants using more than 1,360,800 kg (3,000,000 lb) of carbon annually. Multiple hearth reactivation systems are more expensive than either infrared or fluid bed systems. Figure 6-12 compares the costs of carbon replacement and two methods of carbon regeneration for several carbon consumption rates.

Operating costs do not correlate strongly with system size, but are dependent on carbon usage rates, which rely on the nature of the contaminants in the influent. Other operating and maintenance cost components are carbon type, labor, fuel, steam, power, maintenance, and laboratory analysis.

6.3 Packed Column Aeration

Aeration, also called air-stripping, mixes water with air to volatilize contaminants. The volatilized contaminants are either released directly to the atmosphere or are treated and then released. Aeration is used primarily to remove VOCs.

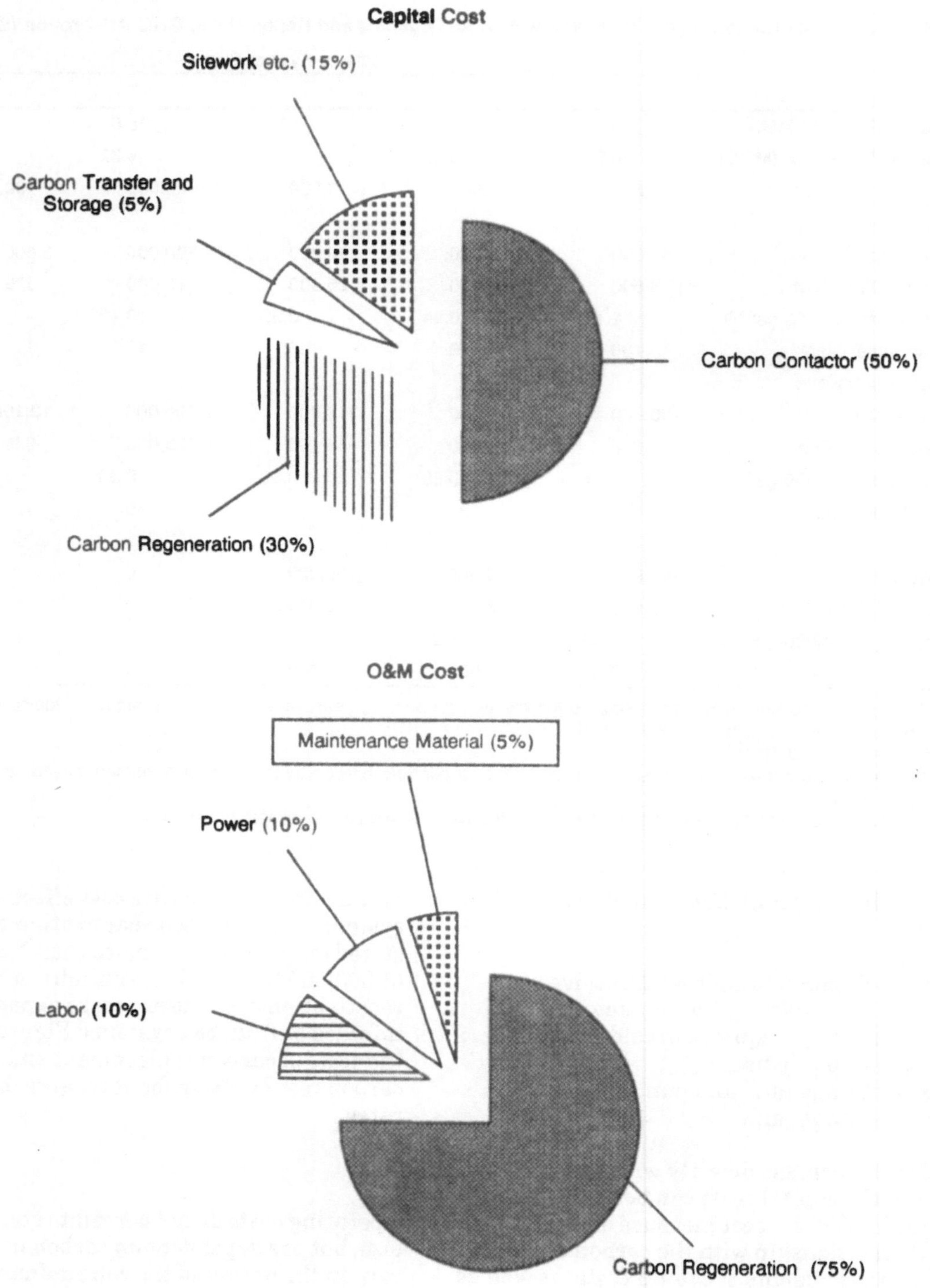

Figure 6-11. GAC facility cost components.

The aeration process is based on the principles of mass transfer from liquids to gases. This transfer is expressed in the following equation:

$$M = (K1)(a)(\Phi P)$$

where:

M = the mass of the substance transferred from water to air expressed in $lb/hr/ft^3$

Kl = the coefficient of mass transfer expressed in $lb/hr/ft^2$

a = the effective area in ft^2

ΦP = the concentration difference or driving force (The driving force is the difference between conditions in the aeration unit and equilibrium conditions for the substance between gas and liquid phases.)

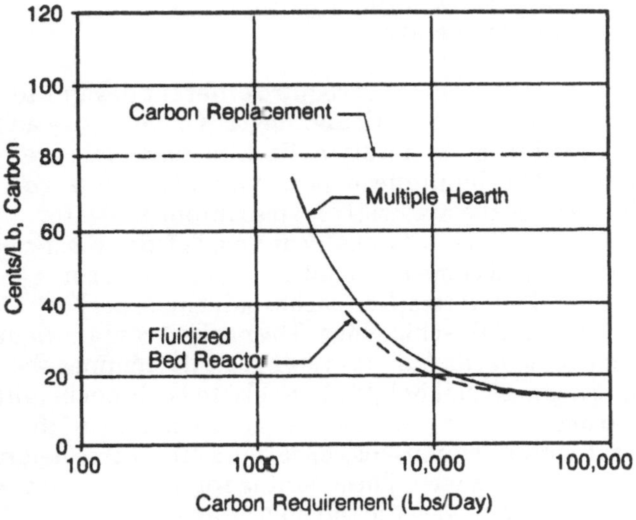

Figure 6-12. Cost comparison of carbon regeneration/replacement options.
Source: Dyksen et al. (1987).

The Henry's Law constant indicates a contaminant's volatility and its affinity for the aeration process. Substances with high Henry's Law constants are easily aerated, while those with low constants are difficult to remove with aeration. Henry's Law constant is calculated with the following formula:

$$H = [(16.04)*P*M]/(T*S)$$

where:

H = the Henry's Law constant in dimensionless units
P = the vapor pressure expressed in mm
M = the molecular weight of the solute
T = the temperature in °K
S = the solubility in mg/L

Table 6-5 lists several compounds and their Henry's Law constants. As the table indicates, vinyl chloride has an extremely large Henry's Law constant relative to any other VOC.

The two basic types of aeration systems are diffused air and waterfall units The diffused air system, discussed in Section 6.5, involves adding air to a volume of water. Waterfall aeration units are more common and are discussed in the remainder of this section.

Packed tower aeration (PTA), or packed column aeration (PCA), is a waterfall aeration process that trickles raw water over a medium within a cylinder to mix water with air. The medium is designed to break the water into tiny droplets, a process enhanced by the introduction of air blown from

Table 6-5. Henry's Law Constants for Nine Organic Chemicals

Type of Organic Chemical	Henry's Law Constant[a] (dimensionless units)
VOCs	
Vinyl chloride	265.00[a]
Trichloroethylene	0.41[a]
Tetrachloroethylene	0.82[a]
cis-1,2-Dichloroethylene	0.32[b]
Pesticides	
Aldicarb	0.00000017[b]
Chlordane	0.004[b]
Dibromochloropropane	0.01[b]
Chlorinated Aromatics	
Polychlorinated biphenols	0.059[b]
Dichlorobenzene	0.081[b]

Note: Constants estimated at about 20°C.
[a] AWWA Research Foundation and KIWA (1983).
[b] U.S. EPA (1988b).

underneath the medium in the tower or column. A typical PTA unit is shown in Figure 6-13.

The major process elements of PTA are the column (or tower), packing medium, blower, booster pump, and instrumentation. Columns can be constructed from fiberglass-reinforced plastic, aluminum, stainless steel, or concrete. Within the column are mist eliminators to prevent water from escaping in the vents, packing material, support grids for the packing material, and liquid distributors to separate the influent into many smaller streams. The four primary designs for liquid distributors are orifice plate (see Figure 6-14), trough-type distributor (see Figure 6-14), orifice headers, and spray nozzles. Adding PTA to an existing plan will require (1) changes in the staging of the well pumps and (2) repumping treated water to the distribution system.

Housing the tower usually is not necessary because the water temperature remains fairly constant throughout the PTA-treatment process. Consequently, water rarely freezes during the process.

PTA systems vary: in some, water cascading over spillways imparts the necessary turbulence; in others, several layers of slats mix the water with air. Innovations in PTA are reflected in the newest additions to the latter type of aeration system. Emerging aeration techniques include the catenary grid and Higee systems which are discussed in Sections 6.7.4 and 6.7.5.

The rest of this section profiles aeration technology including design considerations (Section 6.3.1), operation and maintenance (Section 6.3.4), system performance (Section 6.3.5), and system costs

Fiberglass-reinforced plastic packed column with two carbon vapor phase treatment units; Battle Creek, MI.

(Section 6.3.6). In addition, this section reviews pilot testing of PTA systems (Section 6.3.2) and managing of VOC emissions from aeration (Section 6.3.3).

6.3.1 System Design Considerations

Aeration provides a fixed percentage of contaminant removal regardless of influent concentration. To compensate for uncertainty, aeration systems can be designed to incorporate safety factors of two or three times the expected influent contaminant concentrations to ensure compliance with regulatory standards.

Aeration system performance is affected primarily by column size and airflow. Increases in airflow and column height improve removal efficiencies. Typical design parameters are provided for 13 common VOCs in Table 6-6. Design considerations include:

- Type of organic contaminant(s)
- Concentration of contaminant(s)
- Type of packing material
- Height of packing material
- Air-to-water ratio
- Water loading rate
- Water temperature

Different contaminants require different designs to accommodate the particular degrees of volatility and, thus, affinities for aeration. Packing materials are designed to simultaneously provide a low pressure drop across the material and maximum air-water contact area. They typically are made from plastic or ceramic, and come in the following forms: super intalox, Tellerettes, Tri-packs, pall rings, berl saddles, and Raschig rings. The desired contaminant removal level and air-to-water ratio determines the packing material height. The selected column height and design of air intake louvers must comply with local zoning regulations concerning structural height and noise nuisance. These zoning regulations are the most basic site-specific considerations.

The air-to-water ratio in a column is a function of the water temperature and the desired level of contaminant removal. This ratio determines the size of the system's blower, which is the primary component of operating costs for PTA systems. Air-to-water ratios typically range from 30:1 to 100:1. The water loading rate, the amount of water passing through the column, routinely ranges from 16.9 to 20.4 L/sec/m^2 (25 to 30 GPM/ft^2). The column diameter specification is derived to accommodate the desired water loading on the column.

All of these factors affect aeration's removal efficiency. Figure 6-15 illustrates the dramatically different removal rates achieved by the same packed material depth for three different contaminants. PTA removal effectiveness usually increases as water temperature increases. The decrease in column height required to achieve the same level of contaminant removal as the water temperature increases is shown in Figure 6-16. In most cases, heating influent is not cost effective, but in temporary situations of low flow, it has been used effectively.

Pretreatment to remove iron, solids, and biological growth may be necessary to prevent clogging of the packing material and, thus, the entire system. Posttreatment also may be necessary to reduce corrosive properties that develop in the water due to the increased presence of dissolved oxygen during the aeration process. One solution is the use of a corrosion inhibitor.

6.3.2 Pilot Testing PCA

Figure 6-17 is a schematic diagram of a typical pilot column. Pilot columns are used to test various water loading rates, airflow rates, air-to-water ratios, and packing materials. Typically, 8 to 12 pilot column runs are used to test various combinations of design factors. A test run lasts for about 30 minutes to

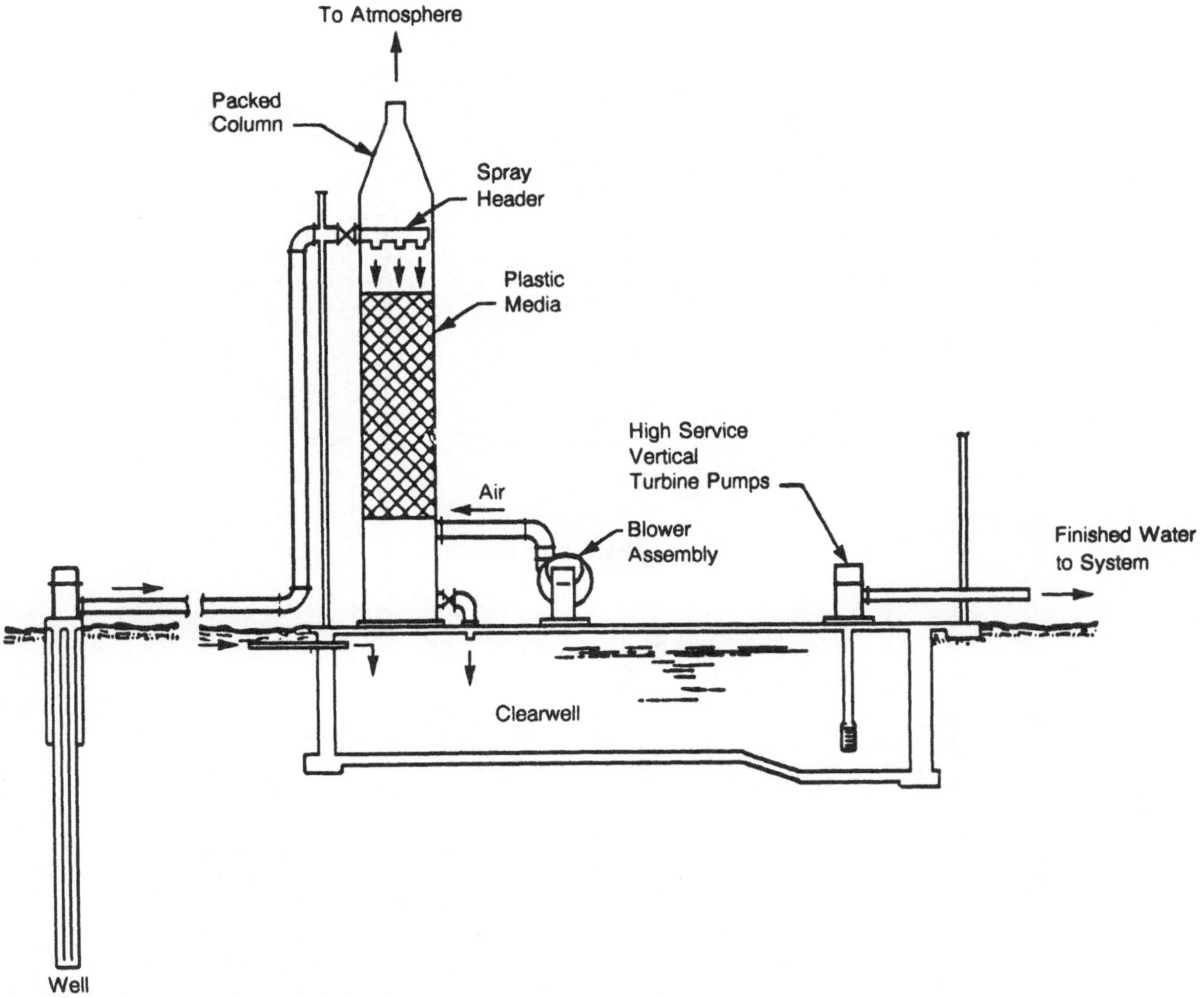

Figure 6-13. Packed tower aeration system.

achieve steady state. Influent and effluent samples are collected for each test run.

The following precautions should be taken to ensure accurate pilot test results:

- Carefully place packing material into the column to avoid channeling and vacant pockets.
- Level the pilot column to avoid channeling and wall effects.
- Select representative performance points for sampling.
- Collect and analyze duplicate samples to check results.
- Verify laboratory results.

6.3.3 VOC Emission Control

To meet air emission regulations, PTA exhaust gas may require treatment. VOC emissions for PTA units are calculated with the following formula:

$$\text{Emission rate (in lb/hour)} = (C1\text{-}C2)*(V)*(5/10^7)$$

where:

C1 = Influent concentration of the VOC expressed in µg/L
C2 = Effluent concentration of the VOC expressed in µg/L
V = Water flow rate in GPM

The emission rate must be evaluated in the context of applicable air quality regulations and other site-specific factors. These factors include proximity to human habitation, treatment plant worker exposure, local air quality, and local meteorological conditions.

Air emission regulations are expressed either in terms of permissible emission rates (lb/day or lb/hr) or projected ground level concentrations (mg/m^3). If the plant's emission rate is unacceptable, then column or plant process design may be changed to bring the plant into compliance.

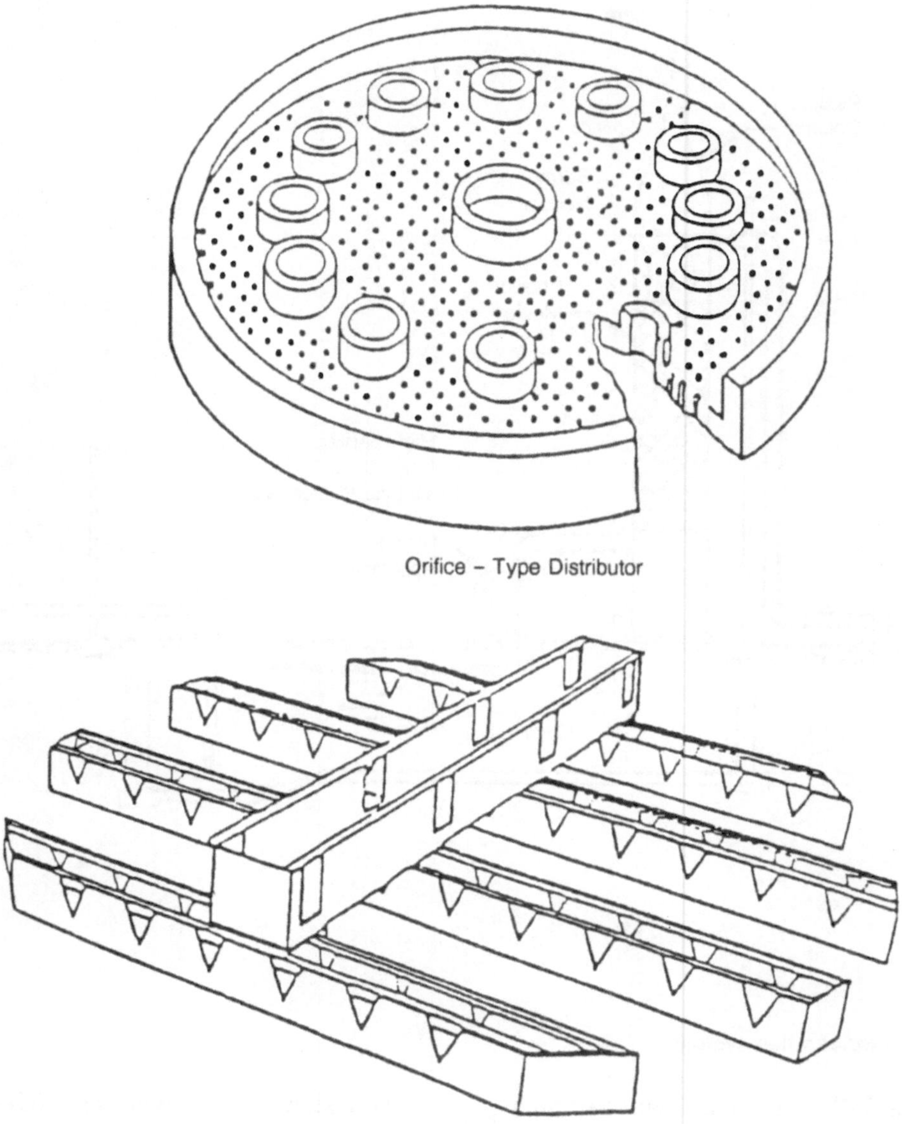

Figure 6-14. Distributor types.

Modifying PTA plants to dilute emissions to an acceptable level is the least costly method of achieving compliance with air regulations. Such plant modifications include increases in tower height, airflow rate, and exhaust gas velocity. If these steps are insufficient to achieve compliance, then a vapor phase treatment component may be required.

The four methods used for vapor phase treatment of VOC emissions are (1) thermal destruction, (2) catalytic incineration, (3) ozone destruction, and (4) carbon adsorption. The first three methods are not used extensively. Thermal destruction is effective but very expensive because of the high energy requirements. Catalytic incineration, depicted in Figure 6-18, lowers the energy requirements of the thermal destruction process. However, catalytic incineration is not used widely because it is not effective in eliminating low levels of chlorinated organic compounds. To date, ozone destruction used in conjunction with ultraviolet radiation has only been evaluated on a pilot scale.

Carbon adsorption for control of VOC off-gases from PTA is accomplished with a vapor phase GAC unit. Currently, GAC is the most frequently used approach to controlling these VOC emissions. Figure 6-19 shows a schematic diagram of a typical gas phase unit. The heating element eliminates humidity, which competes with organic compounds for the available GAC. The specifications of a gas phase unit depend on individual PTA system requirements,

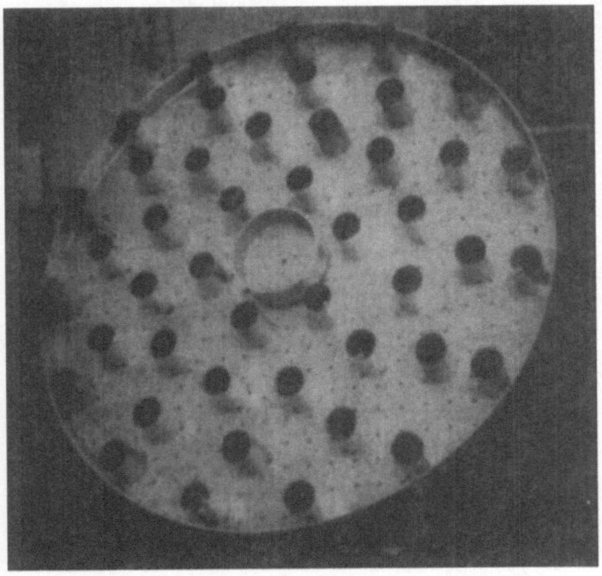

Orifice-type distributor plate for a packed column aerator.

including air-to-water ratio, influent concentration of the VOC, and acceptable level of VOC emissions.

Carbon usage for gas phase control of VOCs from aeration units is less than equivalent VOC controls for the liquid phase GAC units. In addition, since gas phase adsorption kinetics are faster than liquid phase kinetics, the carbon bed size is reduced. Carbon exhaustion is estimated either by using a mass balance approach or monitoring with gas chromatography or mass spectrometry.

6.3.4 Operation and Maintenance

Typically, packed columns are operated automatically. Daily visits assure that all equipment is running satisfactorily. Maintenance requirements generally involve several hours per month to service pump and blower motors and to replace air filters on

he blowers, if necessary. Some packed column installations have reported operational difficulties from the plugging of the packing material and inadequate liquid distribution. However, most installations report no unusual operational problemsthe blowers, if necessary. Some packed column installations have reported operational difficulties from the plugging of the packing material and inadequate liquid distribution. However, most installations report no unusual operational problems.

6.3.5 System Performance

PTA is an effective and practical removal process for several organic compounds applicable to water treatment plants of all sizes. It is typically used for systems drawing ground water. Larger plants using PTA for radon removal have the added task of properly managing the associated air emissions.

PTA effectively removes from drinking water most compounds with high Henry's Law constants, which includes most VOCs. PTA is the BAT for vinyl chloride, which has an extremely high Henry's Law constant; GAC and PTA are BAT for other regulated VOCs. Table 6-7 lists examples of aeration's removal effectiveness for trichloroethylene, vinyl chloride, VOCs, and aldicarb.

6.3.6 System Costs

The cost of removing many organic compounds with PTA, by itself, is well within the range of the cost of delivering treated drinking water in the United States, which typically costs from \$0.26 to \$0.40/m^3 (\$1.00 to \$1.50/1,000 gal of water). The cost of controlling air emissions from PTA units significantly increases total system costs.

Table 6-8 presents estimated costs of removing 95 percent of several VOCs and radon using PTA. As the table indicates, costs in terms of cents per 1,000 gal of treated water drop significantly as system size increases. Of the VOCs listed in the table, the most expensive VOC to remove with PTA is dibromochloropropane; the least expensive is vinyl chloride. Both vinyl chloride and radon are removed equally as well using PTA, therefore the costs for removing these contaminants are similar. Approximate costs per home per year are presented in the table for each of the contaminants.

The capital cost elements for PTA systems include the tower or column, internal column parts, packing material, blower(s), clearwell, booster pump(s), and any associated piping. Site-specific costs may include a raw water holding tank, restaged well pump, blower building, chemical facility, noise control installation, and air emission control.

The cost of vapor phase controls to manage air emissions from PTA operations can strongly influence plant design and total costs. EPA research indicates that adding a vapor phase carbon adsorption unit will double the costs of PTA. Carbon adsorption units have significant capital and operation and maintenance cost components, including costs for carbon contactors, initial activated carbon material, gas heaters, and installation.

Table 6-6. **Typical Air Stripping Design Parameters for Removal of 13 Commonly Occurring Volatile Organic Chemicals[a]**

Compound	Henry's Law Constant	Air-to-Water Ratio	Air Stripper Height (feet)[b]	Diameter of Packed Column (feet)
Trichloroethylene	0.116	29.9	38.03 38.03	8.10
Tetrachloroethylene	0.295	11.8	43.77	5.97
Carbon tetrachloride	0.556	6.2	44.88	4.95
1,1,1-Trichloroethane	0.172	20.1	40.06	7.07
1,2-Dichloroethylene	0.023	150.6	33.47	14.89
Dichloromethane	0.048	71.59	28.61	11.12
cis-1,2-Dichloroethylene	0.093	37.10	34.88	8.73
Vinyl chloride	265.000	0.013[c]	59.58	1.90
Benzene	0.106	32.69	36.25	8.37
Toluene	0.117	29.62	39.04	8.07
m-Xylene	0.093	37.26	40.49	18.34
Chlorobenzene	0.069	50.29	37.60	22.74
1,2-Dichlorobenzene	0.090	38.67	40.45	8.86

[a] Water flow rate - 2.16 MGD (8.17 ml/day), inlet water concentration - 100.0 µg/L, water treatment objective - 1.0 µg/L, air-stripper temperature - 50 °F (10°C), air-stripper packing pressure drop - 50.0 (N/m²)/m packing, air- stripper packing - 3-inch plastic saddles

[b] 1 ft = 0.3 m.

[c] Theoretical calculation based on the extremely high Henry's Law constant.

Source: Crittenden et al. (1988).

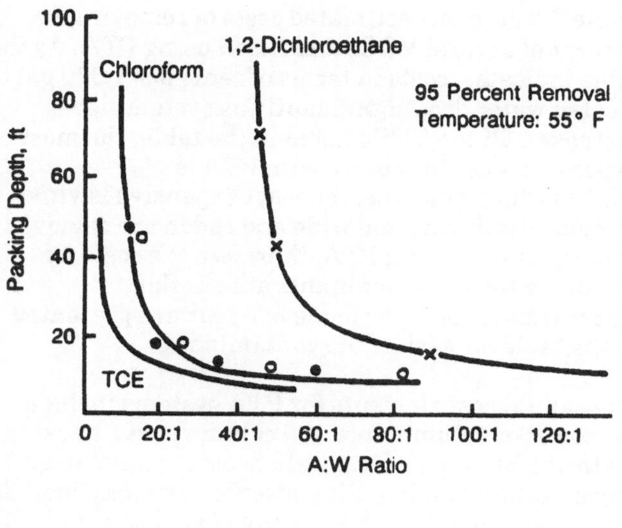

Figure 6-15. Effect of compound on packed-column design.

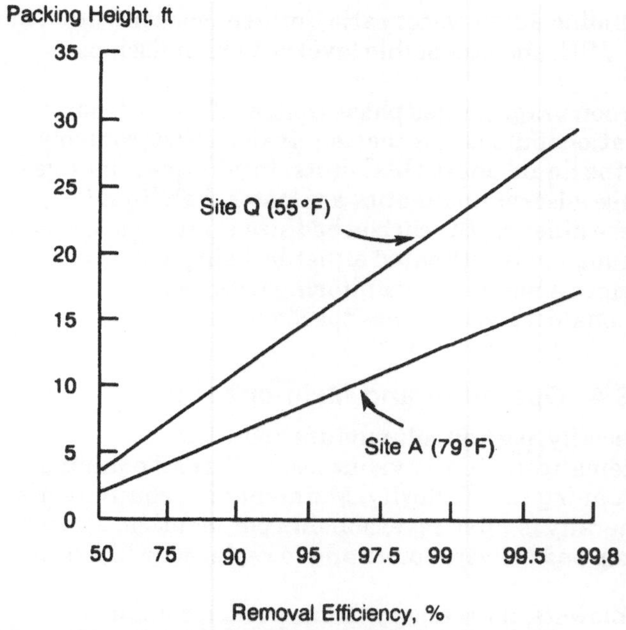

Figure 6-16. Packing height vs. removal efficiency for trichloroethylene.

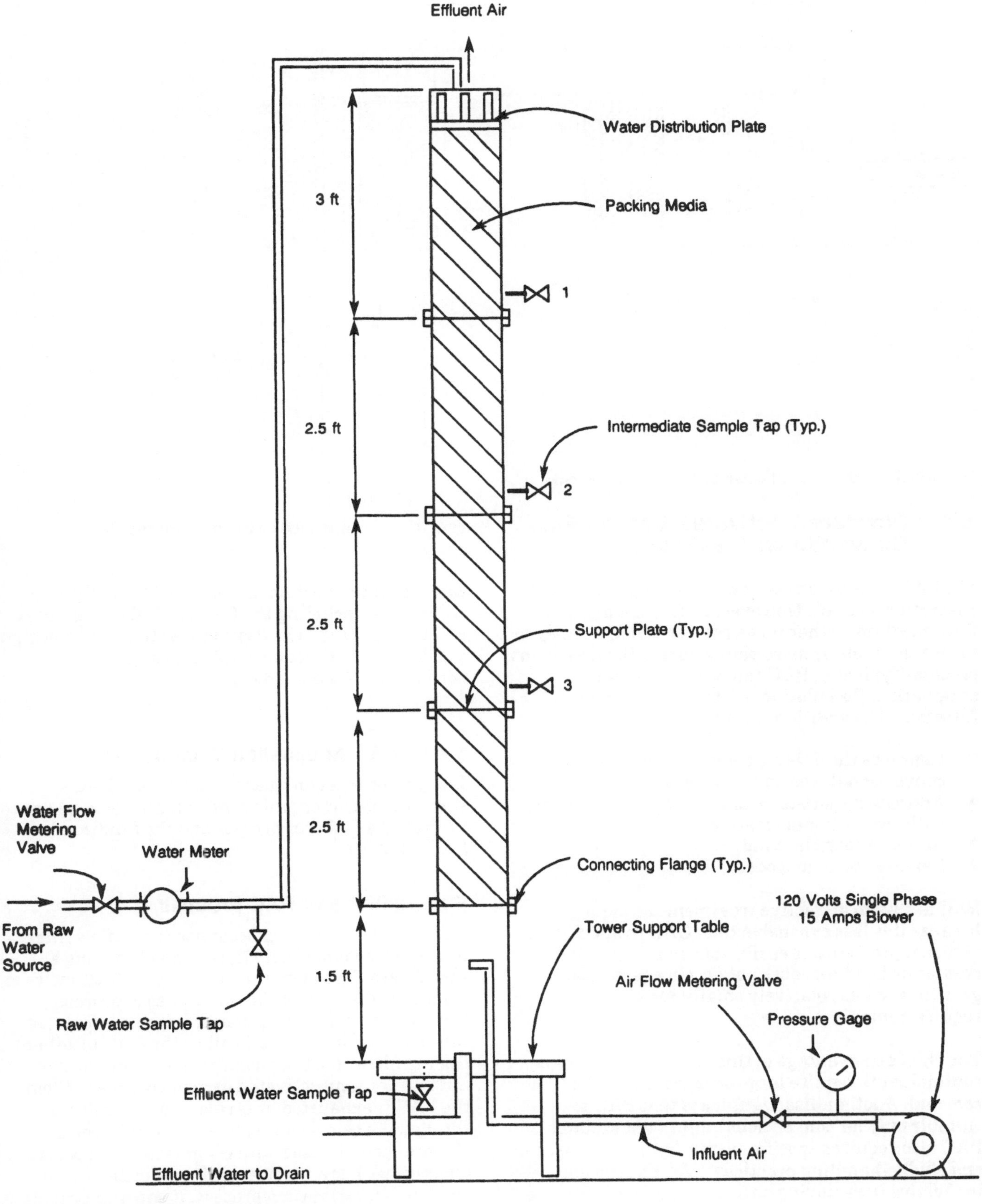

Figure 6-17. Schematic of pilot aeration column.

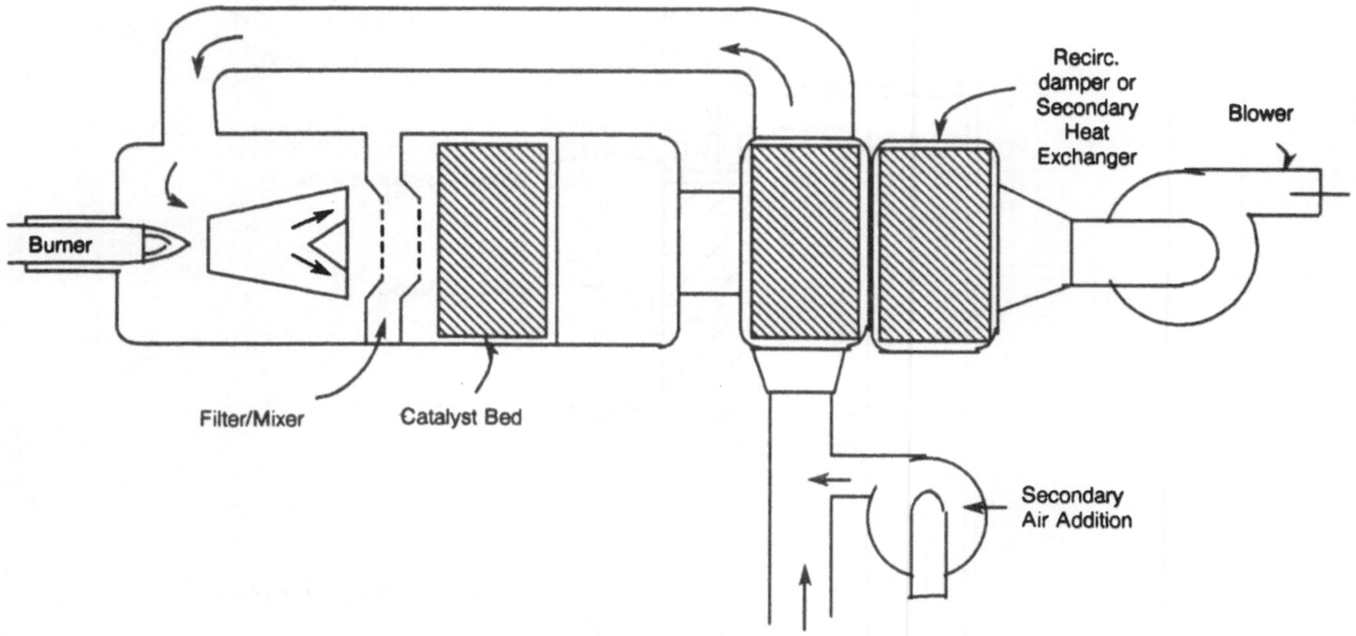

Figure 6-18. Schematic of catalytic incineration process.

6.4 Powdered Activated Carbon Plus Conventional Treatment

The PAC process is based on the same principles of adsorption as GAC. However, PAC is not a granular filter medium; rather it is a powder added directly to the water at one or more points during the treatment process. Typically, PAC can be added during coagulation, flocculation, sedimentation, and filtration. The addition of PAC:

- Improves the organic removal effectiveness of conventional treatment processes
- Addresses short-term and emergency problems with conventional treatment systems
- Acts as a coagulant aid
- Removes taste and odor

PAC is also an attractive treatment technology because it is less expensive than GAC in addressing seasonal problems, is easily started and stopped, creates no headloss, does not encourage microbial growth, and has relatively small capital requirements.

The chief disadvantage of this process is that some contaminants require large dosages of PAC for removal. Another disadvantage is that PAC is suitable only for conventional treatment systems. PAC also requires specific system hydraulics, space, and sludge-handling practices. PAC has proven ineffective in removing natural organic matter, due to the competition from other contaminants for surface adsorption and the limited contact time between the water and PAC. In addition, PAC adsorption is not amenable to basin-mixed flow reactions (as opposed to column-mixed flow reactions).

The rest of this section discusses PAC application techniques, including the Roberts-Haberer process and the fluidized-bed adsorber (Section 6.4.1), design considerations (Section 6.4.2), and system performance (Section 6.4.3).

6.4.1 PAC Application Techniques

Two potential techniques for improving PAC's effectiveness for organic contaminant removal are the Roberts-Haberer process and the fluidized-bed PAC adsorber.

Roberts-Haberer PAC Process

The Roberts-Haberer process uses an upflow filter composed of foamed polystyrene beads having a specific gravity of less than 0.1. Figure 6-20 shows a simplified diagram of the three-phased process. During the conditioning phase, water recirculates through the filter media to allow the PAC to adhere. During filtration, the media act as a combination upflow filter and activated carbon adsorber. When the PAC is exhausted, it is released to the drain through the backwashing process. The Roberts-Haberer process used with coagulation is especially effective with low turbidity influent or combined with the sedimentation process. It can also provide flocculation for a system.

The reported advantages of the Roberts-Haberer process include:

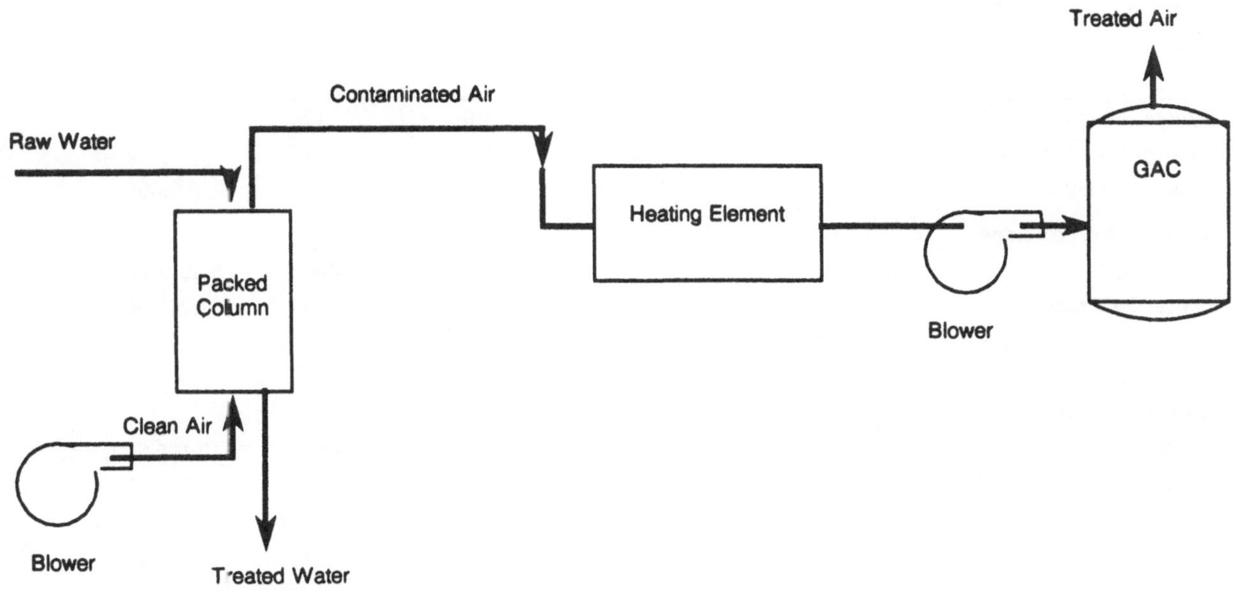

Figure 6-19. Vapor-phase carbon system for treatment of aeration exhaust air.

Table 6-7. Examples of Removal Efficiencies for PTA Systems

TCE

- A full-scale redwood slate tray aeration plant with a 3.8 MGD capacity at an air-to-water ratio of 30:1 achieved 50 to 60 percent reductions from initial TCE influent concentrations of 8.3 to 39.5 µg/L.
- A full-scale multiple tray aeration unit with a 6 MGD capacity achieved 50 percent reductions from initial TCE influent concentrations of 150 µg/L.
- A full-scale packed tower aeration column plant using ground water at an air-to-water ratio of 25:1 achieved 97 to 99 percent reductions from initial TCE influent concentrations ranging from 1,500 to 2,000 µg/L.

Vinyl Chloride

- A pilot packed tower aerator, with 9°C influent, achieved up to 99.27 percent removal of vinyl chloride.
- A spray tower aeration unit removed vinyl chloride from ground water with VOC concentrations of 100 to 200 µg/L.
- An in-well aeration unit with an air-lift pump achieved 97 percent removal of vinyl chloride.

Aldicarb

- Aeration was found to be ineffective in reducing levels of aldicarb because of its low Henry's Law constant.

VOCs

- A four-stage aeration design with four shower heads and a pressure drop of 10 psi achieved 99.9 percent VOC removal.

- Reduced PAC use because of the increased contact time
- Quicker adsorption and increased utilization of carbon, relative to the GAC process, due to the small size of the carbon particles
- Decreased carbon use relative to the GAC process
- Increased SOC removal if used after coagulation, due to the removal of high molecular weight substances by coagulation
- Improved filtration efficiency by removing THM precursors prior to filtration and allowing prefiltration chlorination, which reduces the microbial presence on the filter
- Minimal bacterial growth because of the frequency of backwashes
- Increased ability to accommodate multiple objective water treatments
- Eased recovery or modification of PAC application

Fluidized-Bed PAC Adsorber

This experimental method processes water through flocculated carbon, which extends the contact period. The strength of the flocculated carbon particles increases the carbon's longevity for treatment purposes. Floc strength must be balanced with particle breakup, which increases carbon surface area and, therefore, adsorption of contaminants. Because fluidized-bed PAC adsorbers are relatively new, more research is required for a full evaluation.

6.4.2 System Design Considerations

The primary design considerations for instituting PAC are dosage, contact time, and points of application. Dosages commonly are less than 100 mg/L but can range as high as 300 mg/L, and the minimum contact time is usually 0.25 hour. PAC points of application customarily are (1) before the rapid mix process, (2) after rapid mixing but before flocculation, or (3) after flocculation and before

Table 6-8. Cost for 99 Percent Removal of Several VOCs and Radon Using Packed Tower Aeration ($1989)

	Costs by System Size				
System Capacity (MGD)	0.1	0.5	1.0	10.0	50.0
Average Daily Flow (MGD)	0.032	0.22	0.4	4.32	22.65
Population Served	500	1,500	3,000	22,000	100,000
Dibromochloropropane					
Capital Cost	106,000	420,000	636,000	5,700,000	5,600,000
Annual O&M Cost	4,500	25,000	50,000	460,000	329,000
Total Cost ($/1,000 gal)[a,b]	1.90	0.90	0.85	0.60	0.12
Cost/Home/Year[c]	175	165	160	155	30
Trichloroethylene					
Capital Cost	85,000	210,000	318,000	2,100,000	10,100,000
Annual O&M Cost	2,700	10,800	21,200	201,000	930,000
Total Cost ($/1,000 gal)	1.10	0.45	0.40	0.30	0.25
Cost/Home/Year	75	70	65	60	60
Vinyl/Chloride/Radon					
Capital Cost	54,000	148,000	210,000	1,484,000	5,830,000
Annual O&M Cost	1,600	6,800	16,700	106,000	477,000
Total Cost ($/1,000 gal)	0.70	0.30	0.25	0.20	0.15
Cost/Home/Year	50	45	40	35	30

[a] Total cost is calculated based on amortizing the capital cost over 20 years at 10 percent interest rate, adding the annual O&M cost, and dividing by the total annual flow.
[b] 1,000 gallons = 3.78 m^3.
[c] Cost per home per year is calculated based on total cost per year divided by the number of homes served using 3.0 people per home.

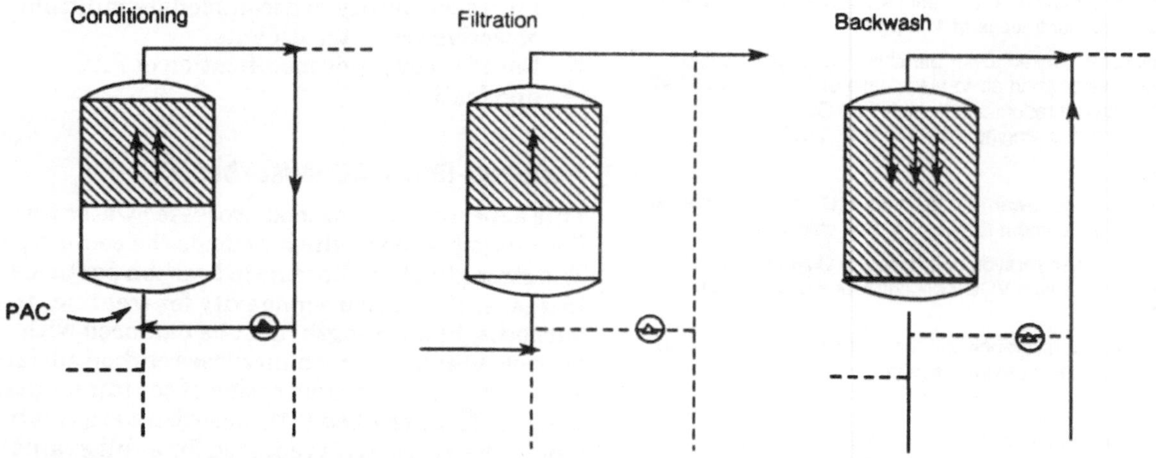

Figure 6-20. Schematic of the Roberts-Haberer process.

sedimentation. Figure 6-21 is a schematic diagram of PAC usage.

6.4.3 System Performance

Adding PAC to conventional treatment systems can greatly improve their performance in removing certain organic chemicals. Table 6-9 compares removal efficiencies of the conventional treatment process with and without PAC for 15 organic compounds, including two VOCs and 13 SOCs.

6.5 Diffused Aeration

The diffused aeration system bubbles air through a contact chamber for aeration; the diffuser is usually located near the bottom of the chamber. The air introduced through the diffuser, usually under pressure, produces fine bubbles that impart water-air mixing turbulence as they rise through the chamber. Diffused aeration units are designed to serve either point-of-use or plant situations. Figure 6-22 depicts a plant-scale system. Figure 6-23 shows a home-scale aeration system.

Table 6-9. Typical Performance of Conventional Treatment Processes without and with PAC

Compound	Conventional Treatment without PAC	Conventional Treatment with PAC	
	Percent Removal	Dosage (mg/L)	Percent Removal
VOCs			
Carbon tetrachloride	–[a]	9.6-30.0	0-25
1,1,1-Trichloroethane	–[a]	7	40-65
SOCs			
Acrylamide	5	8	13
Alachlor	<50	4-34	36-100
Carbofuran	54-79	9-25	45-75
o-Dichlorobenzene	–[a]	8-27	38-95
2,4-D	0-3	11-306	69-100
Ethylbenzene	–[a]	8-27	33->99
Heptachlor	64	11-97	53-97
Lindane	10-20	2-34	82-97
Monochlorobenzene	–[a]	8-27	14->99
Toluene	–[a]	8-27	0-67
2,4,5-TP	63	1.5-17.0	82-99
Toxaphene	–[a]	1-44	40-99
Xylenes	–[a]	8-27	60->99

[a] Information not available.
Source: Miltner and Fronk (1985).

The main advantage of diffused aeration systems is that they can be created from existing structures, such as storage tanks. This type of aeration, however, is less effective than PCA and is generally used only in systems with adaptable existing structures.

6.5.1 System Design Considerations

The critical process design considerations for diffused aeration units are:

- Diffuser type and air bubble size
- Chamber depth: 1.5 to 3 m (5 to 10 ft)
- Air-to-water ratio: 5:1 to 15:1
- Detention time: 10 to 15 minutes
- Chamber hydraulics

One of the most important diffuser design considerations is the air introduction method and the resultant bubble size. Air diffusers use porous plates or tubes, or perforated pipes placed along the bottom or sides of the chamber. Chamber hydraulics affect the uniformity with which the aeration process takes place and, therefore, the completeness of removal. Baffling that achieves plug flow conditions followed by mixing is the general method of assuring proper chamber hydraulics.

6.5.2 System Performance

Table 6-10 presents removal effectiveness data of diffused aeration for five VOCs and nine SOCs. The removal rates, ranging from 11 to 95 percent, are based on diffused aeration units with air-to-water ratios of 5:1 to 15:1 and contact times of 10 to 15 minutes.

6.6 Multiple Tray Aeration

Multiple tray aeration directs water through a series of trays made of slats, perforations, or wire mesh. Air is introduced from underneath the trays, either with or without added pressure. Figure 6-24 is a diagram of a redwood slat tray aerator.

Multiple tray aeration units have less surface area susceptible to clogging from iron and manganese precipitation than PTA. However, this type of aeration is not as effective as PTA and can experience clogging problems, in addition to biological growth and corrosion problems. Multiple tray aeration units are generally available as package systems.

6.6.1 System Design

The principal design considerations for multiple tray aeration are tray type, tray height, pressurized or unpressurized air flow, and air-to-water ratio. Trays are usually made from wood or plastic and range in stack height from 3.6 to 4.8 m (12 to 16 ft). Pressurized air flow is used to increase the air-to-water ratio, with the typical ratio being 30:1.

6.6.2 System Performance

Slat tray aeration with an air-to-water ratio of 30:1 and a tray height of 3.6 to 4.8 m (12 to 16 ft) has achieved 30 to 90 percent reductions of trichloroethylene and 20 to 85 percent reductions of tetrachloroethylene.

6.7 Emerging Applications of Treatment Technologies for Organic Contaminants

This section discusses seven emerging water treatment technologies that hold promise of becoming BATs for removing organic contamination, including:

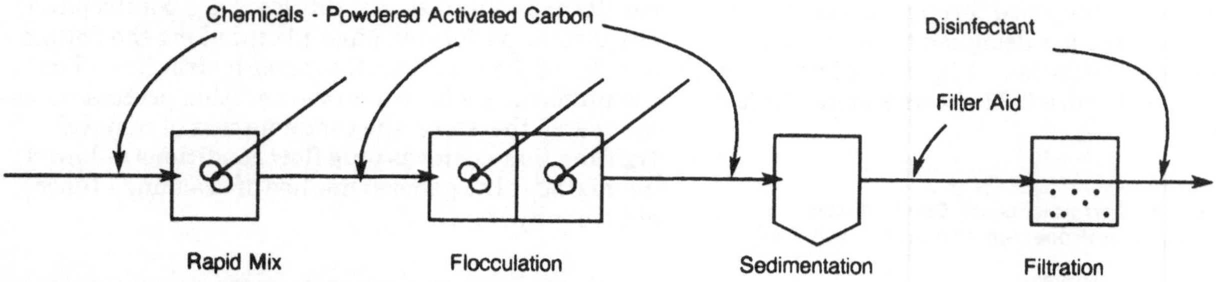

Figure 6-21. Schematic of PAC adsorption process.

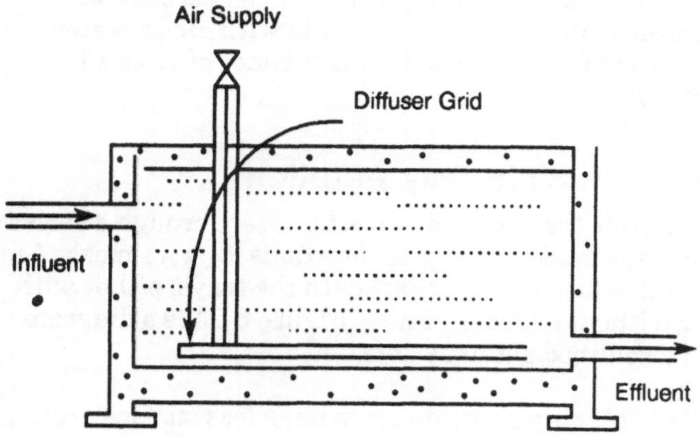

Figure 6-22. Schematic of a plant-scale diffused aeration process.

- Oxidation including ozone and advanced oxidation processes
- Reverse osmosis
- Mechanical aeration
- Catenary grid aeration
- Higee aeration
- Resins

All of these technologies require extensive laboratory and field testing before becoming BATs.

6.7.1 Oxidation Including Ozone

Oxidation is usually accomplished with either chlorine, chlorine dioxide, ozone, or potassium permanganate. Complete oxidation reactions also destroy organic contaminants. Incomplete oxidation reactions, however, produce by-products, most of which are biodegradable and some of which may pose health risks.

Chlorination, the most widely used disinfection, is used routinely for oxidation (e.g., breakpoint chlorination). Unfortunately, THMs and other halogenated by-products are known to be formed during chlorination. Other oxidation agents and processes that do not pose this problem are being studied as agents to reduce levels of halogenated organic contaminants.

Ozone is widely used in Europe and is gaining substantial interest in the United States. A 600-MGD direct filtration plant in Los Angeles, California, is employing preozonation to enhance turbidity removal and filter longevity. Three advanced oxidation processes for treating organic contamination are currently being tested: (1) ozone with high pH levels, (2) ozone with hydrogen peroxide, and (3) ozone with ultraviolet radiation. Each of these processes forms the hydroxyl free radical, which has an oxidation potential about 30 percent higher than molecular ozone. These processes combine organic compound removal with disinfection, and taste and odor control.

Ozone with High pH Levels

Ozone, at low pH levels (less than 7), reacts primarily as the O_3 molecule by selective and sometimes relatively slow reactions. Ozone at elevated pH (above 8) rapidly decomposes into hydroxyl free radicals, which react very quickly. Many organic compounds that are slow to oxidize with ozone, oxidize rapidly with hydroxyl free radicals.

The alkalinity of the water is a key parameter in advanced oxidation processes. This is because bicarbonate and carbonate ions are excellent scavengers for free radicals. Consequently, advanced oxidation processes are incompatible with highly alkaline water. In addition, carbonate ions are 20 to 30 times more effective in scavenging for hydroxyl free radicals than bicarbonate ions. Therefore, ozonation at high pH should be conducted below 10.3 at which level all bicarbonate ions convert to carbonate ions.

Ozone with Hydrogen Peroxide (the Peroxide Process)

The combination of ozone with hydrogen peroxide much more effectively reduces levels of trichloroethylene (TCE) and tetrachloroethylene (PCE) than ozone alone. Table 6-11 shows design criteria and assumptions for a full-scale ozone/hydrogen peroxide plant.

A significant advantage of the peroxide process over GAC and PTA is the absence of vapor controls

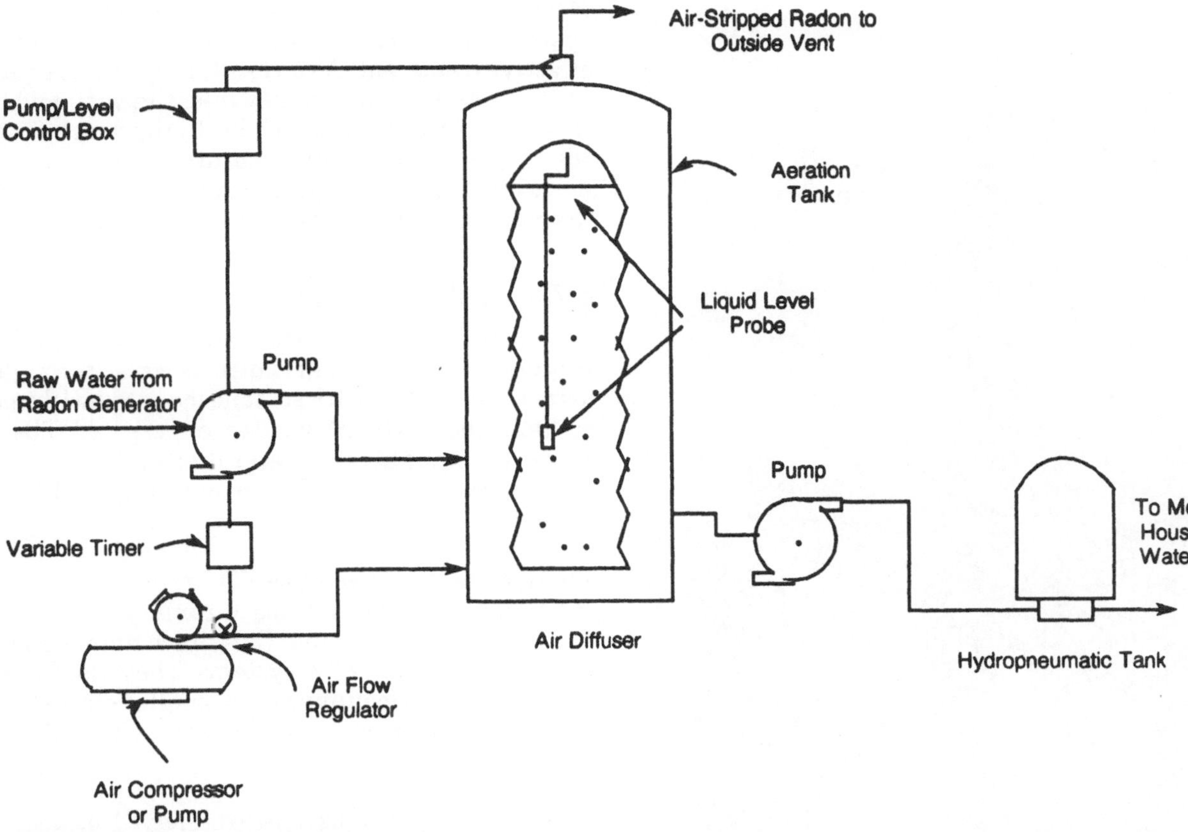

Figure 6-23. Home diffused aeration system.

Table 6-10. Typical Performance of Diffused Aeration

Compound	% Removal
VOCs	
Trichloroethylene	53-95
Tetrachloroethylene	73-95
1,2-Dichloroethane	42-77
1,1-Dichloroethylene	97
1,1,1-Trichloroethane	58-90
SOCs	
Carbofuran	11-20
1,2-Dichloropropane	12-79
cis-1,2-Dichloroethylene	32-85
trans-1,2-Dichloroethylene	37-96
o-Dichlorobenzene	14-72
Ethylbenzene	24-89
Monochlorobenzene	14-85
Toluene	22-89
Xylenes	18-89

Source: Miltner and Fronk (1985).

because the contaminants are destroyed, not merely removed from the water. Table 6-12 compares annual costs for four system types: aeration, aeration with gas-phase GAC, liquid-phase GAC, and an advanced oxidation process.

Ozone/Ultraviolet

In ozone/ultraviolet (UV) treatment, ozone catalyzed by UV oxidizes organic substances. This process breaks down the saturated bonds of the contaminant molecules. Typical contact time is 0.25 hours. Table 6-13 presents system removal efficiencies with 0.25-hour contact times and varying ozone dosages for 13 SOCs. A major advantage of this system is that it does not produce any THMs. These systems also do not require waste disposal because the contaminants are destroyed.

There is some concern about the completeness of the ozone/UV oxidation process and the intermediate breakdown products. If oxidation is incomplete, some of the compounds produced in the intermediate reactions may still be available to form THMs. The influent contaminant profile also affects the performance of these systems. However, if oxidation is followed by a biological filtration step, particularly GAC on sand or GAC adsorber, these oxidation products are mineralized into carbon dioxide and

Multiple tray aerator; Norwalk, CT.

water. Consequently, THM formation potential and TOX formation potential are lowered.

6.7.2 Reverse Osmosis

Reverse osmosis is a proven technology for the removal of inorganic compounds. The process is fully described in Section 7.3.3. Reverse osmosis also is effective in removing THM humic and fulvic acid precursors, pesticides, and microbiological contaminants (viruses, bacteria, and protozoa). This treatment has been shown to remove VOCs with low molecular weights. Table 6-14 presents performance data for reverse osmosis units operating at about 200 psi for 7 VOCs and 16 SOCs.

6.7.3 Mechanical Aeration

Mechanical aeration systems use surface or subsurface mechanical stirring mechanisms to create turbulence to mix air with the water. These systems effectively remove VOCs but are generally used for wastewater treatment systems. Surface and subsurface aerator designs are shown in Figure 6-25 (Roberts and Dandliker, 1982).

Mechanical aeration units consume large amounts of space because they demand long detention times for effective treatment. As a result, they often require open-air designs, which may freeze in very cold climates. These units also have high energy requirements. Mechanical aeration systems, however, are easy to operate and are less susceptible to clogging from biological growth than PTA.

6.7.4 Catenary Grid

Catenary grid systems are a variation of the PTA process. The catenary grid directs water through a series of parabolic wire screens mounted within the column, above which turbulence is created. The screens mix the air and water in the same way as the packing materials in PTA systems. Figure 6-26 depicts a sample unit.

These systems can achieve VOC removal rates comparable to PTA systems. Catenary grid units require more airflow and, thus, have higher energy requirements than PTA systems. They also have shorter aeration columns with smaller diameters. Their more compact design lowers their capital cost relative to PTA.

Catenary grid systems, however, have limitations. Limited data are available concerning this system's removal effectiveness for a wide variety of organic compounds. Also, the procedure for scaling systems up from pilot plants to full-scale operations is not fully developed.

The principal design considerations for catenary grid systems are air-to-water ratio, number of screens, and hydraulic loading rate. Removal efficiency improves with increases in air-to-water ratios and increasing number of screens in the column.

6.7.5 Higee Aeration

Higee aeration systems are another variation of the PTA process. These systems pump water into the center of a spinning disc of packing material to achieve the necessary air and water mix (Glitsch, Inc.). By design, the packing material has a large surface area per unit volume. Air is pumped countercurrently toward the center from the outside of the spinning disc. Simultaneously, water flows from the center of the disc and mixes with the air (see Figure 6-27).

Higee units require less packing material than PTA units to attain equivalent removal efficiencies. They require smaller air volumes and can process high water flows in a compact space. The Higee unit's compact size permits its application within constrained spaces and heights. Current Higee systems are best suited for temporary applications of less than 1 year with capacities up to 6.3 L/sec (100

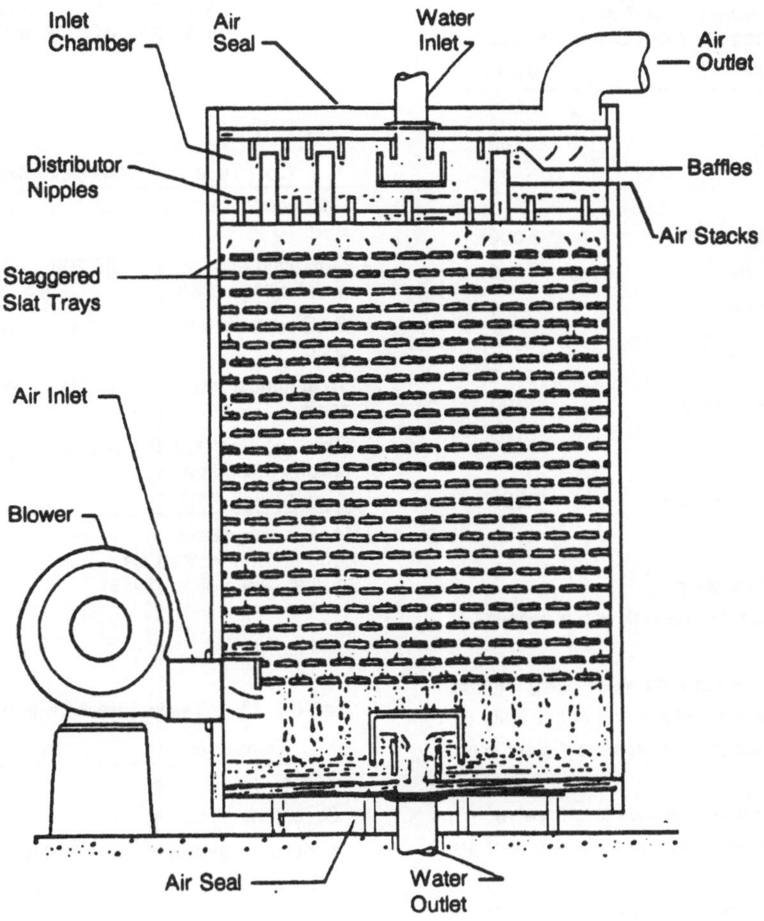

Figure 6-24. Schematic of a redwood slat tray aerator.

GPM). Few data are available concerning the organic compound removal efficiencies of Higee systems.

6.7.6 Resins

In this process, synthetic resins are used in place of GAC to remove organic compounds by adsorption. Their performance varies with resin type, EBCT, and regeneration frequency.

The advantages of resins include shorter EBCT requirements and longer operational life, relative to GAC. Also, resins can be regenerated on site with steam. However, wastewater from this process can be difficult to manage properly because of its high concentrations of hazardous constituents (Ruggiero and Ausubel, 1982).

Resins are more costly than GAC, costing up to \$0.02/g (\$10/lb), compared to less than \$0.01/g (\$0.80/lb) for GAC.

Table 6-11. Design Parameters for Peroxide-Ozone Treatment Plant and Assumptions for Cost Comparisons

Parameter	Value
Plant flow-GPM	2,000
TCE concentration - μg/L	200
PCE concentration - μg/L	20
Reaction tank capacity - gal	6,000
Hydraulic detention time - min	3
Reaction tank stages - number	1
Ozone dosage - mg/L	4
Ozone generator capacity - lb/day	100
Peroxide dosage - mg/L	2
Peroxide storage - gal at 50 percent concentration	1,000

Assumptions

The treatment plant is to be built for one well with a capacity of 2,000 GPM (126 L/s).

- The TCE concentration of the ground water being pumped remains constant for 10 years. At the end of this time, all the TCE has been extracted and the treatment plant is no longer needed.
- The rate of interest remains at 8 percent for 10 years.
- The capital cost of the aqueous-phase GAC treatment plant is $1.3 million.
- The capital cost of the air-stripping treatment plant alone is $325,000.
- The capital cost of the treatment plant for air stripping plus gas-phase GAC adsorption is $725,000.
- The annual operating cost of each system is the power requirement plus one quarter of a person-year for maintenance.
- The cost of power is $1.08/lb.
- The cost of hydrogen peroxide is $1.08/lb.

1 pound = 453.6 grams
1 gallon = 3.785 liters
Source: Aieta et al. (1988a).

Table 6-12. Comparison of Annual Treatment Costs for Removal of PCE and TCE from Ground Water

Cost Type	Aeration	Aeration with Gas-Phase GAC Adsorption	Liquid Phase GAC Adsorption	Peroxide Ozone Advanced Oxidation Process
Capital cost (annualized, 10 years, 8%)	48,400	108,000	192,500	35,000
Operating and maintenance costs	30,200	78,300	13,900	63,900
GAC replacement costs	NA	105,100	210,400	NA
Total annualized costs	78,600	291,400	416,800	98,900
Dollar cost per 1,000 gallons (based on 2,000-gallon flow)	(0.075)	(0.277)	(0.397)	(0.094)

NA = not available.
1,000 gallons = 3.78 m^3
Source: Aieta et al. (1988).

Table 6-13. Typical Performance of Ozonation Process

Compound	Dosage (mg/L)	Percent Removal
SOCs		
Carbofuran	9	100
1,2-Dichloropropane	0.9-6.0	8-22
cis-1,2 Dichloroethylene	2-10	87-93
trans-1,2 Dichloroethylene	0.9-6.0	100
o-Dichlorobenzene	9	88
Ethylene dibromide	0.9-6.0	8-9
Ethylbenzene	1.5-9.0	47-95
Heptachlor	17	100[a]
Heptachlor epoxide	17	26
Lindane	0.4-150.0	0-100
Monochlorobenzene	0.4-6.0	86-98
Toluene	1.5-12.0	49-98
Xylenes	1.5-12.0	54-98

[a] Oxidation of heptachlor produces heptachlor epoxide, which is relatively stable to further oxidation.
Source: Miltner and Fronk (1985).

Table 6-14. Typical Performance of Reverse Osmosis Process

Compound	Percent Removal
VOCs	
1,2-Dichloroethane	15-70
1,1,1-Trichloroethane	15-100
Carbon tetrachloride	95
Trichloroethylene	0-75
Tetrachloroethylene	70-90
Benzene	2-18
p-Dichlorobenzene	0-10
SOCs	
Acrylamide	0-97
Aldicarb	94-99
Alachlor	100
Carbofuran	86-99
1,2-Dichloropropane	10-90
cis-1,2-Dichloroethylene	0-30
trans-1,2-Dichloroethylene	0-30
2,4-D	1-65
o-Dichlorobenzene	65
Ethylbenzene	30
Ethylene Dibromide	37-84
Lindane	50-75
Methoxychlor	>90
Monochlorobenzene	50-100
PCBs	95
Xylenes	10-85

Source: Miltner and Fronk (1985).

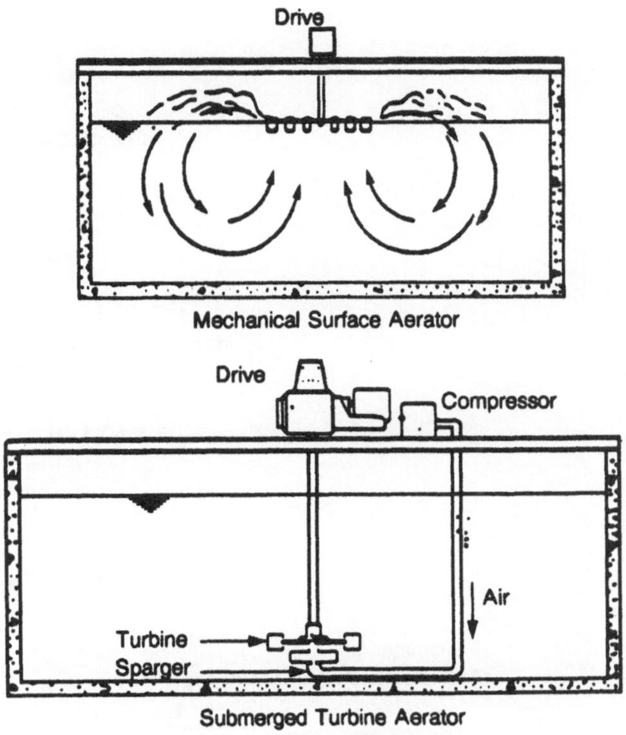

Source: Roberts and Dandliker (1982).

Figure 6-25. Schematic of mechanical aeration process.

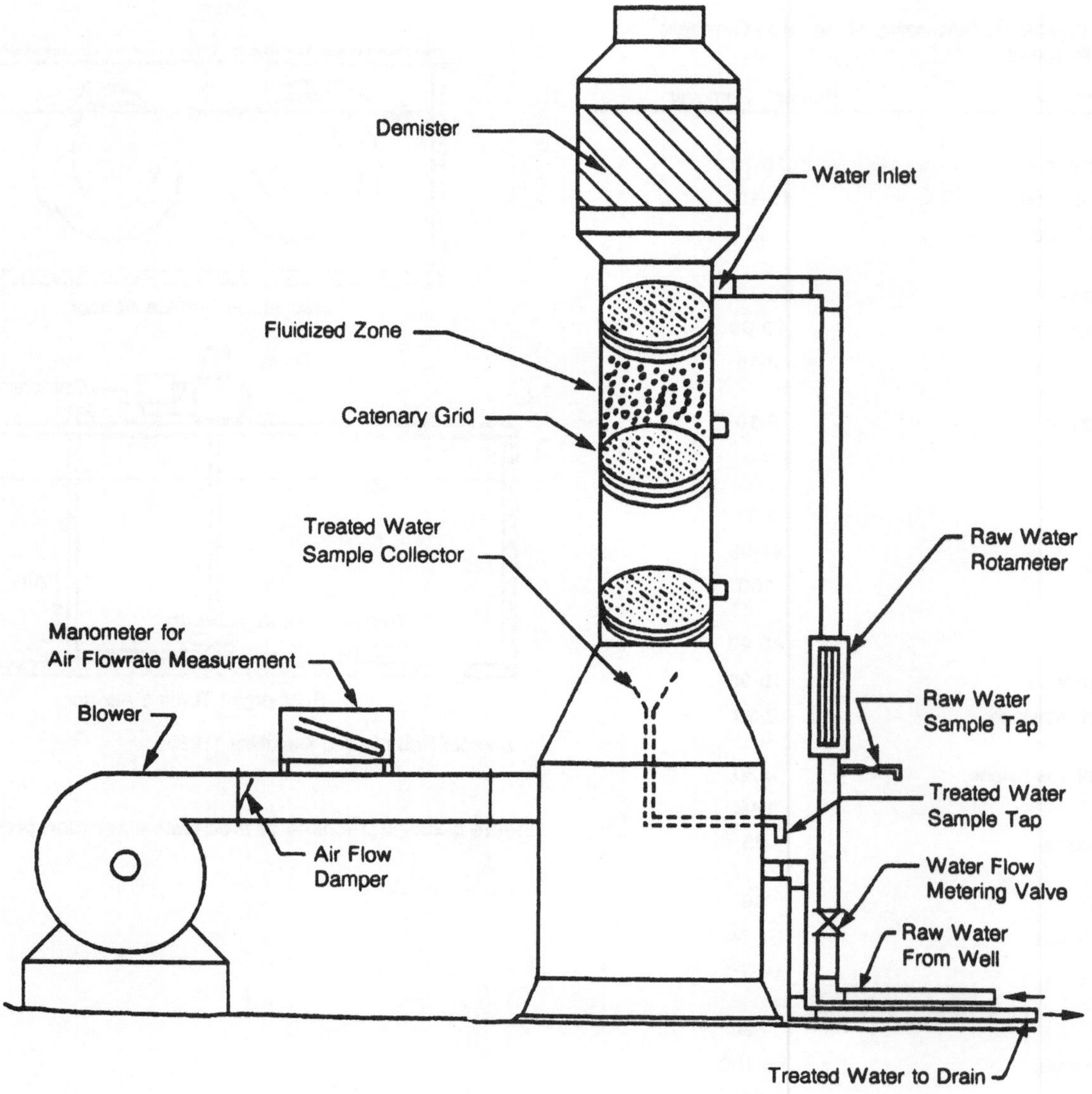

Figure 6-26. Catenary grid system.

Pilot catenary grid aerator unit.

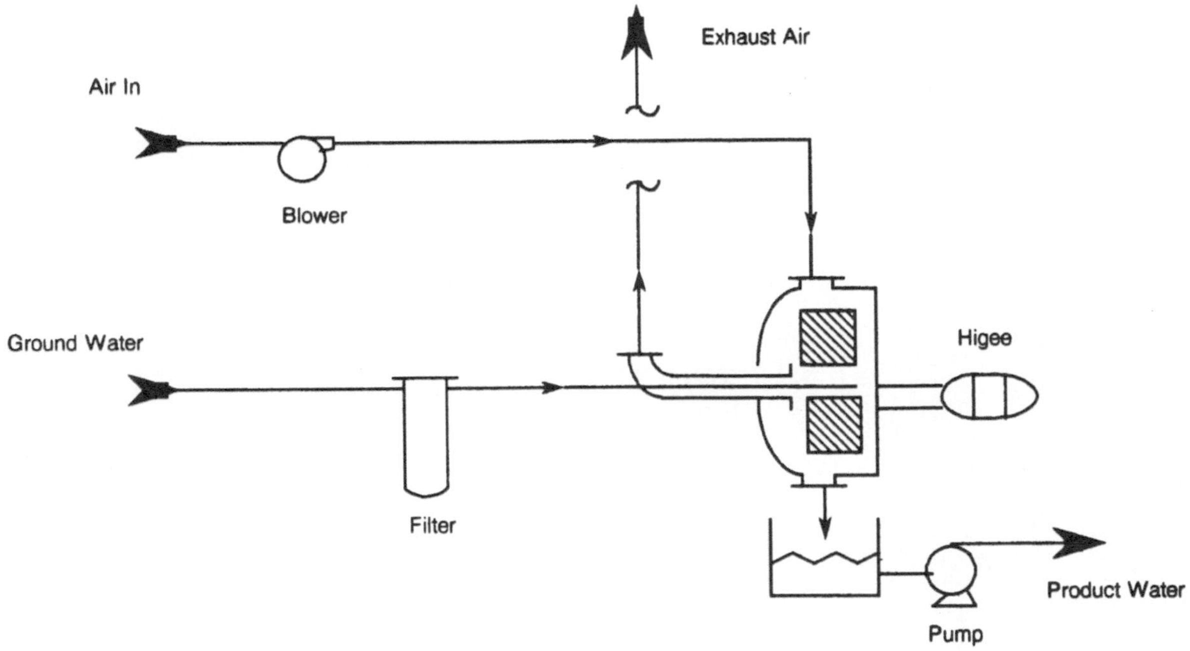

Figure 6-27. Schematic of Higee system.

Chapter 7

Treatments for Inorganic Contaminants

This chapter describes the technologies available for removing inorganic contaminants from drinking water. Inorganic contaminants tend to come from natural sources and, therefore, are more predictable in behavior and composition than organic contaminants that come from manmade sources. Inorganic contaminants are classified as cationic, anionic, or neutral and occur as atoms, ions, or molecules. The two most significant properties affecting the removal of inorganic contaminants from water are valence and solubility in water.

The most common treatments for inorganic contaminants are:

- Precipitation
- Coprecipitation
- Adsorption
- Ion exchange
- Membrane separation by reverse osmosis or electrodialysis
- A combination of two or more of the above five technologies

Other treatments for removing inorganic contaminants include:

- Distillation
- Evaporation
- Oxidation/Reduction
- Air stripping
- Biological treatment

EPA has specified best available technologies (BAT) for three inorganic contaminants, including:

- Corrosion control for lead
- Ion exchange or reverse osmosis for nitrate
- Ion exchange and reverse osmosis for radium

It is expected that aeration will be designated as BAT for radon. Corrosion control targets lead as a corrosion by-product. Lead, when found in source water, requires conventional or other treatment such as reverse osmosis and ion exchange.

Twelve inorganic contaminants are presently regulated under SDWA: lead, radium, nitrate, arsenic, selenium, barium, fluoride, cadmium, chromium (total), mercury, silver, and strontium-90. Lead is not typically found in source water, but rather at the consumer's tap as a result of corrosion of the plumbing or distribution system. The other contaminants are found regionally for a variety of reasons. Radionuclides are of particular concern because few utilities currently test for other than radium and EPA has found them to be more prevalent than previously determined.

This chapter presents two distinct approaches to water treatment for inorganic contamination: (1) preventing inorganic contamination of finished water and/or (2) removing inorganic contaminants from raw water. Section 7.1 discusses corrosion control, the primary method for minimizing inorganic contamination of finished water. The major focus of this subsection is lead. Section 7.2 describes treatment options for removing inorganic contaminants from raw water. Appendix D provides case histories on corrosion control and treatment methods to remove inorganics.

7.1 Techniques for Controlling Corrosion

Lead is found in drinking water chiefly because the metals used in water distribution systems, home plumbing, and appliances, such as water coolers, corrode. These metals generally are not found in significant amounts in source waters, but rather at the final point of use. Zinc, copper, and iron also are commonly present in water as corrosion products. Cadmium has been found in water systems with new galvanized pipe. Lead and cadmium are toxic at low concentrations; copper and zinc are toxic only at much higher levels. Because of its ubiquity and toxicity, lead is the corrosion product of most concern. Cadmium also has significant health implications, but it is not as widespread in drinking water as lead.

Lead levels in drinking water are minimized with corrosion controls. The current lead MCL applies to finished water from a treatment plant. New lead regulations may include an MCL at the consumer's tap. While corrosion is recognized as a critical factor for proper management of any water system, the lack of a standard measure of corrosion has thwarted development of uniform controls.

7.1.1 The Problem of Corrosion

Corrosion occurs because metals tend to oxidize when in contact with potable water and form stable solids on metal surfaces. All metals in contact with water will corrode to some extent. Corrosion has implications for health, costs, and aesthetics. Drinking water contaminated with metals adversely affects human health. Corrosion reduces the useful life of water distribution systems and household plumbing, and is thus responsible for higher costs due to problems with:

- Pumping caused by narrowed pipe diameters resulting from corrosion deposits
- Pumping and water production caused by corrosion holes, which reduce water pressure and increase the amount of finished water required to deliver a gallon of water to the point of consumption
- Water damage caused by corrosion-related pipe failures
- Replacement frequency of hot water heaters, radiators, valves, pipes, and meters
- Customer complaints of color, staining, and taste problems
- Repairs for pipe leaks and breaks

Lastly, corrosion can produce conditions that promote microorganisms that cause disagreeable tastes, odors, slimes, and further corrosion.

All water is corrosive to some degree, but water that is acidic will have faster corrosion rates. Many naturally occurring acidic waters are also soft. Soft water is generally defined as water with less than 100 mg/L of calcium as calcium carbonate ($CaCO_3$); acidic water has a pH of less than 7.0.

The degree of corrosion is determined primarily by the characteristics of the metal and water, and the nature and duration of the contact between the two. Table 7-1 summarizes the factors affecting drinking water corrosivity. Water treatment processes can change water quality characteristics that significantly affect the water's corrosion potential. For example, pH is lowered with the use of coagulants or disinfectants. Other treatment processes affect water chemistry parameters such as disinfectant residual, hardness, and alkalinity.

Table 7-1. Factors Affecting the Corrosivity of Drinking Water

Factor	Effect on Corrosivity
pH	Low pHs generally accelerate corrosion.
Dissolved oxygen	Dissolved oxygen in water induces active corrosion, particularly of ferrous and copper materials.
Free chlorine residual	The presence of free chlorine in water promotes corrosion of ferrous metals and copper.
Low buffering capacity	There is insufficient alkalinity to limit corrosion activity.
High halogen and sulfate-alkalinity ratio	A molar ratio of strong mineral acids much above 0.5 results in conditions favorable to pitting corrosion (mostly in iron and copper pipe).
Total dissolved solids	Higher concentrations of dissolved salts increase conductivity and may increase corrosiveness. Conductivity measurements may be used to estimate total dissolved solids.
Calcium	Calcium can reduce corrosion by forming protective films with dissolved carbonate, particularly with steel, iron, or galvanized pipe.
Tannins	Tannins may form protective organic films over metals.
Flow rates	Turbulence at high flow rates allows oxygen to reach the surface more easily, removes protective films, and causes higher corrosion rates.
Metal ions	Certain ions, such as copper, can aggravate corrosion of downstream materials. For example, copper ions may increase the corrosion of galvanized pipe.
Temperature Rates	High temperature increases corrosion reaction rates. High temperature also lowers the solubility of calcium carbonate, magnesium silicates, and calcium sulfate and thus may cause scale formation in hot-water heaters and pipes.

Source: Adapted from Williams (1986).

The type of corrosion products present depends on the metals composing the solder, pipes, valves, meters, and faucets in distribution and plumbing systems. The most common metals used are steel, iron, galvanized steel, copper, lead, brass, and bronze. Zinc, cadmium, and some lead are present in galvanized coatings. The most frequent sources of lead are brass faucets, meters, pipe solder, and valves. Lead is no longer generally used for pipes, but in older cities service connections between houses and water mains still contain lead pipe. Many cities have initiated lead service line replacement programs to address this source.

While lead pipes have long been recognized as hazardous, lead-based solder was used in the United

States until it was banned in the 1986 SDWA Amendments. This ban prohibits the use of solder containing lead and the associated high lead levels often found in newly constructed homes and in new plumbing in existing homes. Precise estimates of lead levels in drinking water resulting from lead-based solders and fluxes vary. This stems from the difficulty in consistently measuring lead levels from the tap, especially if there is a brass faucet, which contributes to lead levels as well. The method of solder application affects the amount of lead imparted to the water. Improper application, which is difficult to determine in retrospect, allows solder to flow on to the inner portion of the pipe, thus increasing its area of contact with the water.

7.1.2 Diagnosing and Evaluating the Problem

There are dozens of types of corrosion, but the two broad categories of particular concern in water treatment are uniform and nonuniform corrosion. Nonuniform corrosion shortens the useful life of pipes and plumbing more quickly than uniform corrosion, and is thus more of a problem for water systems. The five most common types of nonuniform corrosion to affect water systems are:

- *Galvanic* - Occurs when two different metals are joined together; the more electrochemically active of the joined metals will corrode. (See Table 7-2 for relative activity of various metals and Section 7.1.3.1 for discussion of this characteristic.)

Table 7-2. Galvanic Series - Order of Electrochemical Activity of Common Metals Used in Water Distribution Systems

Metal	Activity
Zinc	More Active
Mild steel	↑
Cast iron	
Lead	
Brass	
Copper	
Stainless steel	Less Active

Source: U.S. EPA (1984).

- *Pitting* - Occurs as uneven pits or holes in the pipe surface, which are undetectable until the pipe fails. Pitting is usually caused by pockets of corrosion initiated by tiny imperfections, scratches, or surface deposits in the pipe.

- *Crevice* - Occurs locally around gaskets, lap joints, rivets, and surface deposits. It is caused by changes in acidity, oxygen concentrations, dissolved ions, and the absence of corrosion inhibitors.

- *Erosion* – Occurs due to the removal of protective coatings through high water velocities, turbulence, sudden changes in flow direction, and abrasive action of suspended solids or gases.

- *Biological* – Occurs in mechanical crevices or accumulations of corroded materials due to the interaction between the metal and bacteria, algae, or fungi.

Most lead corrosion problems are diagnosed indirectly. Risk factors that indicate potentially high lead levels at the tap are:

- The water distribution system or structure's plumbing is made of lead.
- The structure's plumbing has solder containing lead.
- The structure is less than 5 years old.
- The tap water is soft and acidic.
- The water stays in the plumbing for 6 or more hours.
- The structure's electrical system is grounded to the plumbing system.

The presence of any of these factors justifies further investigation.

One direct method of measuring lead corrosivity of water is to sample standing water from the consumer's tap. In addition, consumer complaints, corrosion indices, sampling and chemical analysis, pipe scale examination, and measurement of the corrosion rate over time are all proxies for corrosion contaminants.

7.1.2.1 Consumer Complaints

Many times a consumer complaint is the first indication of a corrosion problem. Table 7-3 lists the most common complaints and their causes. In investigating the extent of the corrosion, complaints can be plotted on a map of a water service area. Then investigators can examine the construction materials used in the water distribution system and in the plumbing of the complaint areas. Random sample surveys are commonly used to confirm the extent of the corrosion problems flagged by these complaints.

7.1.2.2 Corrosion Indices

Corrosion related to calcium carbonate deposition can be estimated using indices derived from common water quality measures. The Langelier Saturation Index (LSI) is the most commonly used and is equal to the water pH minus the saturation pH. The saturation pH refers to the pH at the water's calcium carbonate saturation point, the point where calcium

Table 7-3. Typical Customer Complaints Due to Corrosion

Customer Complaint	Possible Cause
Red water or reddish-brown staining of fixtures and laundry	Corrosion of iron pipes or presence of natural iron in raw water
Bluish stains on fixtures	Corrosion of copper lines
Black water	Sulfide corrosion of copper or iron lines or precipitations of natural manganese
Foul taste and/or odors	By-products from microbial activity
Loss of pressure	Excessive scaling, tubercle build-up from pitting corrosion, leak in system from pitting or other type of corrosion
Lack of hot water	Buildup of mineral deposits in hot water system (can be reduced by setting thermostats to under 60°C [140°F])
Short service life of household plumbing	Rapid deterioration of pipes from pitting or other types of corrosion

Source: Adapted from U.S. EPA (1984).

carbonate is neither deposited nor dissolved. Calcium carbonate may precipitate and form a protective layer on metals. The saturation pH is related to the water's calcium ion concentration, alkalinity, temperature, pH, and presence of other dissolved solids, such as chlorides and sulfates.

The Aggressive Index (AI) is a simplification of the LSI that only approximates the solubility of calcium carbonate and may not be useful. The Ryznar Stability Index (RSI), also a modification of the LSI, uses visual inspections. McCauley's Driving Force Index (DFI) estimates the amount of calcium carbonate that will precipitate based on the same factors as the LSI.

Riddick's Corrosion Index (CI) is distinct from the LSI. Its empirical equation incorporates different factors to predict corrosion, such as dissolved oxygen, chloride ion, noncarbonate hardness, and silica. Appendix E provides the equations, descriptions of test parameters, and interpretations of each index.

7.1.2.3 Sampling and Chemical Analysis

Corrosion can also be assessed by conducting a chemical sampling program. Water with a low pH (less than 6) is more corrosive. Temperature and total dissolved solids can be important indicators of corrosivity, although this varies case by case.

Proper sampling and analysis methods are essential to obtain accurate and meaningful test results. Determining whether lead is from a service pipe or internal plumbing requires sampling at multiple locations or getting accurate and precise samples representing water in prolonged contact with the suspected section of pipe (e.g., service line, soldered joints). Experiments show that sample volume also affects the amount of lead detected in tap water. Other critical sampling elements include sampling location, amount of water in each sample, flow rate of sample, and the contact time of the water with the metal. The amount of time the water remains in the pipes significantly affects lead levels. Also, the time of day is a critical factor in accurately assessing sampling results. Early morning sample results will reflect water that has been held in the pipes overnight, while evening water samples may assess water intermittently drawn by daily activities.

The age of the pipe solder has been shown to affect lead levels in water. One study found that 4 to 5 weeks after an application of lead-based solder, lead levels declined by 93 percent from initial measurements. The effects of solder age and the length of time the water stands in the pipes are illustrated in Table 7-4, Table 7-5, and Table 7-6 for low, medium, and high water pH levels, respectively.

Table 7-4. Percentage of Test Sites with Lead in Drinking Water Greater than 20 µg/L at Low pH (6.4 and less)

Age of Test Site (Years)	First Draw	10 Sec	20 Sec	30 Sec	45 Sec	60 Sec	90 Sec	120 Sec
0-1	100%	100%	100%	100%	100%	86%	86%	88%
1-2	100	71	86	57	57	29	43	14
2-3	86	86	57	57	43	43	43	29
3-4	100	86	100	71	71	71	29	29
4-5	86	57	29	43	43	43	14	0
6-7	78	44	33	33	11	11	11	0
9-10	71	29	14	14	14	14	0	0
15-16	57	14	14	14	14	14	14	14
20 and older	86	27	29	0	14	0	14	0

7.1.2.4 Scale or Pipe Surface Examination

Pipe scale and the inner surface of pipes can be examined by optical or microscopic observation, x-rays, wet chemical analysis, and Raman and infrared spectroscopy. Obviously the most practical and economic method is simple observation. The other methods require expensive equipment and highly skilled personnel. Even simple observation, however, requires taking sections of pipe out of service. Chemical examinations can determine the composition of pipe scale, such as the proportion of calcium carbonate present that shields pipes from dissolved oxygen and thus reduces corrosion.

Table 7-5. Percentage of Test Sites with Lead in Drinking Water Greater than 20 µg/L at Medium pH (7.0-7.4)

Age of Test Site (Years)	First Draw	10 Sec	20 Sec	30 Sec	45 Sec	60 Sec	90 Sec	120 Sec
0-1	100%	90%	90%	60%	30%	20%	10%	10%
1-2	80	60	30	10	20	0	10	0
2-3	40	20	10	10	10	0	0	0
3-4	50	20	20	30	20	30	30	20
4-5	30	10	10	0	10	0	0	0
6-7	10	0	0	0	0	0	0	0
9-10	20	0	0	0	0	0	0	0
15-16	40	20	20	10	0	0	0	0
20 and older	20	0	0	0	10	0	0	0

Table 7-6. Percentage of Test Sites with Lead in Drinking Water Greater than 20 µg/L at High pH (8.0 and Greater)

Age of Test Site (Years)	First Draw	10 Sec	20 Sec	30 Sec	45 Sec	60 Sec	90 Sec	120 Sec
0-1	100%	100%	60%	10%	20%	10%	20%	0%
1-2	67	22	11	11	11	0	11	0
2-3	30	10	10	0	0	0	0	0
3-4	25	0	0	0	0	0	0	13
4-5	30	10	0	0	0	0	0	0
6-7	20	0	0	0	0	0	0	0
9-10	10	0	10	0	0	10	0	10
15-16	33	22	11	11	0	0	0	0
20 and older	20	0	0	0	0	0	0	0

Chemical analyses and microscope assistance then should be used to augment these findings.

7.1.2.5 Rate Measurements

The corrosion rate is usually expressed in thousands of an inch (mils) per year (MPY). The three methods for calculating this rate are:

- Coupon weight-loss
- Loop system concentration increase
- Electrochemical rate measurement

The first two methods are much less expensive than the third.

The coupon weight-loss method uses flat metal pieces or test sections of pipes called coupons or inserts. The flat coupons are commonly placed in the middle of a pipe in the system. The coupons are weighed before and after they are put in place and the recorded weight loss is translated into a uniform corrosion rate with the following formula:

$$\text{Rate of Corrosion (MPY)} = \frac{534 \times W}{D \times A \times T}$$

where:

W = weight loss in mg
D = density of the specimen in g/cm^3
A = surface area of the specimen in sq in
T = exposure time in hours

This method can be used to monitor corrosion progress over time or to spot check the corrosion rate.

The loop system weight-loss method is similar to the flat coupon method, except that it uses actual sections of pipe in the system instead of flat coupons. Loop system corrosion rates are usually determined by analyzing corrosion over a period of time.

Electrochemical rate measurement requires expensive and sophisticated equipment beyond the means of most smaller systems. In this method, two or three electrodes are placed in the corrosive environment, and instrumentation measures the corrosion rate in MPY.

7.1.3 Corrosion Controls

If any of the tests described above reveal unacceptably high levels of lead, then immediate measures should be taken to minimize human exposure until a long-term action plan is developed, approved, and implemented. Some short-term measures for water consumers are:

- Running the water for 1 to 3 minutes before each use
- Using only cold tap water for drinking or cooking
- Using a home-scale reverse osmosis unit or other treatment process, in extreme cases
- Using bottled water

Longer term solutions fall into six categories:

- Distribution and plumbing system design considerations
- Water quality modification
- Corrosion inhibitors
- Coatings and linings
- Use of only nonlead based solder for construction and repairs
- Replacement of lead pipes

7.1.3.1 Distribution and Plumbing System Design Considerations

Many distribution and plumbing system design considerations will reduce corrosion. For example, water distribution systems designed to operate with lower flow rates will have reduced turbulence and decreased erosion of protective layers. Other measures that minimize corrosion include:

- Using only lead-free pipes, fittings, and components
- Selecting appropriate system shape and geometry
- Avoiding sharp turns and elbows
- Avoiding dead ends and stagnant areas
- Eliminating shielded areas
- Providing adequate drainage
- Selecting appropriate metal thickness
- Using welded ends, instead of rivets
- Reducing mechanical stresses
- Avoiding uneven heat distribution
- Providing adequate insulation
- Providing easy access for inspection, maintenance, and replacement
- Eliminating the grounding of electrical circuits to the system

To implement these measures effectively, local plumbing codes may need to be modified.

Distribution system designers should base their materials selection criteria on system water characteristics. For example, water with low pH levels and high dissolved oxygen levels corrodes metals quickly. Concrete may be more appropriate than metal in this case because, although concrete dissolves under low pH conditions, the rise in dissolved calcium in the water is not objectionable and since concrete pipes are thick they have a longer lifetime. Table 7-7 presents the corrosion properties of common distribution system materials.

Table 7-7. Corrosion Properties of Frequently Used Materials in Water Distribution Systems

Distribution Material	Corrosion Resistance	Associated Potential Contaminants
Copper, iron	Good overall corrosion resistance; subject to corrosive attack from high velocities, soft water, chlorine, dissolved oxygen, and low pH	Copper and possibly zinc, tin, arsenic, cadmium, and lead from associated pipes and solder
Lead	Corrodes in water with low pH and high alkalinities	Lead (can be well above MCL[a]), arsenic, and cadmium
Mild steel	Subject to uniform corrosion; affected primarily by high dissolved oxygen levels	Iron, resulting in turbidity and red-water complaints
Cast or ductile iron (unlined)	Can be subject to surface erosion by aggressive waters	Iron, resulting in turbidity and red-water complaints
Galvanized iron	Subject to galvanic corrosion of zinc by aggressive waters; corrosion is accelerated by contact with copper materials;corrosion is accelerated at higher temperatures as in hot-water systems	Zinc and iron; cadmium and lead (impurities in galvanizing process may exceed primary MCLs)
Asbestos-cement	Good corrosion resistance; immune to electrolysis; aggressive waters can leach calcium from cement	Asbestos fibers
Plastic	Resistant to corrosion	
Brass	Fairly good; subject to dezincification depending on the water quality and alloy	Lead, copper, zinc, arsenic

[a]MCL = Maximum contaminant level.
Source: Adapted from U.S. EPA (1984).

Metal electrochemical activity is the measure of a metal's tendency to oxidize. When placed together, metals with different activities create galvanic corrosion cells. Therefore, if different metals must be placed together, minimizing the differences in their activity will lessen corrosion. Galvanic corrosion is also avoided by placing dielectric insulating couplings between the dissimilar metals.

While many corrosion control programs address the lead content of pipes, solder, and other plumbing components, they do not target lead-bearing plumbing fixtures, such as brass faucets or valves.

Changing the composition of solder used in plumbing is considered the most important factor in any lead corrosion control strategy. In the past, solder used in plumbing has been 50 percent tin and 50 percent lead. Alternative lead-free solders are available at a higher cost.

Two alternative solders are made from 95 percent tin and 5 percent antimony or silver. The 1984 costs per gram of standard lead-; antimony-; and silver-based solders were $9.92, $17.64, and $44.09, respectively ($4.50, $8.00, and $20.00/lb, respectively). Plumbers estimate that plumbing in an average residence requires less than 1 lb of solder, which makes solder costs relatively insignificant. However, these alternative solders do not perform exactly the same as lead-based solder. For instance, both silver-tin and antimony-tin solders are more difficult to work with than lead solder because of higher and more narrow melting ranges.

7.1.3.2 Water Quality Modifications

Measures that change water quality to reduce corrosion contamination, especially lead corrosion, include pH and alkalinity adjustment, lime-soda softening, and adjustment of dissolved oxygen levels, although altering oxygen levels is not a common method of control. Any corrosion adjustment program should include a monitoring component that enables dosage modification in response to changing water characteristics over time.

pH Adjustment

Frequently corrosion is mistakenly associated merely with water acidity. The pH level plays a central role in determining the corrosion rate and is relatively inexpensive to control. Generally, water pH less than 6.5 is associated with uniform corrosion, while pHs between 6.5 and 8.0 can be associated with pitting corrosion. Systems using only pH to control corrosion should maintain their water above 9.0 pH to reduce the availability of hydrogen ions as electron acceptors. However, pH is not the only factor in the corrosion equation; carbonate and alkalinity levels may affect corrosion as well.

By adjusting pH, alkalinity, and calcium levels, operators can promote the precipitation of a protective calcium carbonate (scale) coating onto the metal surface of plumbing. While these protective coatings build up over time, they are disturbed by:

- Water turbulence due to high velocities and flow rates
- Stagnation of water causing prolonged periods of contact
- Variations in the control of velocities, flow rates, and stagnation

Calcium carbonate scaling occurs when water is oversaturated with calcium carbonate. Below the saturation point, calcium carbonate will redissolve and at saturation, calcium carbonate is neither precipitated nor dissolved. The saturation point of any particular water source depends on the concentration of calcium ions, alkalinity, temperature, and pH, and the presence of other dissolved materials, such as phosphates, sulfates, chlorides, sulfides, and some trace metals. A coating is usually attained through simple pH control, given adequate concentrations of alkalinity and carbonate. Adding lime or alkaline substances also promotes the formation of calcium carbonate scale in most systems. The Langelier or CCAP Indices are useful for initiating this corrosion inhibiting strategy because they estimate calcium carbonate saturation. Further pH and/or carbonate and/or calcium adjustments may be required to reach equilibrium, as established by system monitoring.

Lime Softening

Lime softening, which is sometimes known as lime-soda softening when soda ash is required in addition to lime, affects lead's solubility by changing the water's pH and carbonate levels. Increasing pH levels reduces the presence of hydrogen ions and increases the presence of hydroxide ions. Hydroxide ions decrease lead solubility by promoting the formation of solid basic lead carbonates that passivate the surface of the pipe. Similarly, elevating carbonate to a certain level increases the presence of carbonate ions, which, in turn, boost the presence of basic lead carbonate or lead carbonate. However, continuing to increase carbonate levels at a pH above about 7.5 will increase lead solubility (see Figure 7-1). If carbonate levels are the only other factors under consideration, then the optimal pH range to avoid lead corrosion is between 9.0 and 10.0.

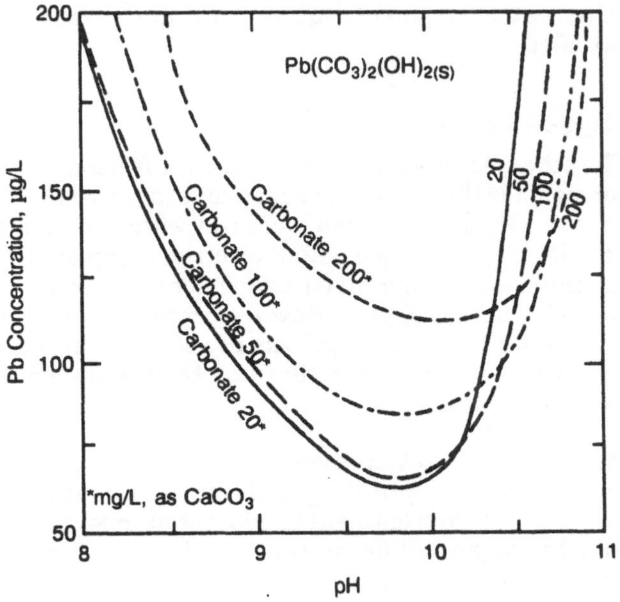

Figure 7-1. Solubility of lead as a function of pH and carbonate.
Source: Adapted from Schock and Wagner (1985).

Usually water carbonate levels are directly related to water alkalinity because alkalinity refers to the water's ability to neutralize acids and the most common base is dissolved inorganic carbon (carbonate species). At drinking water treatment plants, neutralization is performed primarily with calcium and sodium hydroxides, although carbonates may also be added. Carbon dioxide and water form bicarbonate as the pH increases; when the pH reaches 10.3, the carbonic species is expressed as carbonate. Coordinating pH level with alkalinity supplementation may reduce lead corrosion. Optimum alkalinity levels are 30 to 60 mg/L as calcium carbonate ($CaCO_3$).

Adjustment of both pH and alkalinity using lime-soda softening is an effective method for controlling lead corrosion. However, optimum water quality for corrosion control may not coincide with optimum hardness reduction. Water hardness is due mostly to the presence of calcium and magnesium ions. Lime-soda softening reduces the presence of these ions by adding hydrated lime ($Ca(OH)_2$), caustic soda ($NaOH$), or soda ash (sodium carbonate) under certain water quality conditions. This technique has been used to address water hardness, rather than corrosion. Comprehensive water treatment, however, must balance water hardness, carbonate, and alkalinity, as well as corrosive potential.

In addition, pH levels that are well suited for corrosion control may not be optimum for other water treatment processes, such as coagulation or disinfection. To avoid this type of conflict, the pH level should be adjusted for corrosion control immediately prior to water distribution, but after the other water treatment requirements have been satisfied.

Oxygen Levels

The other major water quality factor affecting corrosion is the presence of excessive dissolved oxygen. Dissolved oxygen increases water's corrosive activity by providing a potent electron acceptor. The optimum level of dissolved oxygen for corrosion control is 0.5 to 2.0 ppm. However, removing oxygen from water is not practical because of cost. The most reasonable strategy is to minimize the presence of oxygen.

Minimizing dissolved oxygen levels is effective for ground-water supplies. Ground water is sometimes aerated prior to treatment to address high levels of iron, hydrogen sulfide, and carbon dioxide. This aeration step eliminates free carbon dioxide, which, in turn, reduces the amount of lime necessary in lime-soda softening or for pH control; however, aeration increases corrosion by increasing dissolved oxygen. Consequently, excluding the aeration step and increasing lime softening can effectively reduce unnecessary and potentially counterproductive high oxygen concentrations.

Oxygen requirements are also reduced by extending the detention periods for treated water in reservoirs. Longer detention times allow the oxidation of hydrogen sulfide and organic carbon at the water surface. In addition, correct sizing of the water pumps used in the treatment plant minimizes the introduction of air during pumping.

Other Modifications

The potential of other substances to affect lead's solubility has been investigated. Tests show that adding sulfate, chlorine, and nitrate to water has little impact on lead solubility and, thus, corrosion. However, chlorine lowers pH, and chloramines (as opposed to free chlorine) have been shown to increase the solubility of lead-based solder. A few studies indicate that some natural organic compounds increase lead solubility; however, others have shown that tannins reduce corrosion by forming protective coatings on the metal surface.

7.1.3.3 Corrosion Inhibitors

Corrosion inhibitors form, or cause to form, protective coatings on pipes that reduce corrosion, but may not totally arrest it. The success of any corrosion inhibitor hinges on the treatment plant operator's ability to:

- Apply double and triple ultimate dosages of inhibitor during initial applications to build a base protective coat to prevent pitting. Typical initial coatings take several weeks to form.
- Maintain continuous and sufficiently high inhibitor dosages to prevent redissolving.
- Attain a steady water flow over all the system's metal surfaces for a continuous application of the inhibitor onto all exposed surfaces in the system.

There are several hundred commercially available corrosion inhibitors. Among the most common for potable water supply stems are:

- Inorganic phosphates
- Sodium silicates
- Mixtures of phosphates and silicates

These corrosion inhibitors can be applied with normal chemical feed systems.

Inorganic Phosphates

Inorganic phosphate corrosion inhibitors include polyphosphates, orthophosphates, glassy phosphates, and bimetallic phosphates. Zinc, added in conjunction with polyphosphates or orthophosphates, also helps inhibit corrosion in some cases. Phosphates can inhibit excessive calcium carbonate scale formation, form a protective coating with the pipe metal, or prevent aesthetically objectionable corrosion by-products. Water characteristics that affect the efficiency of phosphate corrosion inhibition include:

- Flow velocity
- Phosphate concentration
- Temperature
- pH
- Calcium
- Carbonate levels

The effectiveness of potential phosphate inhibitors should be confirmed by laboratory and field tests.

Glassy phosphates, such as sodium hexametaphosphate, effectively reduce iron corrosion at dosages of 20 to 40 mg/L. At lower dosages, this glassy phosphate may merely mask corrosion by eliminating the red color associated with iron corrosion. Under some circumstances, adding zinc in dosages of about 2 mg/L improves phosphate's corrosion control. Zinc phosphate also has been used effectively to inhibit corrosion.

Some studies show that orthophosphate is an effective lead corrosion inhibitor within specific ranges of carbonate and hydrogen ion concentrations. Polyphosphates have not been shown to be more effective than simple orthophosphates, and adding zinc sulfate has not improved polyphosphates' performance. In addition, polyphosphates can increase lead solubility in the absence of orthophosphate.

Silicates

Sodium silicates have been used for over 50 years to inhibit corrosion, yet the process by which they form a protective layer is not completely understood. The effectiveness of sodium silicates depends on pH and carbonate concentration. Sodium silicates are particularly effective for systems with high water velocities, low hardness, low alkalinity, and pH of less than 8.4. Typical coating maintenance dosages of sodium silicates range from 2 to 12 mg/L. They offer advantages in hot-water systems because of their chemical stability, as contrasted with many phosphates.

While silicates are hypothesized to inhibit lead corrosion, little data has been gathered to evaluate this theory.

7.1.3.4 Cathodic Protection

Cathodic protection is an electrical method of inhibiting corrosion. However, this expensive corrosion control method is not practical or effective for protecting entire water systems. It is used primarily to protect water storage tanks when a minimum metal surface area is exposed.

Metallic corrosion occurs when contact between a metal and an electrically conductive solution produces a flow of electrons (or current) from the metal to the solution. The electrons given up by the metal cause the metal to corrode rather than remain in its pure metallic form. Cathodic protection overcomes this current with an external power source. The electrons provided by the external power source prevent the metal from losing electrons, forcing it to be a cathode. This reaction allows the metal to remain in its less stable metallic form.

Cathodic protection is accomplished either by inserting electrodes into the metal to induce an external current or by placing a sacrificial galvanic anode in the current system. The sacrificial anode then corrodes in place of the protected metal. Galvanizing is a form of cathodic protection.

7.1.3.5 Coatings and Linings

Mechanically applied coatings and linings differ for pipes and water storage tanks. They usually are applied prior to installation, although some pipes can be lined after installation. While coal-tar products for pipes and tanks have been widely used for linings, they are under regulatory scrutiny due to the presence of polynuclear aromatic hydrocarbons in their composition.

The most common pipe linings are coal-tar enamels, epoxy paints, cement mortar, and polyethylene. Table 7-8 summarizes the advantages and disadvantages of these four primary pipe coatings and linings.

The most common types of water storage tank coatings and linings include coal-tar paints and enamels, vinyls, and epoxy. Eleven of these coatings and linings are described in Table 7-9.

7.2 Treatment Technologies for Controlling Inorganic Contaminants, Including Radionuclides

Inorganic contamination of raw drinking water supplies occurs from a wide variety of sources. Contaminants include metal ions, nitrate, fluoride, and radioactive substances; the most common contaminants are fluoride, arsenic, and nitrate. Contaminants that pose significant problems on a regional basis are selenium, barium, and radium-226. Inorganic contaminants are grouped as either naturally occurring or anthropogenic (resulting from human activities). Many naturally occurring inorganics, such as fluoride, arsenic, selenium, and radium-226, are commonly found in ground-water sources. Anthropogenic contaminants are usually found in surface water supplies. For example, nitrates and nitrites are a problem in agricultural areas or areas without sanitary sewer systems.

Contaminant type, valence, and solubility determine the most appropriate removal mechanism. Each removal technology uses a distinct mechanism or series of mechanisms to remove inorganic contaminants. Valence is particularly important to the removal efficiency of lime softening and coagulation. The primary factors in selecting an inorganic removal technology are:

- Contaminant type(s) and valence(s)
- Influent contaminant concentrations, because many technologies remove fixed percentages of contamination
- Desired effluent contaminant concentrations

Table 7-8. Pipe Wall Linings

Material	Use	Advantages	Disadvantages
Hot applied coal-tar enamel	Lining for steel pipes (used in 50 to 80% of pipes in distribution systems)	Long service life (>50 years) Good erosion resistance to silt or sand Resistant to biological attachment	Need to reapply to welded areas Extreme heat may cause cracking Extreme cold may cause brittleness May cause an increase in trace organics in water
Epoxy	Lining for steel and ductile iron pipes (can be applied in the field or in a foundry)	Smooth surface results in reduced pumping costs Formulated from components approved by the Food and Drug Administration	Relatively expensive Less resistant to abrasion than coal-tar enamel Service life <15 years
Cement mortar	Standard lining for ductile iron pipes, sometimes used in steel or cast-iron pipes	Relatively inexpensive Easy to apply (can be applied in place or in pipe manufacturing process) Calcium hydroxide release may protect uncoated metal at pipe joints	Rigidity of lining may lead to cracking or sloughing Thickness of coating reduces cross-sectional area of pipe and reduces carrying capacity
Polyethylene	Lining used in ductile iron and steel pipe (applied at foundry)	Long service life (50 years) Good erosion resistance to abrasives (silt and sand) Good resistance to bacterial corrosion Smooth surface results in reduced pumping costs	Relatively expensive

Source: U.S. EPA (1984).

Table 7-9. Water Storage Tank Linings and Coatings

Material	Comments
Hot applied coal-tar enamel	Most common coal-tar based coating used in water tanks; tends to sag or ripple when applied above the waterline when tank walls are heated.
Coal-tar paints	Most commonly used to reline existing water tanks; those paints containing xylene and naphtha solvents give the water an unpleasant taste and odor and should be used only above the waterline. Other coal tar paints containing no solvent bases can be used below the waterline but should not be exposed to sunlight or ice; service life of 5 to 10 years.
Coal-tar epoxy paints	Less resistant to abrasion than coal tar enamel; can cause taste and odor problems in the water; service life of about 20 years.
Coal-tar emulsion paint	Good adhesive characteristics, odorless, and resists sunlight degradation but not as watertight as other coal-tar paints, which limits use below waterline.
Vinyl	Nonreactive; hard, smooth surface; service life (about 20 years) is reduced by soft water conditions.
Epoxy	Forms hard, smooth surface; low water permeability; good adhesive characteristics if properly formulated and applied.
Hot and cold wax coatings	Applied directly over rust or old paint, short service life (about 5 years).
Metallic-sprayed zinc coating	Relatively expensive process that requires special skills and equipment, good rust inhibition, and service life of up to 50 years.
Zinc-rich paints	Hard surface; resistant to rust and abrasion; relatively expensive.
Chlorinated rubber paints	Used when controlling fumes from application of other linings is difficult, or where their use is specified.
Asphalt-based linings	Use is generally limited to relining existing asphalt-lined tanks.

Source: U.S. EPA (1984).

- Influent levels of dissolved solids and pH
- Pretreatment requirements, such as filtration required before reverse osmosis or ion exchange treatments
- System size, because of the significant savings from economies of scale for most technologies
- Water flow variability; for example, the coagulation process may not respond readily to flow fluctuations, while ion exchange, reverse osmosis, and activated alumina are relatively unaffected by changes in flow
- Cost
- Sludge management

- Adaptability of existing facilities

Clearly, no single treatment is perfectly suited for all inorganic contaminants. Possible treatment technologies include conventional treatment with coagulation, lime softening, cation exchange, anion exchange, reverse osmosis and electrodialysis, PAC, GAC, and activated alumina. Table 7-10 presents the principal applications and removal efficiencies of the eight treatment technologies for inorganics.

Many inorganic treatment decisions are based on the need to reduce the levels of a single contaminant. However, to treat more than one contaminant, multiple treatments in series or a membrane treatment unit, such as reverse osmosis, is most appropriate. Reverse osmosis and electrodialysis are effective for most contaminants; however, their low tolerance for turbidity makes them inappropriate without preliminary treatment for most surface water. The removal effectiveness of the other seven treatments in Table 7-10 varies from poor to excellent, depending on the contaminant. The removal percentages provided in the table are approximations that apply to favorable water quality conditions.

The actual performance of a removal technology depends on the selection factors listed above. Table 7-11 presents a summary of the most effective treatments for 11 common inorganic contaminants. The assumptions in the table include consideration of treatment costs. Table 7-12 summarizes advantages and disadvantages of each treatment technology.

Section 7.2.1 discusses radionuclide contamination. Radionuclides present different health risks and require different analytical evaluation techniques than other inorganic contaminants, but are removed by many of the same technologies that are effective for inorganic and organic contaminants. Sections 7.2.2 through 7.2.5 review the five treatment technologies designated primarily for inorganic contamination.

7.2.1 Removing Radionuclides in Drinking Water

Radionuclides are radioactive atoms that are characterized by the number of protons and neutrons in their nucleus and their energy content. The number that appears with a radionuclide's chemical abbreviation relates to this nuclear composition. The most common radionuclides in drinking water are radium, uranium, and radon.

Many methods are used to measure radionuclides' levels in drinking water. Atoms emit three types of nuclear radiation: alpha, beta, and gamma radiation. The average estimated costs for analyzing these radiation types range from $25 to $100 per sample.

Many of the techniques for removing radionuclides are the same as those for inorganic and organic contaminant removal. Table 7-13 lists the effective treatment processes for radionuclides. The radium and uranium removal processes can achieve up to 96 and 99 percent removal, respectively. GAC and aeration are effective for removing radon. (Complete discussions of GAC and aeration are provided in Sections 6.2 and 6.3.)

7.2.1.1 Costs

EPA has developed preliminary cost estimates for removing radionuclides (see Table 7-14). The costs for aeration are based on experience from two system sizes, those serving 100 to 500 persons and those serving 100,000 persons. Lower costs for larger systems reflect economies of scale for the aeration process.

Tables 7-15 and 7-16 present the costs of removing radon from drinking water. Table 7-15 compares cost data for removing radon with aeration for three plant sizes and Table 7-16 compares packed column aeration and GAC for managing three different levels of radon in effluent. The costs of removing radon from tap water vary with the radon contamination level and the removal method. Aeration is more expensive and effective than GAC for contaminant levels of 30,000 pCi/L. However, GAC is more expensive and effective at higher radon concentrations of 150,000 pCi/L.

7.2.2 Conventional Treatment: Coagulation and Lime Softening

Coagulation and lime softening are traditionally used to control turbidity, hardness, tastes, and odors, but also are effective in removing some inorganic contaminants. Section 4.1 describes the conventional water treatment processes in common practice at many water utilities. However, employing these processes to remove specific inorganic contaminants is relatively new.

Coagulation, discussed in Section 7.2.2.1, is effective in removing most metal ions or colloidally dispersed compounds, but is ineffective in removing nitrate, nitrite, radium, barium, and sulfate. Lime softening, discussed in Section 7.2.2.2, increases water pH, which precipitates the polyvalent cations and anions of calcium, magnesium, carbonate, and phosphate; however, lime softening is ineffective for nitrate removal.

Typically, higher dosages of coagulants are required for effective inorganics removal than for removal of turbidity, color, or hardness. For example, typical

Table 7-10. Removal Effectiveness for Eight Processes by Inorganic Contaminant

	Contaminant																		
Treatment	Ag	As	As^{III}	As^{V}	Ba	Cd	Cr	Cr^{III}	Cr^{VI}	F	Hg	$Hg^{(O)}$	$Hg^{(I)}$	NO_3	Pb	Ra	Se	Se^{IV}	Se^{VI}
Conventional treatment	H	–	M	H	L	H	–	H	H	L	–	M	M	L	H	L	–	M	L
Lime softening	–	–	M	H	H	H	–	H	L	M	–	L	M	L	H	H	–	M	L
Reverse osmosis	H	–	M	H	H	H	H	–	–	H	H	–	–	M	H	H	H	–	–
Cation exchange	–	L	–	–	H	H	–	H	L	L	–	–	–	L	H	H	L	–	–
Anion exchange	–	–	–	–	M	M	–	M	H	–	–	–	–	H	M	M	H	–	–
Activated alumina	–	–	H	–	L	L	–	–	–	H	–	–	–	–	–	L	H	–	–
Powdered activated carbon	L	–	–	–	L	M	–	L	–	L	–	M	M	L	–	L	–	–	–
Granular activated carbon	–	–	–	–	L	M	–	L	–	L	–	H	H	L	–	L	–	–	–

H = High = >80% removal.
M = Medium = 20-80% removal.
L = Low = 20% removal.
– indicates no data were provided.

Table 7-11. Most Effective Treatment Methods for Removal of Inorganic Contaminants

Contaminant	Most Effective Treatment Methods
Arsenic	As V-iron coagulation, pH 6-8; alum coagulation, pH 6-7; excess lime softening; activated alumina, pH 5-6
	As III-oxidation treatment of As III to As V; use same treatment list for As V
Barium	Lime softening, pH 11; ion exchange softening
Cadmium	Iron coagulation, above pH 8; lime softening; excess lime softening
Chromium	Cr III-iron coagulation, pH 6-9; alum coagulation, pH 7-9; excess lime softening
	Cr-ferrous sulfate coagulation, pH 7-9.5
Fluoride	Ion exchange with activated alumina or bone char
Lead	Iron coagulation, pH 6-9; alum coagulation, pH 6-9; lime or excess lime softening
Mercury	Inorganic-ferric sulfate coagulation, pH 7-8; granular activated carbon
	Organic-granular activated carbon
Nitrate	Ion exchange with anion resin
Radium	Lime softening; ion exchange with cation resin
Selenium	Se IV-ferric sulfate coagulation, pH 6-7; ion exchange with anion resin or activated alumina; reverse osmosis
	Se IV-ion exchange with anion resin or activated alumina; reverse osmosis
Silver	Ferric sulfate coagulation, pH 7-9; alum coagulation, pH 6-8; lime or excess lime softening

Source: Sorg (1980).

alum dosages for turbidity control range from 5 to 40 mg/L, while the dosage for arsenic removal is about 100 mg/L. Higher dosages increase costs and create more sludge.

7.2.2.1 Coagulation

The coagulation processes that remove turbidity as well as inorganic contaminants include adsorption and precipitation. Coagulation with iron or aluminum salts also removes trace anions of selenium, arsenic, and fluoride through coprecipitation or sorption onto flocculated particles.

Coagulation is more effective in removing polyvalent, as opposed to monovalent, cations and anions. The effectiveness of coagulation also depends on influent turbidity and/or color levels, the quality of the flocculated particles produced, and the amount of turbidity or color removed.

Table 7-17 presents potential removal efficiencies using alum and ferric salts as coagulants. The optimum pH level varies depending on the target inorganic substance and the coagulant used. The table provides the most suitable pH level for removing each inorganic substance, when available. As indicated, iron salts are more effective over a wider range of pH levels than alum. Ferric salts are effective at pH levels between 4.0 and 12.0, while alum is generally effective at pH 5.5 to 8.0.

Table 7-12. Advantages and Disadvantages of Inorganic Contaminant Removal Processes

Precipitation and Coprecipitation Used in Coagulation/Conventional

Advantages

- Low cost for high volume
- Often improved by high ionic strength
- Reliable process well suited to automatic control

Disadvantages

- Stoichiometric chemical additions required
- High-water-content sludge disposal
- Part-per-billion effluent contaminant levels may require two-stage precipitation
- Not readily applied to small or intermittent flows
- Coprecipitation efficiency depends on initial contaminant concentration and surface area of primary floc

Ion Exchange

Advantages

- Operates on demand
- Relatively insensitive to flow variations
- Essentially zero level of effluent contamination possible
- Large variety of specific resins available
- Beneficial selectivity reversal commonly occurs upon regeneration

Disadvantages

- Potential for chromatographic effluent peaking
- Spent regenerant disposal
- Variable effluent quality with respect to background ions
- Usually not feasible at high levels of total dissolved solids

Activated Alumina

Advantages

- Operates on demand
- Insensitive to flow and total dissolved solids background
- Low effluent contaminant level possible
- Highly selective for fluoride and arsenic

Disadvantages

- Both acid and base are required for regeneration
- Medium tend to dissolve, producing fine particles
- Slow adsorption kinetics
- Spent regenerant disposal

Membranes (Reverse Osmosis and Electrodialysis)

Advantages

- All contaminant ions and most dissolved non-ions are removed
- Relatively insensitive to flow and total dissolved solids level
- Low effluent concentration possible
- Bacteria and particles are removed

Disadvantages

- High capital and operating costs
- High level of pretreatment required
- Membranes are prone to fouling
- Reject stream is 20 to 90% of feed flow

Source: Clifford (1986).

Adjusting the pH and using coagulant aids can optimize inorganic compound removal. Coagulant aids may reduce the higher doses of coagulant needed for inorganic contaminant removal. Optimal dosages for coagulants, coagulant aids, and pH adjusters may be derived using jar tests and pilot studies.

Table 7-13. Treatment Technologies for Removing Radionuclides

Treatment Technology	Radio-nuclide	Reported Approximate Process Efficiency (percent)	Comments
Conventional treatment with coagulation-filtration	Ra	< 25	High pH and Mg required
	U	18-98	High pH (10 +) and high dosages of ferric chloride or alum only accomplished in lab studies with diatomaceous earth filtration
Lime softening	Ra	75-96	Best choice for large plants
		43-92	Plant-scale results
		80	Plant-scale results
	U	85-90	pH 10.6-11.5
		99	High pH, high Mg
Ion exchange	Ra	95 +	Best choice for small plants; cation exchangers
		99	
	U	99	Brine disposal problem
			Anion exchangers; largely experimental but some full-scale plants on line
Adsorption	Ra	90 +	Adsorption on any solids;
		85-90	experimental
	Rn	62-99	Sand adsorption; experimental
			GAC adsorption
Aeration	Rn	20-96	Depends on process
		93 +	Depends on process
Reverse osmosis	Ra	87-96	Plant-scale data
		87-98	Based on eight plants
		95 +	High-volume brine solution for disposal
	U	95 +	High-volume brine solution for disposal

Table 7-14. Radionuclide Process Treatment Costs

Process	Range of Costs of Removal Process (dollars/1,000 gallons of water)
Coagulation/Filtration[a]	0.07 to 0.28[b]
Lime softening[a]	0.10b
Aeration	0.10 to 0.75[c]
Ion exchange (cationic)	0.30 to 0.80[c]
Ion and manganese treatment	0.30 to 1.10[c]
Lime softening (new)	0.50[b]
Ion exchange (anionic)	1.60 to 2.10[b]
Reverse osmosis	1.60 to 3.20[c]

1,000 gallons = 3.78 m^3

[a] Adding to an existing facility.

[b] 1982 dollars.

[c] 1987 dollars.

Table 7-15. Costs for Removing Radon from Drinking Water by Packed Tower Aeration (99% Removal)

	Population Served		
	100-500	3,300-10,000	75,000-100,000
Total capital cost ($1,000)	67	250	2,200
Operations and maintenance cost ($1,000 per year)	1.2	15	230
Cost (cents/1,000 gallons)	75	14	9

1,000 gallons = 3.78 m^3

Table 7-16. Costs of Radon Treatment at the Plant Scale (200 GPD) for GAC Versus Aeration

Influent Concentrations (pCi/L)	Effluent Concentrations (pCi/L)		Capital Costs in Dollars		Operating Costs in Dollars	
	I[a]	II[b]	I	II	I	II
15,000	1,350-3,300	750	430-760	900	20	60
30,000	2,700-6,600	1,500	430-760	900	20	80
150,000	1,200	67,500	1,500	1,000	40	80

[a]I = GAC process.
[b]II = Aeration process.

Table 7-17. Removal Efficiency Potential of Alum Versus Ferric Chloride

Inorganic Contaminant	Removal Efficiency	
	Alum Coagulant	Iron Coagulant
Ag (pH < 8.0)	90%	-
Ag (pH = 8.0)	-	70%
As V	-	90%
As V (pH < 7.5)	90%	-
As V (pH = 7.5)	-	90%
Cd (pH $\geq$ 8.0)	-	-
Cd (pH $\geq$ 8.5)	70%	-
Cr (III)	90%	-
Cr III (pH = 10.5)	-	90%
Cr VI (using Fe II)	-	90%
Hg	70%	-
Pb	90%	-

7.2.2.2 Lime Softening

Lime softening is a reliable and established treatment for "hard" water and corrosion control, and also removes some inorganic contaminants. (See Section 7.1.3.2 for discussion of the effect of lime softening on corrosion.) The pH is an important design consideration for inorganic contaminant removal using lime softening because it affects:

- Contaminant species (soluble or insoluble)
- Coagulant form
- Type of flocculated particle formed

Higher dosages of lime create higher pH levels, which, in turn, increase precipitation of inorganic contaminants and adsorption of both organic and inorganic species. The harder the water, the more effective the lime softening process is in removing inorganic contaminants. This happens because the softening process depends on the enmeshment of ions within the flocculated particles or adsorption on the surface of precipitates.

Table 7-18 presents the removal efficiency of lime for seven inorganic contaminants. Lime softening is effective for cadmium, lead, silver, fluoride, and cation removal in general. It also effectively removes arsenic(V), barium, and chromium(III) at the specific pH levels shown in the table. Lime-soda softening is generally ineffective in removing nitrate, selenium(VI), mercury(0), or chromium(IV).

Table 7-18. Removals Possible with Lime Softening

Inorganic Contaminant	pH	Percent Removal
As (V)	10.0-10.5	70
As (V)	>10.8	90
As (III)	>10.5	70
Cd	NA	90
Cr (III)	>10.5	70-90
Pb	NA	90
Ag	NA	70
Ba	9.5-10.8	

NA = Not available.

Figure 7-2 shows a typical lime-soda softening process. Characteristic elements of such a system include raw water pumps, lime and soda ash addition facilities, sedimentation, recarbonation, filtration, disinfection, storage, and distribution. The recarbonation step is designed to lower the pH to the point of calcium carbonate saturation after the lime has had its effect in the sedimentation basin.

7.2.3 Reverse Osmosis

Reverse osmosis is an expensive process that uses a semipermeable membrane to remove contaminants from solution. Water with different contaminant concentration is placed on each side of the membrane. The water is directed through the membrane by hydrostatic pressure to the side with the lower concentration of contaminants. Since the membrane permits only water, and not dissolved ions, to pass

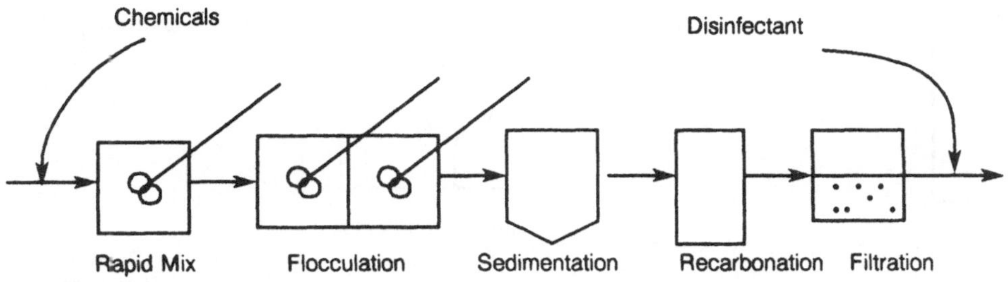

Figure 7-2. Lime softening treatment system.

through its pores, contaminants are left behind in a brine solution. These membranes are available with a variety of pore sizes and characteristics.

Electrodialysis is a process that also uses membranes. However, in this process direct electrical current is used to attract the ions to one side of the treatment chamber.

The membranes adjacent to the influent stream are charged either positively or negatively, and this charge attracts counter-ions toward the membrane. The membranes are designed so as to allow either positively or negatively charged ions to pass through the membrane, thus they move from the product water stream through a membrane to the two reject water streams.

The three essential elements of the system are (1) a source of pressurized water, (2) a direct current power supply, and (3) a pair of selective membranes. The average ion removal varies from 25 to 60 percent per stage. Multistage units can increase the efficiency of removal. Many membrane pairs are "stacked" in the treatment vessel.

Fouling of the membranes may limit the amount of water treated before cleaning is required. Fouling is caused when membrane pores are clogged by salt precipitation or by physical obstruction of suspended particulates. Particulates, suspended in water, can be removed in pretreatment but salts that exceed their solubility product at the membrane surface must be controlled chemically by pH reduction (to reduce carbonate concentration) or chelation of metal ions (by use of phosphate, for example). A recent innovation has been the occasional reversal of the charge on the membranes, a process called electrodialysis reversal (EDR). This helps to flush the attached ions from the membrane surface, thus extending the time between cleanings.

Reverse osmosis is a compact system that is well suited for treating water with high levels of inorganic substances, organic substances, or total dissolved solids. It has effectively treated water with total dissolved solids as high as seawater, at 8,600 mg/L. While reverse osmosis is a proven technology for removing inorganic substances, it has also removed the following effectively:

- Radium
- THM precursors including humic and fulvic acids
- Pesticides
- Microbiological contaminants (viruses, bacteria, and protozoa)

Reverse osmosis systems are particularly effective in series. Water passing through multiple units can achieve near zero effluent contaminant concentrations.

Reverse osmosis is also suitable for small systems with a high degree of seasonal fluctuation in water demand. Reverse osmosis systems are relatively insensitive to changes in flow and operate immediately, without any minimum break-in period. Their operational simplicity and minimal labor requirements make them suitable for small system applications. One disadvantage of reverse osmosis units is high operating and capital costs. Managing the wastewater (brine solution) is also a potential problem.

7.2.3.1 Design Considerations

Typical reverse osmosis units include raw water pumps, pretreatment, membranes, disinfection, storage, and distribution elements. Figure 7-3 is a schematic diagram of a reverse osmosis system. These units are able to process virtually any desired quantity or quality of water by configuring units sequentially to reprocess waste brine from the earlier stages of the process. The principal design considerations for reverse osmosis units are:

- Operating pressure
- Membrane type and pore size
- Pretreatment requirements
- Product conversion rate (the ratio of the influent recovered as waste brine water to the finished water)

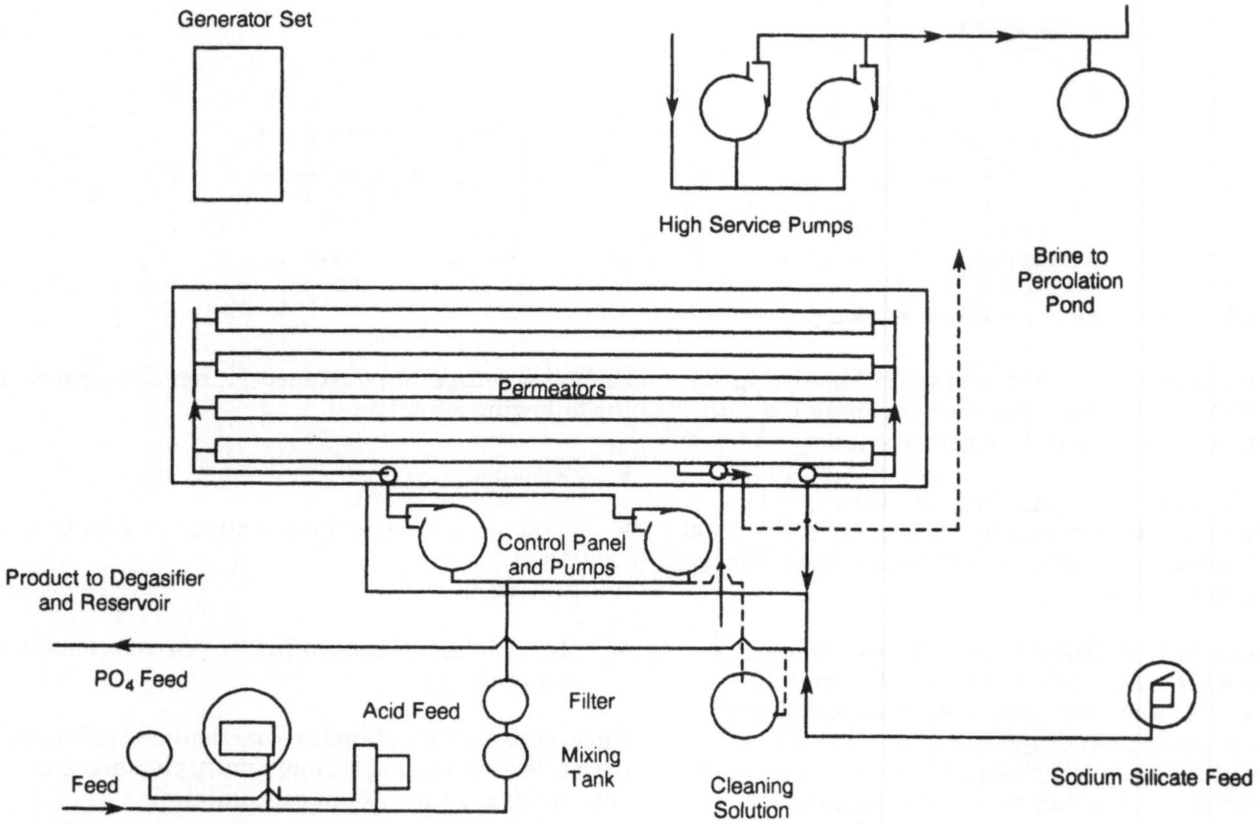

Figure 7-3. Schematic of a reverse osmosis system.

Operating Pressure

Typical operating pressures range from 5.3 to higher than 24.6 kg/cm² (75 to higher than 350 psi). Reverse osmosis systems rated at less than 17.6 kg/cm² (250 psi) are classified as low-pressure units, while those operating above 24.6 kg/cm² (350 psi) are classified as high-pressure units. High water pressure can lead to noise, vibration, and corrosion problems, but, in general, these systems are more effective. Recently developed models, however, are able to perform well at low pressures.

Membrane Type

Two distinct membrane designs for reverse osmosis are illustrated in Figure 7-4. The spiral-wound unit clogs less frequently when processing influent with high solids content. The hollow-fiber membrane, however, has much greater surface area per unit of space than the spiral-wound design. The hollow-fiber unit has about 3,280 m² of membrane surface/m³ of membrane module (1,000 sq ft/cu ft of membrane module), as compared to about 328 m² (100 sq ft) for the spiral-wound unit.

Until the 1970s, membranes were made almost exclusively of cellulose acetate. Now membranes are also made from aromatic polyamide and thin-filmed polymer composites. Different membrane materials will have distinct characteristics, such as hydraulic resistance, pH range, temperature range, chlorine tolerance, and biodegradation tolerance.

Pretreatment Requirements

Assessing pretreatment requirements requires careful consideration of the following factors:

- Influent suspended solid concentration
- Ionic size of the contaminants
- Membrane type

Other influent characteristics such as contaminant concentration, water temperature, and presence of competing ions are also important. Some systems with poor quality influent require extensive pretreatment.

Pretreatment is most commonly used to prevent fouling of the membrane during operation. Fouling is also caused by biological growth during periods of disuse. Typical pretreatment for reverse osmosis includes particle removal by filtration, and sequestering hardness ions by precipitates and pH control to prevent clogging.

A major problem with reverse osmosis units is proper management of the wastewater quantity and quality. The percentage of treated water recovered from these systems typically ranges from 50 to 90 percent of the influent. Systems that produce reject water ranging

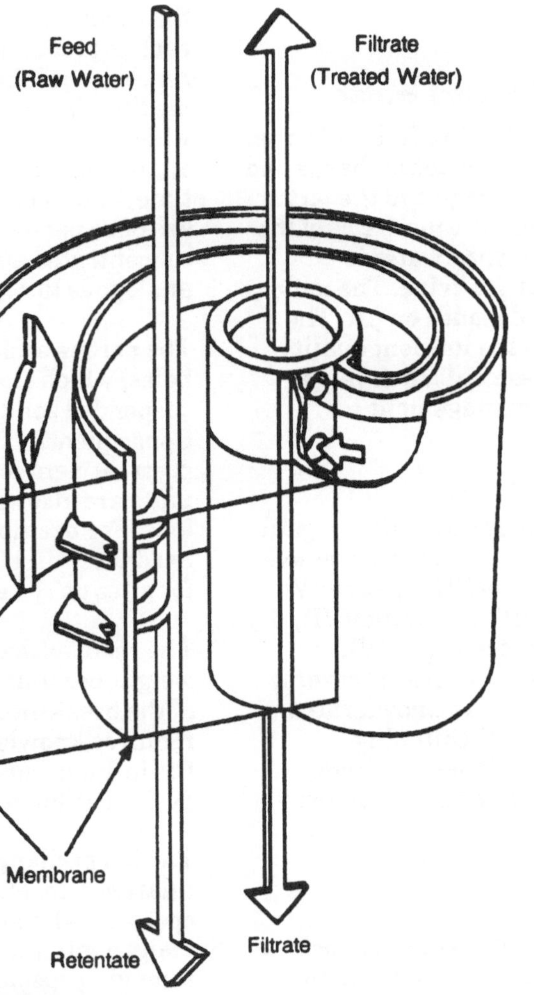

Courtesy of Millipore Corporation

Process Flow Through Spiral-Wound Reverse Osmosis Unit

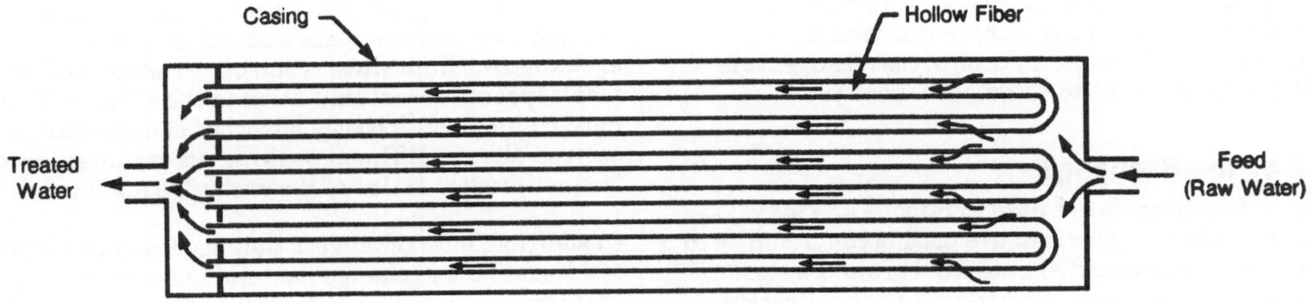

Hollow-Fiber Reverse Osmosis Unit

Figure 7-4. Two types of reverse osmosis membranes.

as high as 90 percent of the feed water are generally uneconomical.

Ratio of Reject Water to Finished Water

The ratio of reject water, or "brine," to finished water depends on several factors primarily ionic charge and ionic size. The higher the ionic charge and the larger the ionic size of the contaminant, the more easily the ion is removed and the more finished water is recovered relative to the amount of brine. The ionic charge for some contaminants depends on pH. The quality of the brine depends on the influent quality and requires site-specific assessments to develop environmentally sound waste management.

7.2.3.2 System Performance

Reverse osmosis can remove essentially all inorganic contaminants from water effectively. It removes over 70 percent of the following: arsenic(III), arsenic(V), barium, cadmium, chromium(III), chromium(VI), fluoride, lead, mercury, nitrite, selenium(IV), selenium(VI), and silver. Properly operated reverse osmosis units will attain 96 percent removal rates, while similarly operated lime-soda softening operations will attain from 75 to 96 percent removal. Comparative ion exchange units achieve between 81 and 97 percent removal.

7.2.3.3 System Costs

The predominant disadvantage of reverse osmosis is its high cost. Operating costs range from \$0.79 to \$1.59/m^3 (\$3 to \$6/1,000 gal) of treated water for a reverse osmosis plant of less than 0.04 m^3/sec (1 MGD). For plants with larger capacities and lower operating pressures, costs may be potentially competitive with other processes. Capital costs are usually high, especially if there is a need for pretreatment. In general, capital costs range from \$264 to \$528/m^3 (\$1 to \$2/gal) of capacity.

7.2.4 Ion Exchange

Ion exchange units are used to remove any ionic substance from water, but are used predominantly to remove hardness and nitrate from ground water. Typical ion exchange units consist of prefiltration, ion exchange, disinfection, storage, and distribution elements (see Figure 7-5).

Inorganics removal is accomplished through adsorption of contaminant ions onto a resin exchange medium. As the name implies, one ion is substituted for another on the charged surface of the medium, which is a resin, usually a synthetic plastic. This resin surface is designed as either cationic or anionic.

The exchange medium is saturated with the exchangeable ion before treatment operations. During ion exchange, the contaminant ions replace the regenerant ions because they are preferred by the exchange medium. After the exchange medium reaches equilibrium with the contaminant ions, the medium is regenerated with a suitable solution, which then resaturates the medium with the appropriate ions. Because of the required "down time," the shortest economical regeneration cycles are once per day. Ion exchange waste is highly concentrated and requires careful disposal, analogous to reverse osmosis reject streams.

The porous exchange medium is covered with tiny holes, which clog when significant levels of suspended solids are in the influent stream. Consequently, filtration may be a necessary pretreatment for ion exchange units. Ion exchange units are also sensitive to the presence of competing ions. For example, influent with high levels of hardness will effectively compete with other cations for sites on the exchange medium.

The resin exchange capacity is expressed in terms of weight per unit volume of the resin. The calculation of the breakthrough time for an ion exchange unit requires knowledge of the resin exchange capacity, the influent contaminant concentration, and the desired effluent quality.

The ion exchange process, like reverse osmosis, is relatively insensitive to flow rate and virtually fully operational upon process initiation. Each resin, of the many available from suppliers, is effective in removing specific contaminants. In addition, ion exchange is able to achieve very low contaminant concentrations in finished water.

The primary disadvantage of ion exchange concerns effluent peaking. Effluent peaking occurs when contaminant ions compete with other ions for exchange medium sites, resulting in unacceptable levels (peaks) of contamination in the effluent. It is most common with poorly adsorbed contaminants, such as nitrate. Effluent peaking necessitates more frequent regeneration of the exchange medium.

In addition, ion exchangers that use sodium chloride to saturate the exchange medium may experience problems with sodium residual in the finished water. Sodium is used because of its low cost, but high sodium residual is unacceptable for individuals with salt-restricted diets. This problem may be avoided by using other saturant materials, such as potassium chloride.

7.2.4.1 System Performance

Ion exchange effectively removes greater than 90 percent of barium, cadmium, chromium(III), silver, and radium using cationic resins as the exchange medium. In addition, it achieves greater than 90

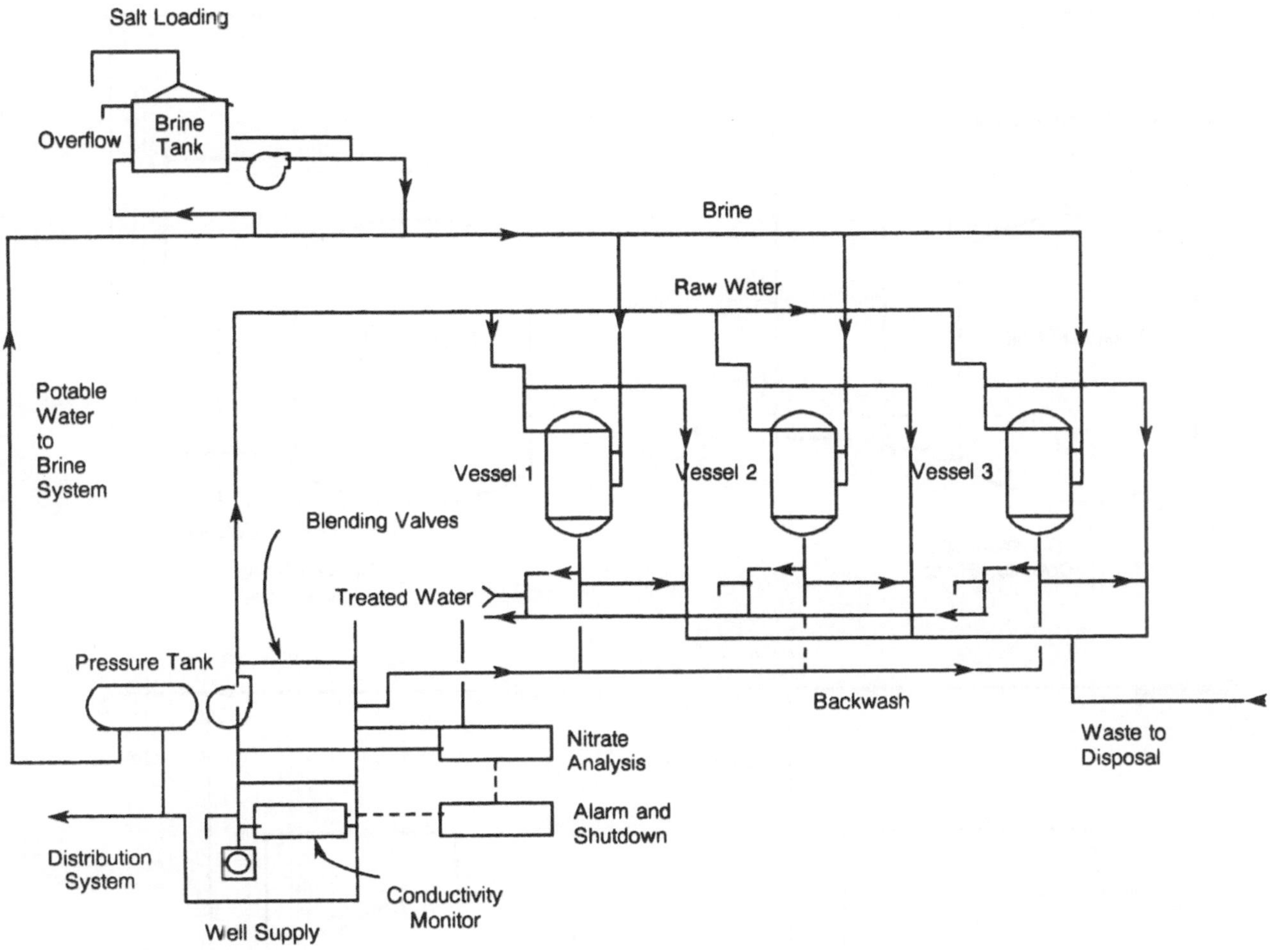

Figure 7-5. Ion exchange treatment system.

Ion exchange facility for nitrate removal; McFarland, CA.

percent removal of nitrites, selenium, arsenic(V), chromium(VI), and nitrate using anionic resins. While it is effective for all of the contaminants listed above, ion exchange is especially well suited to remove barium, radium, nitrate, and selenium.

7.2.5 Activated Alumina

Activated alumina, a commercially available ion exchange medium, is primarily used for fluoride removal from ground water. The activated alumina process is an ion exchange process that consists of the following six basic elements: raw water pumps, pretreatment, activated alumina contact, disinfection, storage, and distribution. Figure 7-6 shows schematic diagrams of the four-phase operational process.

Before processing, the activated alumina medium is saturated with hydroxyl ions from a strong sodium hydroxide solution. During the removal process, activated alumina exchanges the hydroxyl ions for fluoride anions. When the medium becomes saturated, the activated alumina must be regenerated with sodium hydroxide. This increases the pH level of the water in the treatment unit to a point where buffer solutions of 3 percent sulfuric acid

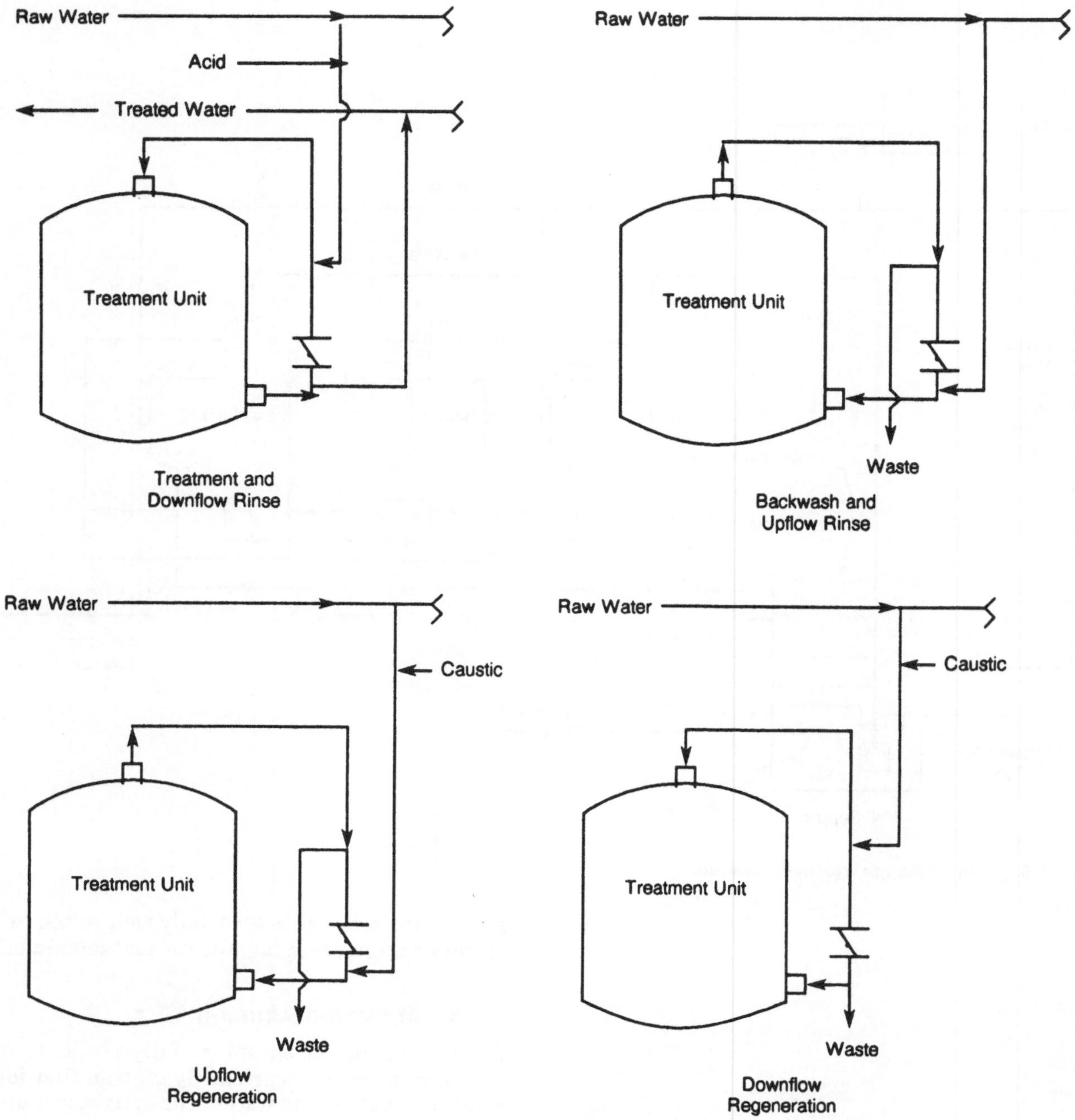

Figure 7-6. Activated alumina systems: Operating mode flow schematics

are required for neutralization, after the water leaves the exchange unit.

While activated alumina requires pretreatment for suspended solids, it can tolerate high levels of total dissolved solids. As with other ion exchange processes, the activated alumina process is sensitive to competing ions. It is also sensitive to pH. Optimum removal efficiency for fluoride occurs below a pH of 8.2.

While activated alumina effectively removes several contaminants, it requires the handling of potentially hazardous strongly acidic and basic solutions. The process of heating the sodium hydroxide solution to maintain it as a liquid in cold climates is particularly hazardous.

Activated alumina's costs are also higher than ion exchange costs. The loss of activated alumina to the sodium hydroxide solution during processing can range up to 20 percent annually. In addition, waste management may also increase costs because of the high contaminant and aluminum concentrations in the waste stream, as well as the high pH.

7.2.5.1 System Performance

Activated alumina is operational immediately because of its relative insensitivity to flow rates. It effectively removes over 90 percent of arsenic(V), fluoride, and selenium(IV); removes 70 percent of selenium(VI); and also effectively removes iron. It is ineffective for barium, cadmium, and radium removal, since these contaminants occur primarily as cations.

Chapter 8

Current and Emerging Research

This chapter describes the activities of EPA's Drinking Water Research Division (DWRD), highlighting the division's recent research efforts and future directions. The impetus for the majority of DWRD's research is to provide technical support for the Agency as it implements the SDWA and its amendments. At present, there are four aspects of the regulatory process influencing current areas of research:

1. Development of Final Surface Water and the anticipated Ground-Water Treatment Rules, which will require disinfection and, in the case of surface waters, filtration of drinking water in communities serving more than 25 persons.
2. The requirement that EPA set Maximum Contaminant Levels (MCLs) for listed contaminants, e.g., volatile organic compounds (VOCs), synthetic organic compounds (SOCs), inorganics, radionuclides, microbiological contaminants, and turbidity according to the timetable specified in the 1986 Amendments to the SDWA.
3. The requirement that every 3 years EPA publish a Drinking Water Priority List of contaminants known or anticipated to occur in public drinking water systems and pose a human health risk; and subsequently regulate at least 25 of the contaminants listed according to the schedule set in the act. The first list of 44 contaminants was already published and, therefore, 25 of those contaminants listed must be regulated by January 1990.
4. The Disinfection By-Products Rule, whose proposal is anticipated along with the Ground-Water Treatment Rule in 1991, will regulate disinfection by-products that have been, and are still being, identified.

In pursuit of these regulatory milestones, research has focused on identifying contaminants that will be regulated and evaluating the removal potential of available technologies. This research includes identifying and characterizing disinfection by-products associated with chlorination and the use of ozone. Also in support of the rules, DWRD is studying various methods of meeting mandatory disinfection requirements; investigating factors affecting the occurrence of lead in drinking water; and evaluating methods of controlling corrosion, exposure of consumers to disinfection by-products, and microbial contamination in distribution systems.

In addition to its research activities, DWRD provides technical assistance to State agencies, EPA regional offices, water utilities, and professional organizations. DWRD also supplies information and provides consultation on the technical aspects of drinking water treatment systems.

8.1 Current Research on Disinfection By-Products

The SDWA and Amendments of 1986 require most public water supplies to disinfect drinking water before distribution (U.S. EPA, 1986a). In the past, chlorine has been the most widely used disinfectant, and while its effectiveness in controlling pathogens is proven, unhealthful chemical by-products have been identified.

Research on these by-products is a very important area of investigation for EPA. A disinfection by-product work group was formed to coordinate research efforts between the EPA Office of Research and Development (ORD) and Office of Drinking Water (ODW). Research efforts are focused on identifying and characterizing the by-products of chlorine disinfection, and evaluating treatment techniques that will control by-product generation. Since a promising strategy for control is to use ozone as an alternate disinfectant, EPA is also researching the by-products and other issues associated with ozone disinfection.

8.1.1 Identifying and Controlling Chlorination By-Products

DWRD has been conducting experiments, mostly in the laboratory, to develop methods of identifying by-products associated with the use of chlorine as a disinfectant. In close association with the Technical Support Division of ODW, DWRD is isolating the various by-products that have been identified in field studies (Stevens et al., 1987).

To date, six compounds (or in some cases groups of compounds) have been identified:

- Trihalomethanes (4 compounds)
- Dihaloacetonitriles (3 compounds)
- Chloroacetic acids (3 compounds)
- Chloral hydrate
- Chloropicrin
- 1,1,1-Trichloropropanone

These by-products comprise 30 to 60 percent of the total organic halogen (TOX) in drinking water. The levels of common chlorination by-products found in drinking water are presented in Table 8-1 (Stevens et al., 1987).

Table 8-1. Common Chlorination By-Products

General Compound Identification	Number of Individual Compounds	Nominal Concentration (μg/l)		
		<10	10-100	>100
Trihalomethanes	4			X
Dihaloacetonitriles	3	X		
Chloroacetic acids	3	X	X	X
Chloral hydrate				X
Chloropicrin		X		
1,1,1-Trichloropropanone		X		

Source: Stevens et al. (1987).

Methods of controlling chlorine disinfection by-products are being studied by DWRD in an onsite pilot plant (Stevens et al., 1989). (Figure 8-1 is a schematic of the pilot plant.) The pilot plant was constructed with two channels: the first was represented by the experiment control; the second was used to test the effects of changing the point of chlorination on the production of by-products.

Through experimentation, it was found that moving chlorination to a point just before filtration minimized the production of total trihalomethanes (TTHMs) while maintaining microbiological quality. The assimilable organic carbon of the finished water, which is a measure of the biological quality, was also found to be reduced. Therefore, the overall quality of the water improved under this scenario.

Although the individual concentrations of the disinfection by-products (DBPs) vary with pH, temperature, and chlorine concentration, pilot studies have shown that the removal of TTHM precursor seems to remove the formation potential for the other individual DBPs. This finding confirms earlier studies and opens up the possibility of precursor removal as an effective means of controlling chlorinated DBP (Lykins et al., 1988a).

An alternative to chlorine is the use of chloramines or chlorine dioxide for minimized DBP. Studies have shown that these approaches can be effective in maintaining microbiological integrity and minimizing by-product formation (Singer, 1988). This issue and the associated technologies will become extremely important if the THM standard is ultimately lowered.

8.1.2 Identifying Ozone By-Products

The DWRD is studying several issues related to using ozone as an alternate disinfectant to chlorine. The broadest issue concerns the identification of by-products formed by ozone reacting with naturally occurring organics in drinking water (Glaze, 1986). The health effects of these by-products are being characterized and compared to the health effects of by-products of chlorination. Also being investigated is the effect of ozone on the precursors of chlorine by-products, an issue if chlorine is used as a secondary disinfectant.

Other research related to ozone as a disinfectant covers its ability to destroy contaminants and its effects on the formation of bromide ion or other brominated organics when bromide ion is present. Some substances formed through the oxidation of bromide ion are known to be harmful to humans.

8.2 Treatment of Organic and Inorganic Contaminants

The SDWA amendments accelerated the schedule for setting MCLs for contaminants in drinking water. Contaminant limits are set according to the removal potential of existing technologies that can be obtained at a reasonable cost. Traditionally, EPA set contaminant limits by studying treatment techniques for each chemical individually. Since there is such a large number of chemicals to be considered for regulation, and the deadlines for promulgating MCLs are quite stringent, EPA is studying the effectiveness of various technologies in removing groups of contaminants using bench- and pilot-scale experiments. Table 8-2 lists the technologies that have been and are being evaluated by DWRD (Clark et al., 1988b; Feige et al., 1987).

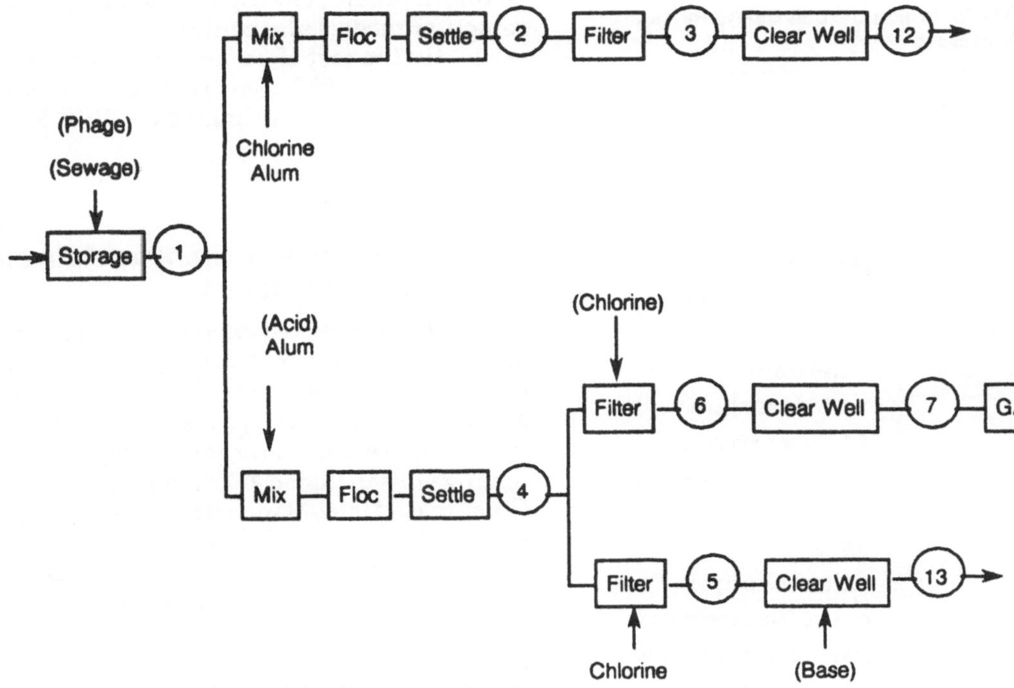

Figure 8-1. Pilot plant schematic for disinfection by-product control.

Table 8-2. Treatment Technologies Evaluated by the Drinking Water Research Division

Contaminant (or Contaminant Classes)	Proven Effective in Field Tests	Proven Effective in Pilot Tests	Being Evaluated as Promising Technologies
Volatile organic compounds	Carbon adsorption, packed tower and diffused aeration	-	Ozone oxidation, reverse osmosis, ultraviolet treatment
Synthetic organic compounds	Carbon adsorption	-	Conventional treatment with powdered activated carbon, ozone oxidation, reverse osmosis, ultraviolet treatment
Nitrates	Ion exchange	Reverse osmosis	
Radium	Reverse osmosis, ion exchange	-	-
Radon	Carbon adsorption	-	Aeration
Uranium	Ion exchange	Reverse osmosis	-

Source: Clark et al. (1988b); Feige et al. (1987).

Granular Activated Carbon (GAC), ozone oxidation, reverse osmosis, ultraviolet (UV) treatment, ultrafiltration, packed tower aeration, and conventional treatment are being studied as methods of controlling organic compounds, including certain disinfection by-products. Ion exchange, reverse osmosis, aeration, and carbon adsorption are being tested for their ability to remove inorganic compounds, including radionuclides. In addition, technologies for controlling secondary sources of pollution from treatment processes are being investigated. This section summarizes the relevant research projects.

8.2.1 Granular Activated Carbon Systems

GAC is the legal feasible technology for treating SOCs. DWRD is currently developing data on the effectiveness of GAC for removing a variety of organic chemicals.

One measure of effectiveness is the carbon usage rate for each chemical. This indicator describes the ability of each compound to be adsorbed by the carbon in the treatment system. It also describes how quickly the carbon bed will be exhausted (adsorption surfaces of the carbon granules will be used) and, therefore, dictates the size of the GAC system needed for optimal operation. EPA is currently developing carbon usage rate data for the 45 chemicals listed in Table 8-3 (Clark et al., 1989a). Carbon usage rates are established by laboratory or field experiments.

8.2.1.1 Establishing Carbon Usage Rates

A mathematical model that uses isotherm results to simulate full-scale carbon adsorption treatment systems is being used (Speth and Miltner, 1989). Pilot testing is a common first step in designing a full-scale treatment system and much data are available correlating pilot-scale test results to actual full-scale operating parameters. Therefore, engineers

Table 8-3. Chemicals for Which Carbon Usage Rates are Being Developed

2,4-D	Dalapon
Silvex	Dinoseb
Lindane	Glyphosate
Methoxychlor	Oxamyl
Toxaphene	N,N-Ethylene-thiourea
Chlordane	Metribuzin
Heptachlor	Hexachlorobenzene
Heptachlor epoxide	*cis*-1,3-dichloropropene
PCS(s)	*o*-chlorotoluene
Pentachlorophenol	1,1-Dichloroethane
Alachlor	1,1-Dichloropropene
Carbofuran	1,1,2,2-Tetrachloroethane
Aldicarb	EDB
Aldicarb sulfone	DBCP
Aldicarb sulfoxide	1,2-Dichloropropane
Tert-Butyl methyl ether	*cis*-1,2-Dichloroethylene
Metolachlor	*trans*-1,2-Dichloroethylene
Tetrachloroethene	Chlorobenzene
1,3,5-Trichlorobenzene	*o*-Dichlorobenzene
Bromobenzene	Toluene
Dibromomethane	Styrene
2,2-Dichloropropane	Ethyl benzene
1,1,1,2-Tetrachloroethane	*o*-Xylene
2,4,5-Trichlorophenoxy acetic acid	*m*-Xylene
Endrin	*p*-Xylene
Phthalate(s)	Cyanazine
Adipate(s)	1,1,2-Trichloroethane
PAH(s)	Isophorone
Hexachlorocyclopentadiene	Trifluralin
Diacamba	Methylene chloride
Simazine	*trans*-1,3-Dichloropropene
Atrazine	*p*-Chlorotoluene
Diquat	1,3-Dichloropropane
Endothall	2,4-Dinitrotoluene
Picloram	1,2,3-Trichloropropane

conducting pilot tests have sufficient data to confidently predict full-scale operations. There is not, however, much information on how well computer models will predict full-scale operations.

DWRD has evaluated microcolumn modeling techniques, focusing on verifying the results of available computer models. DWRD conducted microcolumn tests and used existing models to estimate carbon usage rates and other design parameters for a full-scale operation. The modeled results were then compared to isotherm model predictions and to actual results obtained in the field. The investigations concluded that microcolumn predictions offered no particular advantage over isotherm predictions (Speth and Miltner, 1989). Some computer models also contain a component that estimates the cost of various sized full-scale systems (Adams et al., 1989). Several of these models are in the process of being validated by DWRD.

8.2.1.2 Field Tests of Granular Activated Carbon Systems

DWRD has conducted extensive field studies on the potential of GAC to remove organic contaminants from surface water. It was found that carbon adsorption is an effective means of removing both volatile and synthetic organic compounds. Field testing sites included Jefferson Parish, Louisiana; Cincinnati, Ohio; Manchester, New Hampshire; Evansville, Indiana; Miami, Florida; Huntington, West Virginia; Beaver Falls, Pennsylvania; and Passaic, New Jersey (Lykins et al., 1984).

More recent studies of GAC systems have focused on organics removal from ground water at three sites: Suffolk County, New York; San Joaquin Valley, California; and Wausau, Wisconsin. A brief description of these ground-water treatment studies follows (Clark et al., 1988b).

The purpose of the Suffolk County project was to study the removal of organic compounds, pesticides, and nitrates under low-flow situations (simulating residential usage). GAC in combination with an ion-exchange system, and a reverse osmosis system were operated in parallel at this test site. Data on treatment effectiveness and system costs were collected and analyzed to assist in the design and testing of large public water supply systems. These results will be especially useful to operators of water treatment systems in farming communities where pesticides, nitrates, and other organics are a common problem.

The field study in the San Joaquin Valley examined point-of-use GAC units servicing small treatment systems and private wells of homeowners and farmers. The purpose of the project was to develop cost-effective design and operating guidelines for small GAC systems that are capable of removing pesticides (including dibromochloropropane) from drinking water supplies.

The third site, Wausau, has ground water with high levels of contaminants from a nearby Superfund site. Air stripping was studied as a companion technology to GAC at this site. Modeling techniques were used to predict full-scale design criteria, and the modeled estimates were verified with actual cost and performance data obtained from GAC contractors.

8.2.2 Ozone Oxidation Systems

Ozone oxidation as a method of removing VOCs was studied in DWRD's in-house pilot plant facilities (Fronk, 1987a). To date, tests on 29 VOCs in distilled and ground water revealed that ozone is effective in removing aromatic compounds, alkenes, and certain pesticides. It is not effective, however, in removing alkanes.

Research showed that removal of some alkanes was enhanced by increasing pH. Also, increasing the dosage of ozone improves removal of alkenes and aromatics. For most compounds, the composition of the water (the background matrix) did not significantly affect the removal efficiencies of ozone. In general, it was found that, when treated with ozone, VOCs behaved similarly in aqueous solutions as in the gaseous or liquid phases, as the literature concluded.

In order to further evaluate the effectiveness of using ozone to oxidize VOCs, DWRD funded a cooperative agreement with the Los Angeles Department of Water and Power (LADWP) (Glaze and Kang, 1988). The oxidation of TCE and PCE by ozone and modification of the traditional ozone treatment with the addition of hydrogen peroxide were studied at the bench scale. Results from this project were very promising and indicate that the kinetics of the H_2O_2/O_3 process warrants further research.

8.2.3 Ultraviolet Treatment

Ultraviolet light in combination with ozone may also be an effective means of controlling organic compounds. DWRD is conducting a study with the Los Angeles Water and Power Company to determine whether these two technologies can successfully treat organics in ground water (Glaze and Kang, 1988). The goal of this treatment technique is to oxidize undesirable chemicals into carbon dioxide and water, thus eliminating the need to control off-gases, which are a problem with other oxidation processes.

8.2.4 Reverse Osmosis

DWRD is conducting experiments at the bench and pilot scale to evaluate the potential for controlling volatile and synthetic organic compounds using reverse osmosis. Some field tests have been conducted as well. Preliminary results indicate that, for ground-water treatment systems, certain membranes are very effective in removing a wide range of organic chemicals (Fronk, 1987b).

Reverse osmosis has been proven as an effective treatment technology for radium. Several years ago, EPA conducted field studies in cooperation with Sarasota County, Florida (Sorg et al., 1980). Six different types of reverse osmosis systems varying in size from 3 m^3/day to 0.044 m^3/sec (800 GPD to 1 MGD) were operated in eight locations. Both hollow fiber and spiral wound cellulose acetate membranes were tested. The results of the study indicate that 82 to 96 percent of radium-226 was removed from all systems, thus meeting the EPA MCL of 5 pCi/L.

Reverse osmosis systems have also been tested for removing nitrate. In Suffolk County, where research on removing organic compounds from ground water has been ongoing, seven different commercially available membranes were tested for nitrate removal (Lykins and Baier, 1985). Removals obtained ranged from 75 to 95 percent. Polyamide membranes were efficient in reducing nitrate and SOCs.

A field study in Charlotte Harbor, Florida, tested reverse osmosis under conditions of high pressure (18.6 to 25.2 kg/cm^2 [265 to 359 psig]) (Huxstep, 1981). Removal of nitrate and other inorganic contaminants added to the ground-water influent were monitored. The investigation showed that a high pressure system could obtain better removals of all substances monitored. For nitrate, the high-pressure system attained an 80 percent removal as opposed to the low-pressure system, which attained 6 to 24 percent removal.

8.2.5 Ultrafiltration

Ultrafiltration is being studied as a method of reducing production of TTHMs in small water treatment systems (Taylor et al., 1987). This study involved treating two ground-water systems that contained large amounts of natural organic contaminants that act as TTHM precursors. Both produced more than 400 g/L TTHMs when treated with conventional systems. A low-pressure membrane system to reduce precursors to TTHMs was tested in a pilot plant. During pilot plant tests, the ultrafiltration system produced finished water that easily met the TTHM MCL (100 g/L). Since the tests were successful and the costs of the system were considered reasonable, more extensive testing of this technology is under way.

8.2.6 Packed Tower Aeration

Three field studies were recently completed to study packed tower aeration (PTA) techniques, which have proven effective in removing VOC. In Wausau, Wisconsin, and Baldwin Park, California, research focused on the effectiveness of PTA in removing VOCs from ground water (Clark et al., 1989a). Off-gas control technologies were tested at both sites as well (Crittenden et al., 1988; Lang et al., 1987). In Wausau, it was found that air stripping was effective in removing compounds that normally are expected to be difficult to remove (compounds have lower Henry's Law constants) by air stripping.

At the third site, Brewster, New York, a pilot-scale project evaluating air stripping techniques was completed. Results were used to develop models that consulting engineers could use to estimate costs and performance of full-scale aeration systems (Wallman and Cummins, 1986).

8.2.7 Conventional Treatment

Although conventional treatment is not very effective in removing organic contaminants, field studies were conducted to test the effectiveness of adding powdered activated carbon (PAC) to water prior to conventional treatment (Miltner et al., 1989). These tests were conducted in Tiffin, Ohio, where river water containing seasonally high levels of pesticides from area farms was a problem. The technique was found to effectively remove SOCs.

8.2.8 Ion Exchange

Ion exchange has been studied as a means of removing nitrate and uranium. In field tests, anion exchange has proven effective in removing nitrates. A demonstration nitrate removal plant operated automatically in McFarland, California, for over 3 years (Lauch and Guter, 1986; Guter, 1982). The plant has a capacity of 0.044 m^3/sec (1 MGD) and consists of three anion exchange vessels designed to reduce nitrate-N levels to below 10 mg/L (the current MCL and State standard in California). In this plant, part of the influent stream bypassed the treatment vessels and was blended with the treated water before leaving the plant. The blended water meets the required limits for nitrate and, in addition, meets EPA's Secondary Regulations (40 CFR Part 143) for chloride (250 mg/L) and sulfate (250 mg/L).

Laboratory-, pilot-, and (small) full-scale tests have been conducted to examine the effectiveness of anion exchange in removing uranium (Reid et al., 1985). In the laboratory, more than 9,000 bed volumes were treated before breakthrough was observed on a ground water containing 300 µg/L uranium was passed through the anion exchange resin. In the field, twelve 0.007-m^3 (0.25-ft^3) anion exchange systems were installed at 10 sites (in New Mexico, Colorado, and Arizona) where uranium levels in the ground water exceeded 20 µg/L. The field results actually showed better removals than the laboratory tests; some systems had treated over 50,000 bed volumes of water when they were finally shut down.

A small 0.63-L/sec (10-GPM) full-scale anion exchange system has been shown to be very effective in removing 95 to 98 percent of 47 to 90 µg/L of uranium from a water supply serving a school in Jefferson County, Colorado (Jelinek and Sorg, 1988). The system is regenerated with salt about once every 2 to 3 months and the brine disposed at a school district owned wastewater treatment plant.

Ion exchange systems, using both weak and strong acid resins in the hydrogen form, have been studied as a means of controlling radium from ground water (Clifford et al., 1988). Both types of resins are capable of removing radium to below the MCL of 5 pCi/L. However, both systems require the pH of treated water to be adjusted by either air stripping or the combination of air stripping and caustic addition.

EPA has also sponsored tests on the removal of radium with the DOW Chemical Company's Radium-Selective Complexer (RSC), a synthetic resin with a high affinity for radium. Field tests have shown the RSC to remove radium from more than 50,000 bed volumes of treated water. However, because the RSC cannot be regenerated, it must be disposed when its capacity for radium is reached. For this reason, studies have been conducted to examine its efficiency for removing radium from the brine waste of a cation exchange system. A field study in Colorado showed the RSC to be very effective for concentrating the radium from the brine of an ion exchange system for over 2 years. When its capacity is reached, this material will be disposed at a low radioactive waste disposal site.

Recent EPA studies in Illinois have also shown that activated alumina modified with barium sulfate has a removal capability similar to the RSC. Plain activated alumina also removed radium but its capacity was about a fifth of the modified alumina. The advantage of the plain alumina was that it could be regenerated whereas the modified alumina could not.

The application of preformed manganese dioxide floc for radium removal was studied at the same field site in Illinois and this process (adsorption)showed promising results (Patel and Clifford, 1989). Feeding of 1 to 2 mg/L of manganese oxide as manganese in a contact tank before a multimedia filter showed 80 to 95 percent removal of radium (12 pCi/L) from the ground water. Cost estimates of this process indicate that it should be competitive with ion exchange treatment.

8.2.9 Technologies for Removing Radionuclides

DWRD has funded a project to study several treatment techniques for removing radon gas from drinking water (Kinner et al., 1987a). Packed tower aeration, diffused bubble aeration, and GAC have been evaluated at a field site in New Hampshire. Both the packed tower and the diffused bubble aeration systems were very efficient at removing radon (90 to 99 percent). The GAC system flow rate and influent radon concentration above design loadings were attributed to the lower removals by the GAC. This same study also investigated the removal of radon by several low cost technologies associated with storage. The simple techniques consisted of spraying the water into a storage tank, simple aeration (bubble) in the bottom of the storage tank, and the combination of these methods (Kinner et al., 1987b). These methods showed removals ranging from 50 percent by free fall of the water into the tank to around 90 percent with the combination of free fall and bubble aeration.

8.2.10 Secondary Sources of Pollution

Drinking water treatment technologies are designed to remove pollutants from water. Once removed, however, the pollutants can create a secondary source problem. Research on GAC systems treating surface water found that dioxins were formed as by-products of the reactivation process (Lykins et al., 1988a). Afterburners were installed and operated at 1315.6°C (2,400°F) to eliminate these toxic gases.

Harmful compounds in air stripping of waste gases can also be a problem. The field studies of air stripping technologies at Baldwin Park, California, and Wausau, Wisconsin, included testing of gas-phase carbon adsorption techniques to remove synthetic and volatile organic compounds from waste gases (Lang et al., 1987; Hand et al., 1986). At Baldwin Park a higher waste gas stack is also being used to increase dispersion of waste gases.

8.2.11 Small Systems Technologies

Small systems account for most of the MCL violations under the Safe Drinking Water Act. DWRD has conducted and is continuing to conduct studies of technology that is specifically designed to solve small system problems. A recently completed study demonstrated the feasibility of using point-of-entry devices for controlling arsenic and fluoride in a municipally sound treatment system (Rogers, in press).

8.3 Mandatory Disinfection

The proposed Surface Water Treatment Rule will require communities whose water supply comes from surface water or is influenced by surface water (e.g., ground water that is stored in open reservoirs) and supplies more than 25 persons to filter and disinfect drinking water before distribution. Upon promulgation of the final rule, all affected suppliers must comply within 3 years. The Ground Water Treatment Rule, anticipated in 1991, is expected to apply similar requirements to ground-water drinking sources.

Historically, efficacy of disinfection has been measured by testing for the presence of coliform bacteria. The new regulations, however, specify requirements to control and regulate levels of *Giardia*, viruses, *Legionella*, and heterotrophic bacteria. In view of these requirements, research (extramural and in-house) is being conducted on methods to achieve adequate disinfection of the specified microorganisms. Inactivation studies using chlorine and chloramine have been completed on particle-associated coliforms, and on inactivation of three coliphages using chlorine, chlorine dioxide and monochloramine (Berman et al., 1988; Berman and Sullivan, 1988). Disinfection studies on inactivation of hepatitis A virus and model viruses in water are ongoing; as are studies on inactivation of *Giardia* (Ohio State University) using chlorine, chlorine dioxide, and chloramine; and *Cryptosporidium* (University of Arizona) using chlorine and chloramines (Sobsey et al., 1988). Results of these investigations will provide contact time (CT) values for the various organisms and disinfectants based on temperature, pH, and disinfectant dosage (Hoff, 1986; Clark et al., 1989b). An in-house project will examine the influence of strain variations in *Giardia lamblia* on disinfection inactivation.

A study is being conducted on the inactivation of viruses and *Giardia* using chlorine and chloramines in a flow-through treatment plant (John Hopkins University).

The use of alternative disinfectants, such as ozone applied as a preoxidant, has been shown to increase the level of readily assimilable organic carbon (AOC) in treated water. The trend toward the use of ozone and alternative disinfectants to control THM formation in the United States may result in increased problems of bacterial regrowth, and possible coliform MCL compliance during water storage and distribution. Investigations are under way to examine AOC levels in treated drinking water and the relationship between microbial regrowth (coliforms) and AOC, and a recently completed project examined modifications of a procedure developed in the Netherlands to determine AOC levels in drinking water (Kaplan and Bott, 1989).

Control of *Legionella pneumophilia* in drinking water treatment through prevention of the *Legionella*/free-living amoebae association is also being examined.

8.3.1 Treatment/Distribution Microbiology

Examination of factors involved in microbial water quality have been addressed in other completed studies. Point-of-use or point-of-entry water treatment may have possible applications where central treatment is not feasible for various reasons. Microbiological studies of point-of-entry treatment devices, used primarily for aesthetic purposes (taste and odor removal) have shown that the GAC cartridge devices become bacterial generators, regardless of whether they contain silver as a bacteriostatic agent (Reasoner et al., 1987a,b).

The use of phosphate corrosion control additives on bacterial growth was examined to determine the role of added phosphate on coliform occurrence problems (Rosenzweig, 1987). GAC fines were shown to be a means by which particle-associated bacteria, including coliforms, can pass from treatment into the distribution system (McFeters et al., 1987). Particle-associated coliforms have also been shown to be less susceptible to disinfection than those that are not particle-associated.

8.3.2 Bacterial Detection/Monitoring

A variety of studies, completed and ongoing, have examined various aspects of the bacteriological quality of drinking water. Studies of detection methods for bacteria in potable water, and for monitoring coliform bacteria in potable water include the development and evaluation of a composite sampling protocol for coliforms in treated and distribution water, comparison of media for recovering total coliforms, detection of *Klebsiellain* water systems, the presence-absence coliform test for monitor-drinking water quality, a radiometric method for detection of fecal coliforms in water, and occurrence of pigmented bacteria in treated and distributed drinking water. An overview that summarizes much of drinking water microbiology research during the past decade is available (Reasoner, 1988).

8.4 Prohibition of Lead Materials

The amendments to SDWA prohibit the use of lead in pipes, pipe solder, or any other apparatus that comes in contact with drinking water. On August 18, 1988, the Office of Drinking Water published in the Federal Register a proposed new regulation for controlling lead and copper in drinking water (U.S. EPA, 1988a). This document has drawn a considerable response, and a lot of work and manpower is being devoted to technical issues surrounding this legislation. On October 31, 1988, the Lead Contamination Control Act (LCCA) was enacted as an amendment to SDWA (Pub. Law 92-339). The LCCA added Section 1464, which requires EPA to publish and distribute to the States a guidance document and testing protocol to assist schools in determining the source of lead contamination in school drinking water.

The Drinking Water Research Division has funded a major study in Long Island on the impact of lead and other metallic solders on water quality. Ninety test sites were selected to provide ten sites in each of nine age groups ranging from new construction to construction more than 20 years old. During the study, the pH was adjusted to three levels. The results of this study indicated that leaching of lead into drinking water is greatly affected by pH and the age of a home. The leaching was the highest for new homes using low pH water. Furthermore, first draw samples after overnight detention times had the highest amounts of lead. A project report will summarize the results of the study.

In-house research includes studying lead contamination from kitchen faucets and in water coolers. Of the 12 faucets used in this study, the tree cast brass faucets contributed significant amounts of lead to the drinking water. Traces of lead were obtained from the other metallic faucets and a plastic faucet was lead free. The results of this study were presented at the Water Quality Technology Conference in November 1988, and have been accepted for publication in the July 1989, issue of the *Journal of the American Water Works Association* (Gardels and Sorg, 1989).

Twenty-two water coolers supplied by the U.S. Navy and two from a Portland, Maine, School District were used in the lead studies. Several coolers contributed lead to the drinking water in amounts considerably above the present standard. At the end of the leaching period, the reservoirs were cut open and several had lead-lined tanks. The results of this study appeared in the April 10, 1989, Federal Register entitled "Drinking Water Coolers that Are Not Lead Free" (U.S. EPA, 1989c). This document and the one entitled "Lead Contamination in School Drinking Water Supplies" were issued pursuant to the Lead Contamination Control Act (U.S. EPA, 1989d). Some chemical analyses and x-ray diffraction examinations of internal deposits were performed on two of the reservoirs. The results will be presented as part of a review of lead corrosion research at the AWWA Annual Conference in June 1989.

Initiation of new in-house research into the effect of orthophosphate on lead solubility under different water qualities, determination of corrosion products and inhibitor films in lead pipes from field sites, integrating new data into corrosion models, and the effect of water softening on corrosivity towards lead and copper is planned. DWRD is also participating in a study to gather baseline data and look at the impact of municipal softening on water corrosivity in Oakwood, Ohio, where there are still many lead service lines.

8.5 Systems and Cost Modeling Studies

The DWRD has been modeling the deterioration of water quality in distribution systems. Subsequent to water leaving the treatment plant, conditions in the distribution system sometimes lead to the introduction of contaminants to the water or to the occurrence of other types of water quality decline. The research in this area is focused on computer modeling to predict and assist in alleviating distribution system problems (Clark et al., 1988c,d).

In addition to field and laboratory research, DWRD is developing a computer model that will predict water quality based on the known characteristics of the distribution system. For example, TTHM levels can be projected throughout the distribution system based on the time of travel and the mixing of water from various sources (Eilers and Clark, 1988).

Other modeling efforts include a project being conducted in cooperation with French scientists investigating the effects of hydraulics on the propagation of contaminants in distribution systems. Case studies have been conducted to develop models that analyze distribution system characteristics and identify factors that contribute to the potential for water main breaks (Clark et al., 1988a). Some models estimate costs of repairs and other system renovations.

DWRD has conducted cost modeling for many years. Cost data have been collected from design studies and merged with performance data to develop cost and performance models. These models are very useful in estimating the behavior of full-scale systems (Adams et al., 1989).

8.6 Future Directions

As the Safe Drinking Water Act matures, areas of major research will be to complete activities in the areas mentioned to this point. In addition, increasing emphasis will be placed in the following areas:

- Use of ozone as a disinfectant and oxidant. Particular emphasis will be placed on the formation and control of disinfection by-products and the maintenance of microbiological integrity of distribution systems.
- Effects of the use of corrosion inhibitors as a means of controlling lead leaching from household plumbing.
- Characterization and treatment of residuals from drinking water unit processes.
- Development of increasingly sophisticated cost and performance models.

Chapter 9

References

Abdel-Rahman, M.S., D. Couri, and J.D. Jones. 1980 Chlorine dioxide metabolism in rats. J. Environ. Pathol. and Toxicol. 3 421-430.

Adams, J.Q., and R.M. Clark. 1989. Cost estimates for GAC treatment systems. JAWWA. 81(1):35-42. January.

Adams, J.Q., R.M. Clark, and R.J. Miltner. 1989. Controlling organics with GAC: a cost and performance analysis. JAWWA. 81(4):132-140. April.

Aieta, E.M., and J.D. Berg. 1986. A review of chlorine dioxide in drinking water treatment. J. Am. Water Works Assoc. 78:62-72.

Aieta, E.M., K.M. Reagan, J.S. Lang, L. McReynolds, J.-W. Kang, and W.H. Glaze. 1988. Advanced oxidation processes for treating ground water contaminated with TCE and PCE: pilot-scale evaluations. JAWWA. 80(5):64-72. May.

Aieta, E.M., and P.V. Roberts. 1986. Application of mass-transfer theory to the kinetics of a fast gas-liquid reaction: chlorine hydrolysis. Environ. Sci. Technol. 20:44-50.

Akin, E. et al. 1987. The U.S. EPA Office of Research and Development Health Research Program on Drinking Water Disinfectants and their By-Products. Issue paper prepared for a Science Advisory Board Program Review.

American Water Works Association. 1973. Water Chlorination Principles and Practice. AWWA Publication No. M20, p. 12.

American Water Works Association Research Foundation and KIWA. 1983. Occurrence and removal of volatile organic chemicals from drinking water.

Augenstein, H.W. 1974. Use of chlorine dioxide to disinfect water supplies. J. Am. Water Works Assoc. 66:716-717.

Berman, D., E.W. Rice, and J.C. Hoff. 1988. Inactivation of particle-associated coliforms by chlorine and monochloramine. Appl. Environ. Microbiol. 54(2):507-512.

Berman, D. and R. Sullivan. 1988. Inactivation of three coliiphages by chlorine, chlorine dioxide, and monochloramine at 5°C. Project Report WERL-C1-0807. EPA/2-88/039. U.S. EPA Risk Reduction Engineering Laboratory. Cincinnati, OH.

Bryck, J., B. Walker, and G. Mills. 1987. *Giardia* removal by slow sand filtration - pilot to full scale. Paper presented at the AWWA National Conference. June.

Chrobak, R.S., D.L. Kelleher, and I.H. Suffet. 1985. Full-scale GAC adsorption performance compared to pilot plant predictions. Paper presented at the 1985 Annual Conference of the American Water Works Association, Washington, DC. June 27.

Clark, R.M., R.J. Miltner, C.A. Fronk, and T.F. Speth. 1989a. Unit process research for removing VOCs from drinking water: an overview in significance and treatment of volatile organic compounds in water supplies. Lewis Publishers, Inc., Chelsea, Michigan.

Clark, R.M., E.J. Read, and J.C. Hoff. 1989b. Analysis of inactivation of *Giardia lamblia* by chlorine. Journal of Environ. Eng. ASCE. 115(1):80-90. February.

Clark, R.M., R.G. Eilers, and J.A. Goodrich. 1988a. Distribution system: cost of repair and replacement. *In*: Proceedings of the Conference on Pipeline Infrastructure, ASCE, Boston, MA. June 6-7.

Clark, R.M., C.A. Fronk, and B.W. Lykins, Jr. 1988b. Removing organic contaminants from ground water. Environ. Sci. Tech. 22(10):1126-1130.

Clark, R.M., W.M. Grayman, and R.M. Males. 1988c. Contaminant propagation in distribution systems. J. of Environ. Eng. ASCE. 114(4):929-943.

Clark, R.M., W.M. Grayman, R.M. Males, and J.A. Coyle. 1988d. Modeling contaminant propagation in drinking water distribution systems. Aqua. No. 3, pp. 137-151.

Clifford, D., W. Jijjeswaropu, and S. Subramonian. 1988. Evaluating various absorbents and membranes for removing radium from ground water. JAWWA. 80(7):94.

Clifford, D., S. Subramonian, and T. Sorg. 1986. Removing dissolved inorganic contaminants from water. Environ. Sci. and Tech. 20(11):1072-1080. November.

Cline, G.C., T.J. Lane, and M. Saldamando. 1985. Packed column aeration for trichlorethylene removal at Scottsdale, Arizona. Paper presented at the 1985 Annual Conference of the AWWA held in Washington, DC. June.

Collins, F. and C.Y. Shieh. 1971. More water from the same plant. The American City. p. 96. October.

Conley, W.R. and S.P. Hansen. 1987. Trident pilot study: Lakeport, California. Unpublished Report. HDR Engineering. March.

Crittenden, J.C., R.D. Cartright, B. Rick, S.R. Tang, and D. Perram. 1988. Using GAC to remove VOCs from air stripper off-gas. JAWWA. 80(5):64-72.

Cryer, E.T. 1986. Preozonation at Kennewick, Washington, Case history. Paper presented at Intl. Ozone Assoc. Workshop on Drinking Water Treatment with Ozone, Perrysburg, OH, April.

Culp, R.L. 1976. Increasing water treatment capacity with minimum additions. Public Works. August.

DeMarco, J. 1983. Experiences in operating a full-scale granular activated carbon system with on-site reactivation. *In*: I.H. Suffet and M.J. McGuire (ed.), Treatment of Water by Granular Activated Carbon. Advances in Chemistry Series, American Chemical Society Books.

DeWalle, F.B., J. Engeset, and W. Lawrence. 1984. Removal of *Giardia lamblia* cysts by drinking water plants. Report No. EPA/600/S2-84-069, U.S. EPA, MERL, Cincinnati, OH. May.

Dobbs, R.A. and J.M. Cohen. 1980. Carbon adsorption isotherms for toxic organics. Report No. EPA 600/8-80-023. Office of Research and Development, MERL, Cincinnati, OH.

Dyksen, J.E., K. Raman, R.F. Raczko, and R.M. Clark. 1987. GAC treatment costs - minimizing them. Paper presented at the AWWA Annual Conference, Kansas City, MO. June 14-18.

Eilers, R.G. and R.M. Clark. 1988. Flow separation conditions at pipe walls of water distribution mains. *In*: Proceedings of the AWWA Annual Conference, Orlando, FL. June 19-23. pp. 1619-1624.

Feige, W.A., R.M. Clark, B.W. Lykins, Jr., and C.A. Fronk. 1987. Treatment of water from contaminated wells. *In*: Rural Groundwater Contamination. Lewis Publishers, Inc.

Fronk, C.A. 1987a. Destruction of volatile organic contaminants in drinking water by ozone treatment. Ozone Science and Engineering. 9(3):265-287.

Fronk, C.A. 1987b. Removal of low molecular weight organic contaminants from drinking water using reverse osmosis membranes. *In*: Proceedings of the AWWA Annual Conference, Kansas City, MO/Denver, CO.

Gardels, M.C. and J.J. Sorg. 1989. A laboratory study of lead leaching from drinking water faucets. JAWWA. July.

Glaze, W.H. 1986. Reaction products of ozone: a review. Environ. Hlth Persp. Vol. 69, p. 151.

Glaze, W.H. and J.-W. Kang. 1988. Advanced oxidation processes for treating ground water contaminated with TCE and PCE: laboratory studies. JAWWA. 80(5):57-63. May.

Glaze, W.H., J.-W. Kang, and M. Aieta. 1987. Ozone-Hydrogen peroxide systems for control of organics in municipal water supplies. *In*: D.W. Smith and G.R. Finch (eds), Proc. Second Intl. Conf. on the Role of Ozone in Water and Wastewater Treatment. TekTran International Ltd., Kitchener, Ontario, Canada. pp. 233-244.

Glitsch, Inc. P.O. Box 660053. Dallas, TX.

Guter, Gerald A. 1982. Removal of nitrate for contaminated water supplies for public use. Final Report. Project Summary. Report No. EPA/600/S2-82-042. Cincinnati, OH.

Hand, D.W., J.C. Crittenden, J.L. Gerbin, and B.W. Lykins, Jr. 1986. Design and evaluation of an air stripping tower for removing VOCs from ground water. JAWWA. 78(9):87-97.

Hansen, S.P. 1987. Adsorption clarifier pilot study evaluation. Unpublished Report. Lake Arrowhead Community Services District, September.

Hoff, J.C. 1986. Inactivation of microbial agents by chemical disinfectants. Report No. EPA/600/2-86-067. Risk Reduction Engineering Laboratory, U.S. EPA, Cincinnati, OH.

Howe, E.W., E.M. Aieta, S. Liang, and M.J. McGuire. 1989. Removal of chlorite residuals with granular activated carbon: a case study. Presented at International Chlorine Dioxide Symposium, Denver, CO (Washington, DC: Chemical Manufacturers Association).

High-tech water on the green fields of Iowa. 1985. Water Engineering and Management. January.

Huxstep, Martin R. 1981. Inorganic contaminant removal from drinking water by reverse osmosis. Project Summary. Report No. EPA/600/S2-81-115. WERL, Cincinnati, OH.

Jelinek, R.T. and T.J. Sorg. 1988. Operating a full-scale ion exchange system for uranium. JAWWA. 8(7):79. July.

Kaplan, L.A. and T.L. Bott. 1989. Nutrients for bacterial growth in drinking water: bioassay evaluation. Report No. EPA/600/2-89/030. Risk Reduction Engineering Laboratory. U.S. EPA. Cincinnati, OH.

Kinner, N.E., C.E. Lassard, and G.S. Schell. 1987a. Low cost/low technology aeration techniques for removing radon from drinking water. Report No. EPA/600/M-87/031. U.S. EPA.

Kinner, N.E., C.E. Lassard, and G.S. Schell. 1987b. Radon removal from small community public water supplies using granular activated carbon and low technology/low cost techniques. *In*: Proceedings of the AWWA Seminar on Radionuclides in Drinking Water, Kansas City, MO.

Lang, J.S., A.R. Trussel, M.D. Umphries, and C.H. Tate. 1987. Gas adsorption of air stripping emissions. *In*: Proceedings of AWWA Seminar on Treatment Processes for the Control of Synthetic Organic Chemicals, Kansas City, MO.

Lange, K.P., R. T. Rice, and W. Hemphill. 1985. Integration of contact clarifier and filtration processes, Microfloc Product. Paper presented at the 1985 Rocky Mountain AWWA/WPCA Conference, Albuquerque, NM.

Lange, K.P., W.D. Bellamy, and D.W. Hindricks. 1984. Filtration of *Giardia* cysts and other substances. Vol. 1: Diatomaceous Earth Filtration. Report No. EPA/600/2-84-114. U.S. Environmental Protection Agency, MERL, Cincinnati, OH. June. NTIS: PB-84-212703.

Lauch, R.P. and G.A. Guter. 1986. Ion exchange for the removal of nitrate from well water. JAWWA. 78(5):83-88.

Letterman, R.D. and T.R. Cullen, Jr. 1985. Slow sand filtration maintenance costs and effects on water quality. Report No. EPA/600/2-85-056. U.S. Environmental Protection Agency, Cincinnati, OH. May. NTIS: PB-85-199669/AS.

Lin, S.D. 1989. Proposed lead rules add chores, costs to utility burden. Water Engineering and Management. pp. 35-47. February.

Lowry, J.D. and J.E. Brandown. Removal of radon from ground water supplies using granular activated carbon or diffused aeration. University of Maine, Department of Civil Engineering, Orono, ME.

Lykins, B.W., Jr., R.M. Clark, and J.Q. Adams. 1988a. Granular activated carbon for controlling THMs. JAWWA. pp. 85-92. May.

Lykins, B.W., Jr., R.M. Clark, and D.H. Cleverly. 1988b. Polychlorinated dioxin and furan discharge during carbon reactivation. Journal of Environ. Eng. ASCE. 114(2):300-316. April.

Lykins, B.W., Jr. and M.H. Griese. 1986. Using chlorine dioxide for trihalomethane control. J. Am. Water Works Assoc. 78:88-93.

Lykins, B.W., Jr. and J.A. Baier. 1985. Removal of agricultural contaminants from ground water. In: Proceedings of the AWWA Conference. pp. 1151-1164.

Lykins, B.W., Jr., E.E. Geldreich, J.Q. Adams, J.C. Ireland, and R.M. Clark. 1984. Granular activated carbon for removing nontrihalomethane organics from drinking water. Report No. EPA/600/S2-84-165.

McFeters, G.A., A.K. Camper, M.W. LeChevallier, S.C. Broadway, and D.G. Davies. 1987. Bacteria attached to granular activated carbon in drinking water. Environmental Research Brief. Report No. EPA/600/M-87/003. Risk Reduction Engineering Laboratory, U.S. EPA, Cincinnati, OH.

McKeon, W.R., J.J. Muldowney and B.S. Aptowicz. 1986. The evolution of a modified strategy to reduce trihalomethane formation. *In*: Proc. Annual Meeting, Am. Water Works Assoc., Denver, CO. pp. 967-997.

McKinnon, R.J. and J.E. Dyksen. 1982. Aeration plus carbon adsorption remove organics from Rockaway Township (NJ) ground water supply. Paper presented at the 1982 Annual Convention of the American Society of Civil Engineers, New Orleans, LA. October 25-27.

Miller, G.W., R.G. Rice, C.M. Robson, R.L. Scullin, W. Kuhn and H. Wolf. 1978. An assessment of ozone and chlorine dioxide technologies for treatment of municipal water supplies. EPA Report No. EPA/600/2-78-147.

Miltner, R.J. and C.A. Fronk. 1985. Treatment of synthetic organic contaminants for Phase II Regulations. Internal Report. DWRD, U.S. EPA. December.

Miltner, R.J., T.F. Speth, and C.A. Fronk. 1989. Treatment of seasonal pesticides in surface waters. JAWWA. 81(1):43-52. January.

Morand, J.M. and M.J. Young. 1983. Performance characteristics of package water treatment plants. Project summary. Report No. EPA/600/S2-82-101, Municipal Environmental Research Laboratory, Cincinnati, OH. March.

Morse, D. 1988. Lead: taking it from the tap. Civil Engineering, pp. 71-73.

Myers, G.L., A. Thompson, D.M. Owen and J.M. Baker. 1986. Control of trihalomethanes and taste and odor at Galveston County Water Authority. *In*: Proc. Annual Meeting Am. Water Works Assoc., Denver, CO, 1986, pp. 1667-1675.

National Academy of Sciences. 1980. Drinking Water and Health. Vol. 1, Chapter 2. Washington, D.C., National Academy Press.

O'Brien, R.P., D.M. Jordan, and W.R. Musser. 1981. Trace organics removal from contaminated ground waters with granular activated carbon. Calgon Corporation, Pittsburg, PA. Paper presented at the National American Chemical Society Meeting, Atlanta, GA. March 29 - April 3.

Patel, R. and D. Clifford. 1989. Radium removal from water by manganese dioxide adsorption and diatomaceous earth filtration. Draft Report. U.S. EPA.

Pinsky, D.E. and V.F. Coletti. 1988. Rehabilitating a New England water treatment plant. J. New England Water Works Assoc., pp. 132-157. June.

Pyper, G.R. 1985. Slow sand filter and package treatment plant evaluation operating costs and removal of bacteria, *Giardia*, and trihalomethanes. Report No. EPA/600/2-85-052. U.S. Environmental Protection Agency, Cincinnati, OH. April.

Randtke, S. 1988. Organic contaminant removal by coagulation and related processes. JAWWA. 80(5):40-56. May.

Reasoner, D.J. 1988. Drinking water microbiology research in the United States: an overview of the past decade. Water Sci. Technol. 20(11/12):101107.

Reasoner, D.J., J.C. Blannon, E.E. Geldreich, and J.A. Barnick. 1987a. Microbiological studies of activated carbon point-of-use systems: an interim report. Report No. EPA/600/2-88/004. Risk Reduction Engineering Laboratory, U.S. EPA, Cincinnati, OH.

Reasoner, D.J., J.C. Blannon, and E.E. Geldreich. 1987b. Microbiological characteristics of third-faucet point-of-use devices. JAWWA. 79(10):60-66.

Reid, George N., P. Lassovszky, and S. Hathaway. 1985. Treatment, waste management and cost for removal of radioactivity from drinking water. Health Physics. 48(5):671-694.

Roberts, P.V. and P. Dandliker. 1982. Mass transfer of volatile organic contaminants during surface aeration. Paper presented at the 1982 Annual Conference of the AWWA, Miami Beach, FL. May.

Rogers, K.R. in press. Arsenic and fluoride removal from drinking water at San Ysidro, N.M. Risk Reduction Engineering Laboratory, Office of Research and Development, U.S. EPA, Cincinnati, OH.

Rosenzweig, W.D. 1987. Influence of phosphate corrosion control compounds on bacterial growth. Report No. EPA/600/2-87/045. Risk Reduction Engineering Laboratory, U.S. EPA, Cincinnati, OH.

Routt, J.C. 1989. Chlorine dioxide trihalomethane control and distribution system odor problem: a utility's 5-year experience. Paper presented at the AWWA Annual Meeting, Los Angeles, CA. June.

Roy, D., R.S. Englebrecht, and E.S.K. Chian. 1982. Comparative inactivation of six enteroviruses by ozone. JAWWA. 74(12):660-664.

Ruggiero, D.D. and R. Ausubel. 1982. Removal of organic contaminants from drinking water supply at Glen Cove, New York, Phase II. Report No. EPA/600/2-82/027, U.S. EPA. March.

Ruggiero, D.D. and R. Ausubel. 1980. Removal of organic contaminants from drinking water supply at Glen Cove, New York, Phase I. Report No. EPA600/2-80-198.

Schock, M.R. 1980. Response of lead solubility to dissolved carbonate in drinking water. JAWWA. Vol. 72, p. 695.

Schock, M.R. and I. Wagner. 1985. Corrosion and solubility of lead in drinking water: Internal corrosion of water distribution systems. AWWA Research Foundation/DVGW.

Sequeira, J.A., L. Harry, S.P. Hansen, and R.L. Culp. 1983. Pilot filtration tests at the American River water treatment plant. Public Works. 114:1:36.

Singer, P.S. 1988. Alternative oxidant and disinfectant treatment strategies for controlling trihalomethane formation. Report No. EPA/600/S2-88/044, Water Engineering Research Laboratory, Cincinnati, Ohio. October.

Singer, P.S. 1986. THM control using alternate oxidant and disinfectant strategies: an evaluation. *In*: Proc. 1986 Annual Conference, Am. Water Works Assoc. Denver, CO. pp. 999-1017.

Slootmaekers, B., S. Tachiyashiki, D. Wood, and G. Gordon. 1989. The removal of chlorite ion and chlorate ion from drinking water. Presented at International Chlorine Dioxide Symposium, Denver, CO, November (Washington, DC: Chemical Manufacturers Association).

Sobsey, M.D., T. Fugi, and P.A. Shields. 1988. Inactivation of hepatitis A virus and model viruses in water by free chlorine and monochloramine. *In*: Proceedings of International Conference on Water and Wastewater, Microbiology, Newport Beach, California. 2(64):1-7.

Sobsey, M.D. 1988a. Detection and chlorine disinfection of hepatitus A in water. CR-813-024, EPA Quarterly Report. December 1988.

Sorg, T.J. 1979. Treatment technology to meet the interim primary drinking water regulations for inorganics, Part IV: chromium and mercury. JAWWA. pp. 454-466. August.

Sorg, T.J. 1978. Treatment technology to meet the interim primary drinking water regulations for inorganics, Part I: nitrite and fluoride. JAWWA.pp. 105-112. February.

Sorg, T.J. and G. Logsdon. 1980. Treatment technology to meet the interim primary drinking water regulations for inorganics, Part V: barium and radionuclides. JAWWA. pp. 411-422. July.

Sorg, T.J. and G. Logsdon. 1978. Treatment technology to meet the interim primary drinking water regulations for inorganics, Part II: arsenic and selenium. JAWWA. p. 379-393. July.

Sorg, T.J., R. Forbes, and D. Chambers. 1980. Removal of radiu-226 from Sarasota County, Florida's drinking water by reverse osmosis. JAWWA. pp. 230-237. April.

Sorg, T.J., M. Csanady, and G. Logsdon. 1978. Treatment technology to meet the interim primary drinking water regulations for inorganics, Part III: cadmium, lead, and silver. JAWWA. pp. 680-691. December.

Speth, T.A. and R.J. Miltner. 1989. Effect of preloading on the scale-up of GAC microcolumns. JAWWA. 81(4):141-148. April.

Stevens, A.A., L.A. Moore, and R.J. Miltner. 1989. Formation and control of nontrihalomethane by-products. *In*: Proceedings of the AWWA Water Quality Technology Conference, St. Louis, Missouri.

Stevens, A.A., L.A. Moore, C.J. Slocum, B.L. Smith, D.R. Seeger, and J.C. Ireland. 1987. By-products of chlorination at ten operating utilities. *In*: Proceedings of the Sixth Conference on Water Chlorination: Environmental Impact and Health Effects. Oak Ridge Associated Universities, Oak Ridge, Tennessee. May 3-8.

Tanner, S. 1988. Evaluation of Slow Sand Filters in Northern Idaho. Idaho Division of Environmental Quality. May 9.

Taylor, J.S., D.M. Thompson, J.K. Carswell. 1987. Applying membrane processes to ground water sources for THM control. JAWWA. 79(8):72-82.

U.S. EPA. 1989. National Primary and Secondary Drinking Water Regulations. Proposed Rule. Fed. Reg. 54(97):22064. May 22.

U.S. EPA. 1989a. Guidance manual for compliance with the filtration and disinfection requirements for public water systems using surface water sources. Draft. Science and Technology Branch, Office of Drinking Water, U.S. EPA, Washington, DC, October.

U.S. EPA, 1989b. Drinking water; national primary drinking water regulations; filtration, disinfection, turbidity, *Giardia lamblia*, viruses, *Legionella*, and heterotrophic bacteria. Final Rule. Fed. Reg. 54(124):27485-27541. June 29.

U.S. EPA. 1989c. Drinking water coolers that are not lead free. Proposed List and Request for Comments, Fed. Reg. 54(67):14320-14322. April 10.

U.S. EPA. 1989d. Lead contamination in school drinking water supplies. Guidance Document and Testing Protocol. Notice of Availability and Request for Comments. Fed. Reg. 54(67):14316-14318. April 10.

U.S. EPA. 1988a. Drinking water regulations: Maximum Contaminant Level Goals and Natural Primary Drinking Water Regulations for lead and copper. Proposed Rule. Fed. Reg. 53(160):31516. August 18.

U.S. EPA. 1988b. Technologies and costs for the removal of synthetic organic chemicals from potable water supplies. Draft. U.S. EPA, Office of Drinking Water, Washington, DC. September.

U.S. EPA. 1988c. Drinking water; substitution of contaminants and drinking water priority list of additional substances which may require regulation under the Safe Drinking Water Act. Federal Register 53(14):1891-1902.

U.S. EPA. 1987. Technologies and costs for the removal of microbiological contaminants from potable water supplies. Draft. Office of Drinking Water, Criteria and Standards Branch.

U.S. EPA. 1986a. Safe Drinking Water Act 1986 Amendments. Report No. EPA/570/9-86-002. Office of Drinking Water. August.

U.S. EPA. 1986b. Design manual: municipal wastewater disinfection. Report No. EPA/625/1-86/021 (October 1986). U.S. EPA, Office of Research and Development, Water Engineering Research Laboratory, Center for Environmental Research Information, Cincinnati, OH 45268.

U.S. EPA. 1984. Corrosion manual for internal corrosion of water distribution systems. Report No. EPA/570/9-84-001. Office of Drinking Water. April.

U.S. EPA. 1983. Microorganism removal for small water systems. Report No. EPA/570/9-83-012. U.S. EPA, Office of Drinking Water, Washington, D.C. June.

U.S. EPA. 1979. Estimating water treatment costs. Volumes 1-4. Report No. EPA/600/2-79-162a. Municipal Environmental Research Laboratory, Office of Research and Development, U.S. EPA, Cincinnati, OH. August.

U.S. EPA. 1973. Process design manual for carbon adsorption. Technology Transfer. October.

Wallman, H. and M.C. Cummins. 1986. Design scale-up suitability for air stripping columns. Report No. EPA/600/S2-86-009. U.S. EPA, Drinking Water Research Division, Cincinnati, OH.

Westerhoff, G.P. 1971. Experiences with higher filtration rates. JAWWA. 63:6:376.

Westerhoff, G.P. and R. Miller. 1985. Design of the nation's first major GAC facility for drinking water treatment at Cincinnati, Ohio. Paper presented at the 1985 Annual Conference of the AWWA held in Washington, DC. June.

Williams, R.B. and G.L. Culp. 1986. Handbook of Public Water Systems. New York, New York, Van Nostrand Reinhold Company.

Wolfe, R.L., M.H. Stewart, K.N. Scott and M.J. Mcguire. 1989. Inactivation of *Giardia muris* and indicator organisms seeded in surface water supplies by peroxone and ozone. Env. Sci. Technol. 23(6):744-745.

Experience Modifying Existing Filtration Systems

This Appendix provides brief case histories of modifications resulting in plant expansions (Section A.1) and installation of different filtration systems, including slow sand (Section A.2), package plant (Section A.3), and diatomaceous earth (Section A.4) filters. Experience selecting a filtration system is provided in Section A.5.

A.1 Upgrading Existing Treatment Facilities

A.1.1 Horizontal Flow Basin Example

This example involved a modification of existing facilities. The plant equipment includes a rectangular sedimentation basin and rapid sand filters with a capacity of 0.18 m^3/sec (4 MGD). The basin is 9 meters (30 ft) wide, 40.5 meters (133 ft) long, and averages 4.5 meters (15 ft) in depth. The design surface overflow rate is 40.7 m^3/day/m^2 (1,000 GPD/sq ft). The sedimentation basin has a single overflow weir across the outlet end. Sedimentation is preceded by mechanical flocculation with a 40-minute detention period. Coagulant aids are added during periods of high turbidity and low water temperature conditions. Influent water characteristics include a maximum turbidity ranging from 25 to 30 NTU and a temperature rarely below 10°C (50°F).

The plan was to expand the plant capacity from 0.18 to 0.35 m^3/sec (4 to 8 MGD). At a 0.35 m^3/sec (8-MGD) capacity, the sedimentation basin overflow rate is 81.5 m^3/day/m^2 (2,000 GPD/sq ft) or 0.95 L/sec/m^2 (1.4 GPM/sq ft). This basin overflow rate, coupled with the plant's influent turbidity, are the primary factors used to derive the appropriate overflow rate for the tube modules. According to the recommended overflow rates for systems with influent temperatures generally over 10°C (50°F), an overflow rate of 2.04 L/sec/m^2 (3 GPM/sq ft) is acceptable for the basin area covered by the tube settlers when preceding either dual- or mixed-media filters.

The area of the basin covered by tube modules was calculated using the allowable tube capacity to achieve the desired overflow rate. The two-step calculation is based on acceptable plant capacity and tube rate parameters:

1. $$\text{Quantity of tubes (area)} = \frac{\text{capacity, GPM}}{\text{allowable tube rate, GPM/sq ft}}$$

$$= \frac{8\ \text{MGD} \times 700\ \text{GPM/MGD}}{3\ \text{GPM/sq ft}}$$

$$= 1{,}870\ \text{ft}^2 \text{ or } 173.7\ \text{m}^2$$

2. $$\text{Area} = \text{length} \times \text{width}$$

$$1{,}870 = \text{length} \times 30\ \text{ft}$$

$$\text{length} = 1{,}870/30 = 62.5\ \text{ft or } 19\ \text{m}$$

The tube lengths were rounded off to the nearest 10 ft to permit use of standard tube module dimensions.

The tube modules were installed in an area extending from the outlet end of the basin to an isolation baffle wall separating the tube modules from the area of inlet turbulence. Three new effluent launders extending from the exit end wall were required to ensure uniform flow through the tube modules. The launders were installed at 3-m (10-ft) intervals. The tube modules were submerged 1.2 m (4 ft) because of the 4.8-m (16-ft) depth of the basin.

A.1.2 Upflow Solids Contact Clarifier Example

This example also involves the modification of existing facilities. The plant has two 3.9-m^2 (42-sq ft) upflow clarifiers of 189.2 L/sec (3,000 GPM),each with peripheral collection launders. The total surface area of each clarifier is 163.5 m^2 (1,760 sq ft). The influent center well consumes 18.6 m^2 (200 sq ft) of settling area.

Peak flows reach 1.3 L/sec/m^2 (1.92 GPM/sq ft), which is high enough to impede clarifier performance, especially at colder temperatures. Influent turbidity ranges from 30 to 70 NTU. Influent temperature rarely falls below 10°C (50°F).

The plan is to increase the capacity of each clarifier to 252.3 L/sec (4,000 GPM). At this flow, the clarification rate based on the available overflow rate allowable for tube settlers would be 1.77 L/sec/m^2 (2.6 GPM/sq ft). This rate is within the recommended limits for the expected influent turbidity levels and could be achieved by totally covering the basin with tubes. Total coverage of each basin is the simplest solution because of the ease of supporting the tube modules. This would result in 144.9 m^2 (1,560 sq ft) of tubes, with radial launders added to provide equal flow distribution. At the 1.77 L/sec/m^2 (2.6-GPM/sq ft) clarifier overflow rate, the probable effluent turbidity would fall within 3 to 7 NTU. An effluent with this turbidity level is acceptable for filtration with dual- or mixed-media filters.

A.1.3 Sacramento, California (Sequeira et al., 1983)

This case involves upgrading the capacity of a municipal plant from 2.6 to 4.4 m^3/sec (60 to 101 MGD). This plant, built in 1963, uses the American River as its water source. Originally the plant provided for 25 minutes of flocculation followed by 115 minutes of sedimentation detention. The expansion reduced flocculation detention to 14 minutes and sedimentation time to 65 minutes. Filter flow rates increased from 2.0 to 3.6 L/sec/m^2 (3.0 to 5.3 GPM/sq ft) at the expanded flow.

The feasibility study found that a change in filtration media alone from rapid sand to either dual or mixed media, would compensate for higher turbidities anticipated from the increased plant flow rate. It was concluded that the modification could be accomplished without structural changes.

Extensive pilot tests of various filter media were conducted fromFebruary to May, 1982, to determine the optimal filter medium in terms of efficiency and economics. The tests involved four types of filter media; each pilot filter was monitored for turbidity and headloss.

Figure A-1 shows the pilot filtration equipment setup. Influent came either directly from the river or after coagulation, flocculation, and sedimentation in the plant. Pilot filter influent turbidities ranged from 3 to 33 NTU and pilot filter rates were set at 3.4, 4.4, and 6.8 L/sec/m^2 (5, 6.5, and 10 GPM/sq ft). Alum dosages ranged from 12 to 21 mg/L and polymer dosages ranged from 0 to 0.3 mg/L. The City of Sacramento allows only 0.1 NTU of turbidity in finished water, so the pilot tests were terminated when turbidity levels achieved that figure. Both mixed- and dual-media filters performed well at all filter test flow rates, but mixed-media filters were more reliable in achieving the desired turbidity goal.

The recommendations from the pilot test data endorsed both dual- and mixed-media filters. A polyelectrolyte preparation and feed system were also recommended. During qualifying tests of the new filters, the plant produced effluent turbidities of less than 0.1 NTU (generally between 0.06 and 0.08 NTU).

A.1.4 Erie County, New York (Westerhoff, 1971)

This case involves upgrading a 1.93 m^3/sec (44 MGD) plant to 2.63 m^3/sec (60 MGD). The Erie County Water Authority built this plant in 1961 using Lake Erie as its water source. The steps of the plant treatment train are:

- Aeration
- Chemical addition
- Rapid mixing
- Flocculation and sedimentation
- Filtration
- Chlorination

The Authority decided to add six new mixed-media filters to the existing four rapid sand filters as the first stage of facility expansion. The decision to use mixed-media filters for the expansion was based on a year-long study in which 48 parallel filters were tested. Sand filters were tested at a flow rate of 1.36 L/sec/m^2 (2 GPM/sq ft), while mixed-media filters were tested at rates of 1.36 to 6.79 L/sec/m^2 (2 to 10 GPM/sq ft). Tests were terminated at headlosses of 2.4 m (8 ft). Test influent turbidities ranged from 2 to 4 NTU, clarifier overflows ranged from 25.5 to 38.3 m^3/day/m^2 (625 to 940 GPD/sq ft), and alum dosages ranged from 14 to 16 mg/L.

The test results indicated that:

- Sand filter effluent turbidity was 0.10 NTU or less for greater than 80 percent of the time.

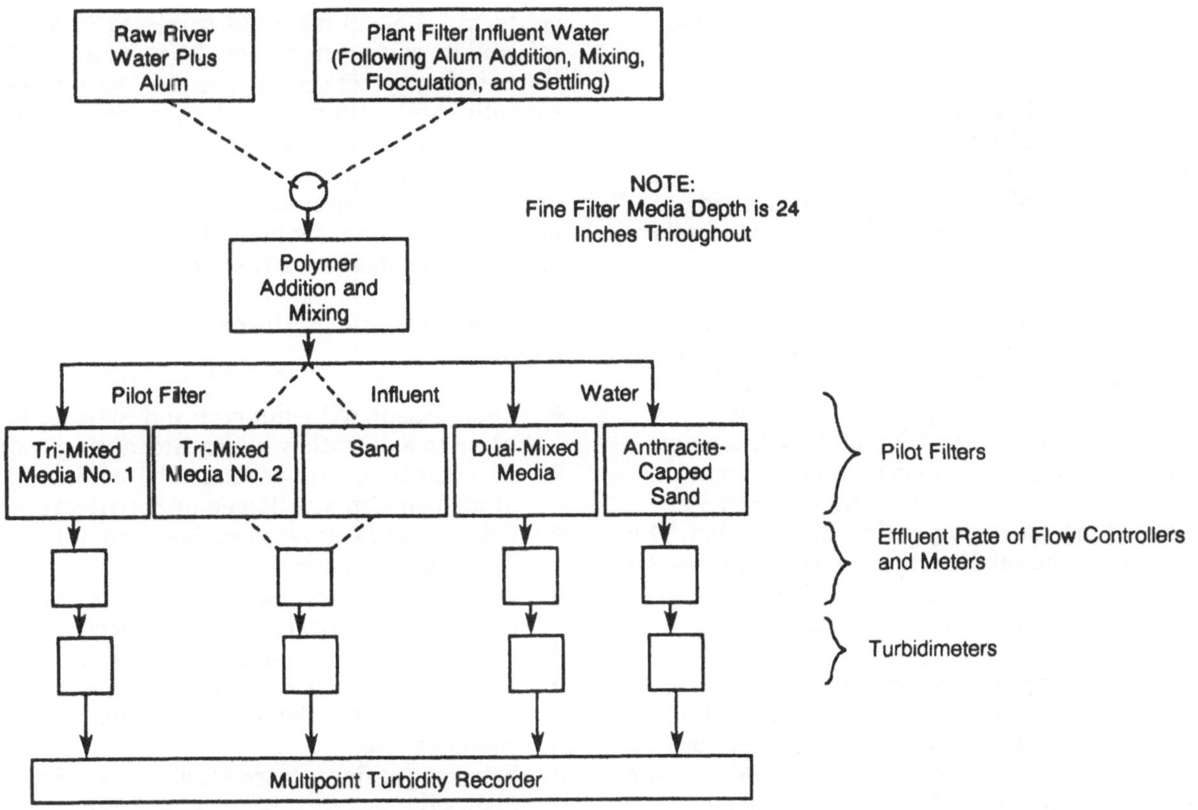

Figure A-1. Flow diagram of the pilot filtration equipment.

- Mixed-media filter effluent turbidity was 0.10 NTU or less for greater than 88 percent of the time.
- Mixed-media filters produced consistently less turbid finished water at flow rates of 1.36 to 6.79 L/sec/m^2 (2 to 10 GPM/sq ft) than did sand filters at flow rates of just 1.36 L/sec/m^2 (2 GPM/sq ft).
- Mixed-media filters had lower microscopic counts of particulate matter at flow rates of up to 4.1 L/sec/m^2 (6 GPM/sq ft) than did sand filters at 1.36 L/sec/m^2 (2 GPM/sq ft).
- Mixed-media filters at flow rates of 3.4 L/sec/m^2 (5 GPM/sq ft) had an average run length of 29 hours.
- Mixed-media filters at flow rates of 4.1 L/sec/m^2 (6 GPM/sq ft) had an average run length of 20 hours.
- Mixed-media filters at flow rates of from 3.4 to 4.1 L/sec/m^2 (5 to 6 GPM/sq ft) used proportionately less wash water than sand filters at a flow rate of 1.36 L/sec/m^2 (2 GPM/sq ft).

The finished water goals for this modified plant, using mixed-media filters at a rate of 4.1 L/sec/m^2 (6 GPM/sq ft), include:

- Average turbidity of less than 0.1 NTU
- Maximum turbidity of less than 0.5 NTU
- Average total microscopic count of less than 200 su/mL
- Maximum total microscopic count of less than 300 su/mL
- Less than 1 unit of color
- Less than detectable odor
- Less than 0.05 mg/L of aluminum
- Less than 0.05 mg/L of iron

A.1.5 Corvallis, Oregon (Collins and Shieh, 1971)

This case involved upgrading an 0.35 m^3/sec (8 MGD) plant to 0.92 m^3/sec (21 MGD). This municipal plant, which uses water from the Willamette River, was planned in phases of four MGD subunits, with the first phase built in 1949. The second phase was built in 1961, but the third and fourth were never built. Instead, the 0.35 m^3/sec (8 MGD) plant was expanded with nonstructural modifications during the early 1970s. Willamette River water has turbidity ranging from 15 to 30 NTU, with surges to 1,000 NTU. The original facility included flocculation and sedimentation basins, with two rapid sand filters per set of flocculation and sedimentation basins.

The expansion involved the replacement of the sand media with mixed media in the filters. It also added shallow-depth sedimentation using tube settlers and coagulation control techniques. Typical chemical dosages used to treat the raw water include:

- 20 to 40 mg/L of alum
- 10 to 20 mg/L of lime
- 2 mg/L of chlorine
- 0.1 to 0.2 mg/L of polymer as coagulant aid
- 5 to 10 mg/L of activated carbon for taste and odor control

Almost all treatment plant piping was enlarged to accept the expanded flows. Settling tubes covering 60 percent of the 325.2 m^2 (3,500 sq ft) rectangular sedimentation basins were positioned at the discharge end of the basins. The tube modules were supported on a simple "I" beam extended across the basin width. New effluent weirs and launders were also installed to ensure proper flow distribution. The overflow rate in the settling basin with tubes was 2.8 L/sec/m^2 (4.2 GPM/sq ft) compared to 0.71 L/sec/m^2 (1.05 GPM/sq ft) prior to the tube installation.

Experience has proved that the modified plant, at full-scale operation, consistently produced water with less than 0.2 NTU turbidity. Also, filter run lengths have increased from 40 to 65 hours due to the improved turbidity removal achieved by tube settling. The filtration rate is now 5.1 L/sec/m^2 (7.5 GPM/sq ft). The plant uses a pilot filter to establish coagulant dosages and has turbidimeters on each filter to continuously monitor effluent turbidity. The increased filter cycle length reduced operating costs through backwash water savings.

The cost of this expansion in 1969 was $430,000. In contrast, the original expansion plan of constructing new flocculation, sedimentation, and filter units to 0.7 m^3/sec (16 MGD) capacity was estimated at $650,000, in 1969 dollars. The nonstructural expansion included:

- New 0.22 m^3/sec (5-MGD) reservoir
- New high service pump station
- A cross-town 40.6-cm (16-in) transmission pipe to the new reservoir
- 219 L/sec (5 MGD) larger capacity than the original expansion

In addition, the expansion resulted in improved finished water quality.

A.1.6 Novato, California (Culp, 1976)

This case involved the expansion of a North Marin County Water District plant from a capacity of 0.16 to 0.27 m^3/sec (3.75 to 6.2 MGD) during 1973 and 1974, without major additions to the plant.

Stafford Lake is the District's water source. The turbidity ranges from 2 to 35 NTU, with the usual turbidity between 10 and 20 NTU. Stafford Lake's coliform MPN values range from 5 to 72,400 per 100 mL and its color ranges from 25 to 50 units. Although the existing system could manage the high raw water turbidity and coliform count, the system had a clarification problem when the water exhibited low turbidity and high color. Also, high plankton populations caused taste and odor problems. The plankton and other biological contaminants significantly shortened filter run length.

The characteristics of the existing treatment system included:

- Disinfection with chlorine and coagulation with alum in a downflow hydraulic mixing chamber
- Flocculation, with additions of recirculated sludge and lime, followed by clarification
- Additions of activated carbon immediately prior to rapid sand filtration
- Dechlorination through the addition of sulfur dioxide in the filtered water clearwell
- Elevated water storage for backwashing
- Discharge of backwash water to a recovery pond and recycling to the plant influent for reprocessing
- Spraying of flocculated sludge on land for disposal

The expansion required three major changes:

1. Modifying the clarifier by adding settling tubes

2. Changing the filter media from sand to mixed media

3. Installing a coagulant control center and turbidity monitoring unit

These changes are discussed below.

Settling tube modules were added to the annular outer settling compartment of the circular clarifier. As a general rule the addition of settling tubes permits processing two to four times as much water with no loss in clarification effectiveness. However, due to the presence of light alum floc, it was decided to increase capacity to only 263 from 164 L/sec (6.0 from 3.75 MGD). This change increased the surface overflow rate on the clarifier from 1.03 to 1.7 L/sec/m^2 (1.52 to 2.5 GPM/sq ft) and decreased the detention time from 1.3 to 0.8 hours. The weir loading increased from 1.47 to 2.3 L/sec/m^2 (7 to 11 GPM/ft).

In addition, capacities of certain plant equipment were increased to accommodate the increased plant flows. In addition to expanded pump capacities, pipelines, meters, chemical feeders and valves, and controls were modified to accommodate the increased plant capacity. The simple hydraulic mixing chamber was equipped with a vertical mechanical rapid mixer. Provisions were made to add polymer to aid flocculation. Also, a system to store and feed sodium hydroxide solution for pH control replaced the lime system.

The filter bottoms were replaced along with replacement of sand media with mixed media. The mixed-media filter was supported by graded silica gravel with an upper layer of coarse garnet gravel. Gate valves were replaced with butterfly valves.

A coagulant control center and turbidity monitoring equipment were installed to ensure compliance with the California Department of Health Services standard of 0.5 NTU.

The changes resulted in achieving a 107 L/sec (2.45 MGD) increase in capacity at a cost of $337,445, in 1974 prices. The expansion took only 8 months to complete. The plant has performed efficiently and reliably at the maximum design capacity, consistently achieving less than 0.15 NTU turbidity levels and effective bacteriological removal.

A.2 Slow Sand Filter Systems

A.2.1 Idaho State (Tanner, 1988)

In a joint 1-year project in cooperation with Idaho State, the EPA and University of Washington's Department of Environmental Health focused on three typical slow sand filter plants. In the course of this study, 13 additional slow sand filter plants were surveyed, and the following plant performance data were collected:

- Turbidity
- Total and fecal coliform removal
- *Giardia* samples
- Microscopic particle evaluation
- Temperature

In addition, the study collected and examined the following operating data:

- Filtration rates
- Sand depth
- Effective sand size
- Cleaning method and frequency
- Problems

From the results of the study, the researchers concluded that:

- Slow sand filters are very effective and reliable, if they are properly designed, operated, and maintained.
- Slow sand filter performance is not reliably determined through examination of turbidity and coliform removal.
- Microscopic particulate analysis (MPA) can be a good indicator of slow sand filter performance.
- The easiest method of projecting slow sand performance is to examine design and operational specifications.
- Raw water quality can have a large impact on slow sand performance.
- Slow sand filters can have significant operational problems.

This study made the following recommendations in considering the slow sand filter option:

- Use pilot filters to guide design and construction.
- Use experienced design professionals.
- Use multiple beds for the filters.
- Use properly educated operators.
- Reduce filtration rates during winter.
- Include proper watershed management in operation and maintenance procedures.

The researchers also recommended the use of routine sanitary surveys, which would include monitoring influent water quality, operating factors, and plant performance. Specifically, water quality should be monitored as to water source, watershed management and status, turbidity, and temperature. Operating factors monitored should include filtration rates, cleaning procedures, and bypassing. Finally, monitoring plant performance should include the recording of turbidity, cleaning frequencies, and freezing frequencies.

A.2.2 New York State (Letterman and Cullen, 1985)

A recent study of seven slow sand filter plants in New York State examined the effects of cleaning procedures on the effluent water quality. The water quality parameters examined included turbidity, total particle count, standard plate count, and total coliforms. Characteristics studied at the seven plants included average operating flow, design filtration rate, average operating filtration rate, filter sand uniformity coefficient, and effective sand size (see Table A-1).

The operational data from these plants are contained in Table A-2. Even during the ripening periods, all but one plant achieved effluent turbidities of less than 0.43 NTU. The one plant that exceeded 0.43 NTU had submicron particle size turbidity, which is ineffectively controlled with slow sand filtration. All seven plants performed well in removing particles greater than 2 μm. The removal rate for particles ranging from 2 to 60 μm was from 90 to 99.8 percent.

Table A-1. Characteristics of Slow Sand Filter Installations in New York

Location	Slow Sand Filtration Average Operating Flow Rate (MGD)	Design Filtration Rate (mgad)	Average Operating Filtration Rate (mgad)	Filter Sand	
				Uniformity Coefficient	Effective Size (mm)
Auburn	6.0	2.83	3.6	2.4	0.45
Geneva	2.5	4.9	4.9	1.9	0.37
Hamilton	~0.3	-	1.0	2.4	0.27
Ilton	1.5	-	4.1-4.6	2.2	0.37
Newark	2.0	4.1	4.1	1.7	0.35
Ogdensburg	3.6	5.1	4.6	1.7	0.35
Waverly	1.2	4.1	4.1	2.4	0.15

1 MGD = 0.044 m^3/sec; 1 MGAD = 0.935 m^3/m^2/day
Source: Letterman and Cullen (1985).

Total plate count removals varied widely from 0 to 50 percent.

A.2.3 McIndoe Falls, Vermont (Pyper, 1985)

This was a 2-year study of a municipal slow sand filter plant with two 37.2-m^2 (400-sq ft) filters, sand depth of 106.7 cm (42 in), effective silica sand size of 0.33 mm, and a design filtration rate of 1.92 m^3/m^2/day (2.05 mgad).

The raw water was obtained from two shallow spring-fed ponds with several beaver dams and lodges. Influent water turbidity ranged from 0.4 to 4.6 NTU, with a seasonal average of 2.1 NTU. The study examined the removal of turbidity, bacteria, coliforms, and *Giardia* cysts.

The evaluation of this slow sand filter plant revealed the following:

- Effluent turbidity of less than 1 NTU was achieved 99.19 percent of the time.
- Effluent turbidity of less than 1 NTU was achieved 99.68 percent of the time, after the first 100 days of operation.
- Effluent turbidity of less than 0.2 NTU was achieved 72 percent of the time (during this time influent turbidity was 1.45 NTU or less).
- Influent total coliform count was reduced from 1,300 per 100 mL or less to 10 per 100 mL or less, 86 percent of the time.

Table A-2. Filter Ripening Data - Summary

Location	Type of Operation Visit[a]	Date of Site Visit	Raw Water Turbidity During Site Visit (NTU)	Water Temperature During Site Visit	Filter Turbidity Approximately 5 Hours After Filter Startup (NTU)		Evidence of Ripening Period	Approximate Length of Ripening Period (days)
					Scraped/Resanded Filter	Control Filter[b]		
Auburn	(1)	Jul 83	1.2-2.0	~19°C	0.43	0.27	Yes	0.25
Auburn	(1)	Jul 83	1.2-2.0	~19°C	0.28	0.27	None	-
Auburn	(1)	Jul 84	2.0-2.8	~18°C	0.22	0.23	None	-
Geneva	(1)	Jul 83	-	-	-	-	-	-
Hamilton	(1)	May 84	1.0-1.5	~12°C	0.28	None	None	-
Ilton	(1)	Jul 83	2.0-4.0	~23°C	0.30	0.40	Minimal (particle count only)	0.5
Newark	(1)	Aug 83	1.2-3.5	~13°C	0.35	0.35	None	-
Newark	(2)	Jan 84	0.6-2.7	~4°C	0.41	0.12	Yes	2
Ogdensburg	(3)	Aug 83	0.3-0.6	~15°C	0.12	0.10	None	-
Ogdensburg	(3)	Feb 84	1.0-1.2	~2°C	0.22	0.24	None	-
Waverly	(1)	Jun 84	6.0-11.0	~15°C	2.3	1.6	Yes	10

[a] (1) Scraping operation.
(2) Resanding operation.
(3) Scraping combined with resanding.
[b] Control filter - Filter on-line at least 1 month, except Ogdensburg where the filter was on-line 1 week.
Source: Letterman and Cullen (1985).

- Influent occurrences of high total coliforms (spikes) and plate count bacteria were removed in water temperatures of from 5 to 10°C.

- Biological removal efficiencies were lower at temperatures of less than 5°C, especially from 0 to 1°C; for example:

  - *Giardia* removal efficiency was lowered to 93.7 percent in one test.

  - At 1°C total coliform count removal was reduced from 98 to 43 percent and the standard plate count bacterial removal was reduced from 98 to 80 percent over a 9-day monitoring period.

- Influent standard plate counts of 500 per mL were reduced to 10 per mL or less, 94 percent of the time.

- Influent total coliform counts were reduced from 440 organisms per 100 mL to 4 per mL.

- Influent heterotrophic organisms were reduced from 520 organisms per 100 mL to 15 organisms per 100 mL, based on the standard plate count.

- Influent *Giardia* levels were reduced by 99.98 percent under warm water conditions.

- Influent *Giardia* levels were reduced by 99.36 to 99.91 percent in water temperatures of less than 7°C.

A.2.4 Village of 100 Mile House, British Columbia, Canada (Bryck et al., 1987)

This example concerns a water supply operation that only chlorinated their water before distribution to a 2,000 person service area. *Giardia* in the raw water intake from beavers and muskrats upstream became a concern. In 1984,the village decided to construct a slow sand filter system to augment their chlorination operation after *Giardia* occurrences in 1981, 1982, and 1983. The new plant included the following components:

- Surface water intake
- Raw and treated water pumping station
- Chlorine equipment
- Contact tank
- Clear well
- Three slow sand filter beds

The three filters used filter media derived from local sand that was washed, dried, and sieved. The resultant effective sand size for their filter media ranged between 0.2 and 0.3 mm, with an average of 0.25 mm. The sand media uniformity coefficient ranged between 3.3 and 3.8, with an average of 3.5.

Each of the filter cells are 43 m (141 ft) long, 6 m (20 ft) wide, 3.75 to 3.9 m (12.5 to 13 ft) deep, and have a total surface area of 262 m^2 (2,820 ft^2). The design filter rate is 0.11 L/sec/m^2 (0.16 GPM/ft^2). All three filters are operated to the accepted maximum unit flow of 2,422.4 m^3/day (640,000 GPD) per filter for a total of 7,255.8 m^3/day (1,917,000 GPD). To protect against freezing, the filters are covered with precast panels. Filter walls have rigid insulation and are also insulated with backfilled soil.

The performance results of the new plant are presented in Figures A-2 and A-3. Influent sample *Giardia* cyst counts for the period of November 1985 through November 1986 are contained in Figure A-2. No cysts were detected in the effluent during that period. Influent and effluent sample turbidity data are contained in Figure A-3. Effluent turbidity ranged from 0.15 to 3.5 NTU,and was lower both when influent turbidities were lower and after the filter ripened.

Operating cycles for the filters ranged from 52 to 215 days, with the longer filter cycles occurring in the winter and spring months. Increased loadings of algae during late spring and summer months contributed to shorter filter cycles during those periods.

The construction cost of the new plant, in 1984 Canadian dollars, was $780,000. The average annual operating costs were $20,700, including costs for chlorination; energy requirements; media replacement; cleaning; and labor for daily inspection, which consumed from 0.5 to 1.0 hours. Cleaning required 16 person hours per filter and cost about $225 per cleaning. The total operating cost was estimated to be $0.25/$m^3$ ($0.96/1,000 gal).

A.3 Package Plants

A.3.1 Conventional Package Plants (Morand and Matthew, 1983)

The EPA surveyed the effectiveness of six conventional package plants, most of which were built during the 1970s. The results of this survey were published in March 1983. The plants were chosen for study because they operated throughout the year, used surface water as their source, and served small populations. Profiles of the six plants are contained in Table A-3; treatment processes are described in Table A-4. Design capacities ranged from 545.0 to 3,028.0 m^3/day (144,000 to 800,000 GPD) and average daily flows ranged from 45.4 to 1,249.0 m^3/day (12,000 to 330,000 GPD).

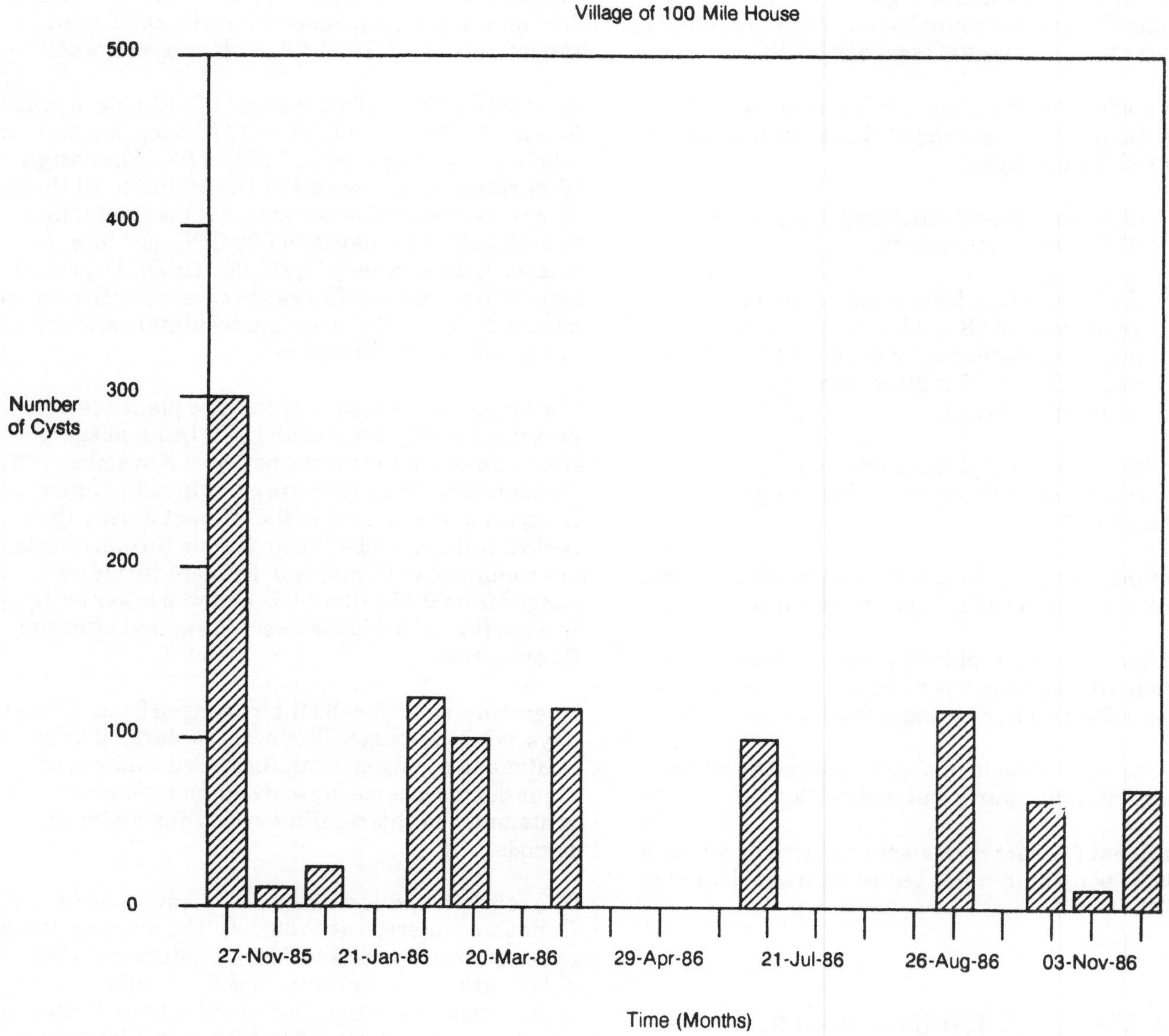

Figure A-2. ***Giardia*** **cysts in the raw water.**

The survey examined samples for turbidity, total coliforms, and chlorine residuals through the collection of grab samples from the influent, effluent, and distribution systems. Influent and effluent turbidities are shown in Table A-5. Plants C, T, and W consistently met the 1 NTU standard, while plants P, V, and M met the 1 NTU standard less than half the time. This discrepancy in the latter plants was due to:

- Inadequate design detention time
- Inadequately trained operators
- Inadequate time allocated for the operation
- Periods of high influent turbidities for plants V and M

After the appropriate adjustments were made to the equipment and operations, plants P, V, and M met the 1 NTU standard.

A.3.2 Adsorption Clarifier Package Plants

This section contains descriptions of six facilities that have used adsorption clarifier package plants. The plant profiles and effluent data are from facilities in Lake Arrowhead, California; Greenfield, Iowa; Lewisburg, Virginia; Philomath, Oregon; Harrisburg, Pennsylvania; and Red Lodge, Montana.

Lake Arrowhead, California (Hansen, 1987)

This case history involves a pilot study where *Giardia* cysts were introduced into a water supply to test the removal effectiveness of an adsorption clarifier. The adsorption clarification/filtration pilot plant had a 1.26 L/sec (20-GPM) capacity. The concentration of *Giardia* introduced to the raw water was 2,100 cysts per liter. The results of the pilot test revealed that filtration removed 100 percent of the cysts. Plant effluent turbidity ranged from 0.05 to

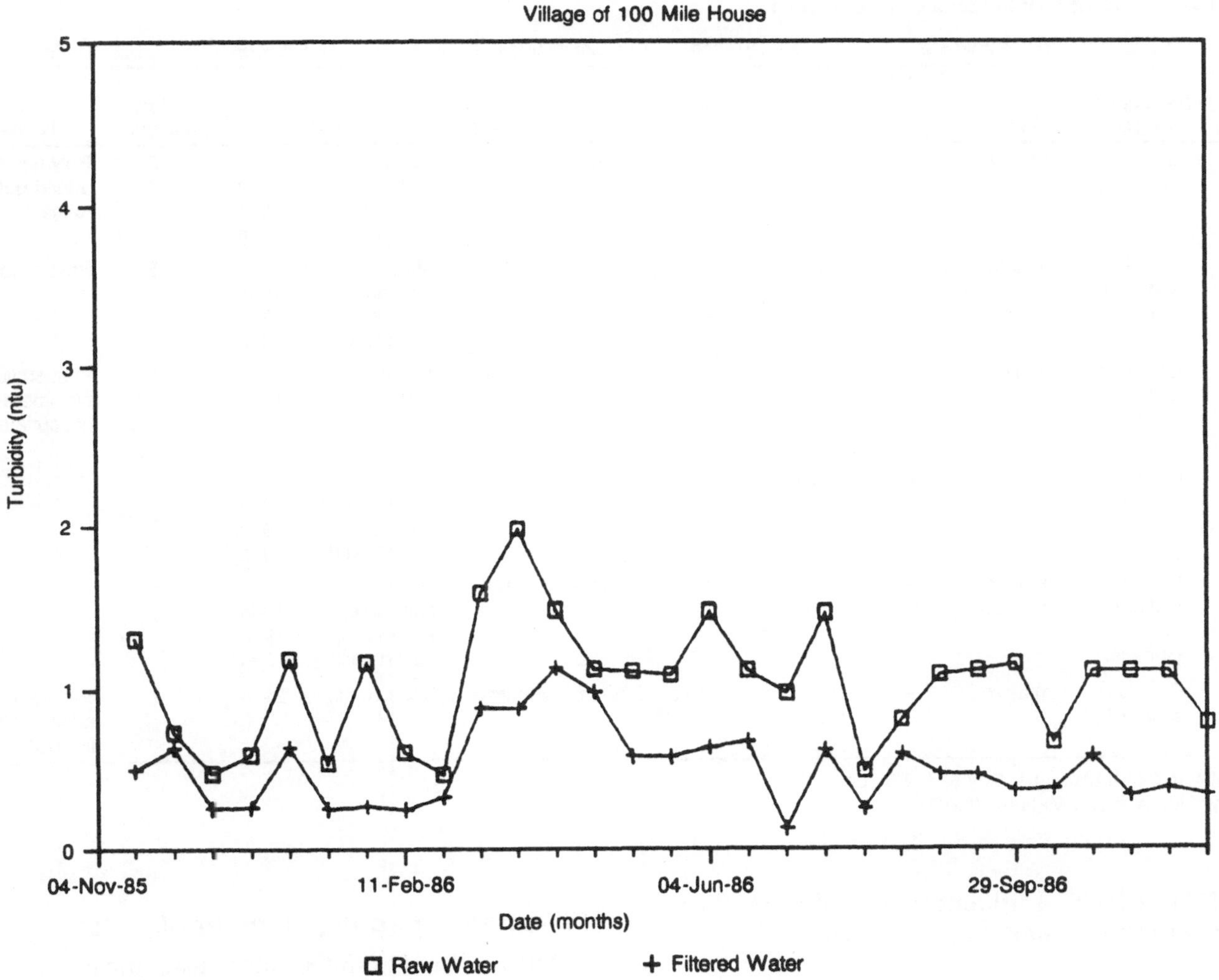

Figure A-3. Average raw and filtered water turbidity.

Table A-3. Water Treatment Facilities Surveyed in Field Study

Site	Manufacturer Model, Year	Design Flow Rate GPM (L/s)	Population Served/ No. of Meters	Average Volume per Day, GPM (mL/d)	Group Served	Type of Distribution Pipe Used	Source
W	Neptune Microfloc AQ-40, 1973	200 (12.6)	1,500/552	110,000 (0.42)	City	PVC	Surface impoundment
T	Neptune Microfloc AQ-40, 1973	200 (12.6)	1,000/360	78,000 (0.30)	City	PVC, cast iron, asbestos cement	Surface impoundment
V	Neptune Microfloc AQ-40, 1976	200 (12.6)	---/423	72,000 (0.27)	PSD[a]	PVC	River
M	Neptune Microfloc AQ-112, 1972	560 (35.3)	---/1,680	330,000 (1.25)	PSD	PVC	River
P	Neptune Microfloc Water Boy, 1972	100 (6.31)	---/411	82,000 (0.31)	PSD	PVC	Surface impoundment
C	Permulti Permajet, 1971	200 (12.6)	State park	57,000 (0.22)	State park	Asbestos cement	River

[a]PSD = Public Service District.
Source: Morand and Matthew (1983).

Table A-4. Treatment Process Characteristics

		Rapid Mix		Flocculation		Sedimentation		Filtration			
Site	Prechemical Addition	Type	Detention Time (sec)	Type	Detention Time (min)	Type	Loading (GPM/ft²)	Media		Rate (GPM/ft²)	Notes
W	C1, alum, soda ash	In pipe	3	Paddle	12.8	Tubes	100	Mixed anthracite, silica sand, garnet sand,	 18 in. 9 in. 3 in.	5	Polymer added before tubes
T	C1, alum, soda ash, poly	In pipe	3	Paddle	12.8	Tubes	100	Mixed anthracite, silica sand, garnet sand,	 18 in. 9 in. 3 in.	5	Post-soda ash
V	C1, alum, soda ash, poly	In pipe	3	Paddle	14	Tubes	100	Mixed anthracite, silica sand, garnet sand,	 18 in. 9 in. 3 in.	5	Post-sodium hexameta-phosphate
M	C1, alum, soda ash, poly	Chamber	30	Paddle	10	Tubes	100	Mixed anthracite, silica sand, garnet sand,	 18 in. 9 in. 3 in.	5	
P	C1, alum, soda ash, carbon (summer)	Chamber not used	–	Paddle	10	Tubes	150	Mixed anthracite, silica sand, garnet sand,	 18 in. 9 in. 3 in.	5	
C	C1, alum, soda ash, poly	In pipe	–	Upflow solids contact 2 hr detention time rise rate – 1 GPM/ft²				Silica Sand,	24 in.	2	Soda ash added before filtration

GPM/ft² = 0.679 L/sec/m²; 1 inch = 2.54 cm
Source: Morand and Matthew (1983).

0.06 NTU. Clarified effluent contained cysts when effluent turbidity ranged up to 0.3 to 0.4 NTU.

Greenfield, Iowa (WEM, 1985)

This case history presents operating performance data from a 0.044 m^3/sec (1.0 MGD) plant between July and December 1984. The plant's influent comes from Lake Greenfield which contains turbidity, taste, color, and iron contaminants. At the Greenfield plant, powdered activated carbon is added to control odor and taste problems. Table A-6 presents the operating data for the plant. Pilot test turbidities were similar to those from regular operational experience, typically ranging from 0.3 to 0.5 NTU. Alum dosages in the full-scale plant range from 7 to 20 mg/L, which are slightly lower than those established in pilot tests.

Lewisburg, West Virginia (Lange et al., 1985)

This case involves a 0.087 m^3/sec (2-MGD) plant that has operated since December 1983. Three months of records on turbidity removal are contained in Table A-7. The clarifier generally achieved 90 percent turbidity removal. While turbidity ranged as high as 50 NTU in May 1985, the plant was able to produce effluent with turbidities of less than 0.5 NTU at all times.

Philomath, Oregon (Lange et al., 1985)

A 0.044 m^3/sec (1.0-MGD) plant utilizing an adsorption clarification unit began operation in February 1986. Seven months of turbidity data and chemical feed data are contained in Table A-8. Influent turbidity ranged from 3.02 NTU to 26.7 NTU in February 1986. The system performed effectively with filter effluent turbidity ranging from 0.17 to 0.26 NTU. Net water production did not fall below 88 percent during this period.

Harrisburg, Pennsylvania (Lange et al., 1985)

This example involves an existing plant that was built in 1964 without sufficient clarifying capacity. The original plant contained a circular upflow sludge blanket clarifier and stored backwash gravity filters. The system was operated in the direct filtration mode at 2.7 L/sec/m^2 (4 GPM/sq ft) because clarification capacity was inadequate.

Three adsorption clarifier packaged units were added to expand the plant to 0.394 m^3/sec (9 MGD) with each adsorption clarifier operating at between 6.79 and 10.19 L/sec/m^2 (10 and 15 GPM/sq ft). Three months of turbidity and chemical feed records are found in Table A-9. Overall turbidity removal ranged from 70 to 90 percent. The length of filter runs doubled from 13 to 26 hours with the addition of the

Table A-5. Turbidity Values (NTU)

Plant C		Plant W		Plant T		Plant V		Plant M		Plant P	
Raw	Clearwell Effluent	Raw	Clearwell Effluent	Raw	Clearwell Effluent	Raw	Clearwell Effluent	Raw	Clearwell Effluent	Raw	Clearwell Effluent
8.5	0.3	–	0.9	10.0	1.9	4.0	1.8	–	0.2	12.0	0.8
6.2	0.2	5.0	0.3	8.0	0.2	12.0	2.8	39.0	3.8	4.4	2.4
1.2	0.3	4.2	0.4	6.0	0.4	–[a]	9.6	40.0	2.6	–	7.0
1.6	0.1	19.0	0.8	3.2	1.1	35.0	1.5	27.0	2.4	3.5	1.5
2.2	0.1	9.2	2.0	3.2	0.2	42.0[a]	2.0	6.0	1.2	2.0	0.1
4.0	0.1	11.5	0.3	3.2	0.2	10.0[a]	2.4	3.8	0.1	1.2	0.5
12.6	0.7	12.0	0.2	5.8	0.2	90.0[a]	8.5	73.0	11.0	15.6	9.7
5.2	0.2	11.0	0.3	10.4	3.2	28.0[a]	5.4	3.6[a]	0.1	3.1	2.2
2.2	0.2	29.7	0.9	3.4	0.7	19.0[a]	0.3	3.8	0.3	177	
.2	1.9			12.8	0.2			47.0[a]	1.2	6.0	0.5
						13.0[a]	0.8	70.0[a]	16.0		
						8.0[a]	0.3	25.0[a]	3.4		
						6.0[a]	0.3	>100.0[a]	55.0		
						>100.0[a]	0.5	>100.0[a]	31.0		
						60.0[a]	0.5	8.5[a]	2.2		
						24.0	1.2	4.3	0.4		
						13.0	0.3	4.0[a]	1.0		
						2.7	1.2	9.6[a]	1.9		
						1.2	1.0	19.1	1.1		
						3.3	0.5	64.0	6.9		
								8.2	1.0		

[a]Averaged values for day.
Source: Morand and Matthew (1983).

Table A-6. Operating Data - Greenfield, IA

Month		Turbidity, NTU			Iron, mg/L		Chemical Feed, mg/L		
		Raw	Clarifier	Filter	Raw	Filter	Alum	Polymer	Chlorine
Jul, 1984	Mean	5.9	-	0.51	0.13	0.03	15.5	0.32	7.0
	Range	2.5-13.0	-	0.32-0.70	0.08-0.22	0.01-0.07	11.1-19.6	0.28-0.37	6.0-8.7
Aug, 1984	Mean	3.9	1.26	0.46	0.18	0.03	12.2	0.32	3.9
	Range	2.5-6.0	-	0.39-0.62	0.07-0.32	0.0-0.04	8.9-23.9	0.16-0.48	2.3-5.6
Sep, 1984	Mean	4.4	1.6	0.39	0.14	0.02	8.4	0.35	7.8
	Range	3.6-5.3	-	0.36-0.60	0.10-0.24	0.01-0.03	6.9-12.9	0.32-0.37	7.4-8.1
Oct, 1984	Mean	6.3	2.3	0.40	0.21	0.02	6.7	0.20	5.9
	Range	4.0-8.2	1.8-2.9	0.26-0.56	0.14-0.29	0.01-0.03	-	0.13-0.26	3.7-7.4
Nov, 1984	Mean	6.4	2.3	0.33	0.20	0.01	13.1	0.18	3.1
	Range	4.2-13.0	2.0-2.5	0.17-0.47	0.14-0.33	0.01-0.02	7.3-22.3	0.15-0.21	2.2-5.1
Dec, 1984	Mean	2.8	1.7	0.35	0.11	0.01	10.2	0.22	2.5
	Range	1.8-7.5	1.6-2.0	0.27-0.48	0.02-0.17	0.00-0.03	8.3-14.6	0.17-0.25	2.1-2.8

adsorption clarifiers, in spite of the 20 to 30 percent increase in flow.

Red Lodge, Montana (Lange et al., 1985)

This case involves installing a new 0.061 m^3/sec (1.4-MGD) plant starting up in January 1984. Prior to construction, the water was only chlorinated because of its high quality. Because the raw water turbidity is so low, the plant is equipped with a bentonite feed system to assist in the treatment of the plant's typically cold, low turbidity (less than 1 NTU) influent. Bentonite is added to the influent before the alum and polymer feed points. The treatment goal is to produce effluent with less than 0.1 NTU turbidity to guard against *Giardia* breakthrough. As Table A-

Table A-7. Operating Data - Lewisburg, WV

Month		Turbidity, NTU Clarifier Influent	Clarifier Effluent	Filter Effluent
Dec, 1984	Mean	5.0	1.2	0.16
	Range	1.8-18	0.6-2.5	0.10-0.20
May, 1985	Mean	9.6	1.0	0.18
	Range	2.0-5.0	0.4-6.0	0.10-0.50
Jun, 1985	Mean	1.7	0.60	0.16
	Range	0.70-5.0	0.1-1.1	0.10-0.50

Table A-8. Operating Data - Philomath, OR

(Monthly Average Values)

Month	Turbidity, NTU Raw	Clarifier	Filter	Chemical Feed, mg/L Alum	Soda Ash	Polymer	Net Production[a]
Feb, 1986	26.7	3.40	-	15.5	21.2	0.29	-
Mar, 1986	6.50	1.67	0.17	11.4	4.5	0.08	91%
Apr, 1986	6.27	1.49	0.20	-	-	-	90%
May, 1986	5.82	1.45	0.23	11.7	1.2	0.08	88%
Jun, 1986	4.25	1.21	0.14	11.3	4.8	0.07	93%
Jul, 1986	3.43	-	0.22	10.0	5.7	0.07	93%
Aug, 1986	3.02	-	0.26	16.6	-	0.07	93%

[a]Net production is the percentage of raw water that was turned into finished water.

10 indicates, the plant achieved its turbidity goal for the entire 6-month period over which data were collected. It should be noted that the turbidity of the clarifier effluent is sometimes higher than that of the raw water influent due to occasional carryover of alum-coagulated bentonite particulates.

A.4 Diatomaceous Earth Filters

A.4.1 Colorado State University Study (Lange et al., 1984)

The study examined *Giardia* inactivation in addition to removal of turbidity, total coliform bacteria, standard plate count bacteria, and particles. Eight operational parameters evaluated for their influence on removal effectiveness were:

- Seven grades of diatomaceous earth
- Hydraulic loading rates of 0.68, 1.36, and 2.72 L/sec/m^2 (1, 2, and 4 GPM/ft^2)
- Influent temperatures, ranging from 5 to 19°C
- Influent concentrations of bacteria, ranging from 100 to 10,000 per 100 mL
- Influent concentrations of *Giardia* cysts, ranging from 50 to 5,000 cysts per liter
- Filter headloss rates
- Filter run times
- Alum coating of the diatomaceous earth

The results of the study indicated that *Giardia* removal exceeded 99 percent for all grades of diatomaceous earth at filter hydraulic loadings of 0.68 to 2.72 L/sec/m^2 (1.0 to 4.0 GPM/sq ft), and at all study temperatures.

In addition, the grade of diatomaceous earth affected the removal rates of other contaminants. The coarsest grade of material achieved removal of 99.9 percent of the *Giardia*, 95 percent of the cyst-sized particles, 20 to 35 percent of the coliform bacteria, 40 to 70 percent of the heterotrophic bacteria, and 12 to 16 percent of the turbidity. The finest grade of diatomaceous earth achieved the removal of 99.9 percent of the bacteria and 98 percent of the turbidity.

A.4.2 McIndoe Falls, Vermont (Pyper, 1985)

This case involves an evaluation of the parallel operation of a diatomaceous earth and a slow sand filter. The diatomaceous earth filter rate ranged from 0.68 to 1.22 L/sec/m^2 (1.0 to 1.8 GPM/sq ft), using a 0.63 to 1.26 L/sec/m^2 (10 to 20 GPM) pressure unit.

The pressure diatomaceous earth filters dependably removed 99.97 percent of the *Giardia* cysts and 86 percent of the coliforms. They also achieved 80 percent standard plate count bacteria removal in 70 percent of the samples. The average influent contained 271 total bacteria count per 100 mL and 30 standard plate count bacteria per mL. The average effluent contained 38 total bacteria count per 100 mL and 6 standard plate count bacteria per mL.

A.5 Selecting a Filtration System

A.5.1 Lake County, California (Conley and Hansen, 1987)

This case study concerns the process of selecting a centralized new treatment facility for an area that was previously served by individual supplies. Clear Lake, the source of raw water, experiences severe seasonal taste and odor problems caused by blue-green algae blooms (*Microcystis* and *Anabaena*).

Clear Lake is classified as eutrophic. The taste and odor problems typically start in May and finish in November with lake water taste and threshold odor numbers (TON) ranging as high as 10 during this period. Clear Lake's turbidity ranges from 3 to 90 NTU. Clear Lake has excellent mineral qualities, as

Table A-9. Operating Data – Harrisburg, PA

Month		Turbidity, NTU Raw	Clarifier	Filter	Chemical Feed mg/L Alum	Polymer	Clarifier Rate GPM/ft²
Jul, 1985	Mean	8.1	1.36	0.17	NA	NA	NA
	Range	5.7-15.1	0.60-3.4	0.13-0.32	-	-	-
Aug, 1985	Mean	8.6	0.94	0.18	11	0.12	14
	Range	4.8-14.3	0.7-1.5	0.13-0.25	8.1-13.4	0.10-0.15	11.9-16.5
Sep, 1985	Mean	9.3	1.2	0.18	9	0.07	13
	Range	5.9-37	0.5-4.9	0.09-0.34	4.9-18.7	0-0.31	11.7-14.9

GPM/ft² = 0.679 L/sec/m²
NA = not available

Table A-10. Operating Data – Red Lodge, MT

Month		Turbidity, NTU Raw	Clarifier	Filter	Chemical Feed mg/L Alum	Polymer	Bentonite	Water Temperature °F
Jan, 1985	Mean	0.20	0.29	0.04	2.5	0.48	0.93	36
	Range	0.15-0.33	0.21-0.35	0.03-0.05	1.4-3.5	0.48-0.48	0.93-0.93	34-38
Feb, 1985	Mean	0.19	0.28	0.05	2.3	0.48	0.93	34
	Range	0.16-0.21	0.22-0.39	0.30-0.07	1.4-3.6	0.48-0.48	0.93-0.93	33-38
Mar, 1985	Mean	0.20	0.29	0.05	3.0	0.48	0.93	36
	Range	0.18-0.22	0.25-0.32	0.03-0.06	1.9-3.9	0.48-0.48	0.93-0.93	35-37
Apr, 1985	Mean	1.1	0.49	0.08	5.0	0.60	0	40
	Range	0.22-1.6	0.27-1.1	0.04-0.14	2.3-7.1	0.48-0.75	-	38-43
May, 1985	Mean	2.1	0.41	0.07	7.3	0.48	0	44
	Range	0.16-4.5	0.16-0.80	0.03-0.14	3.6-13.4	0.48-0.48	-	43-47
Jun, 1985	Mean	3.5	0.41	0.07	10.4	0.60	0	45
	Range	1.5-6.2	0.21-0.75	0.04-0.18	7.4-12.8	0.48-0.75	-	43-49

°C = (°F - 32) × 5/9

shown in Table A-11, but the lake water requires turbidity removal, taste control, and odor control.

Several other jurisdictions in addition to Lake County use Clear Lake for a water supply. Consequently, there were data available from other operations using the same influent. One of these neighboring jurisdictions, the City of Lakeport, installed a plant equipped with both ozone and activated carbon, which effectively manages the odor and taste problems.

A feasibility study recommended a complete treatment system rated at 0.088m³/sec (2 MGD) with preozonation, chemical coagulation, pH control, flocculation, sedimentation, filtration, and postfiltration. A suitable package plant was selected as the most cost-effective choice.

The selected package plant required pilot tests for establishing process feasibility and design criteria. Four utilities were contacted with regard to their experiences before pilot testing commenced. The utilities reported that taste and odor had been increasingly troublesome, and they described rising floc problems due to air bubbles introduced by algae. In addition, algae blooms had interfered with the clarification process. Water quality data from

Table A-11. Clear Lake Water Quality Analysis

Mineral Constituents	Concentrations, mg/L High	Low	Median
Calcium	30	17	23
Magnesium	21	9.8	15
Sodium	14	4	10
Potassium	2.8	0.1	2.0
Bicarbonate	212	96	145
Sulfate	35	5.1	9
Chloride	10	3.2	6
Nitrate	11	0.0	1.6
Fluoride	0.4	0.0	0.1
Boron	1.2	0.1	0.7
Silica	34	0.7	14
Hardness	158	78	115
Turbidity	90 TU[a]	3 TU	15 TU

[a]Turbidity Units

January 1977 to November 1980 for a sampling station on Clear Lake is summarized in Table A-12.

The odor and taste problems had been successfully treated by the other utilities with potassium permanganate and powdered activated carbon. However, the two utilities using ozone received the fewest complaints. Also, some utilities had recently installed postfilter granular activated carbon contactors.

As a result of reviewing the existing data, Lake County decided to install adsorption clarifiers as a prefiltration step. The adsorption clarifier uses an upflow design with a buoyant medium to assist the clarification process.

The manufacturer was requested to perform the pilot tests. The two primary test objectives were to:

- Measure turbidity removal capabilities
- Determine the effectiveness of adding potassium permanganate or powdered activated carbon prior to clarification to remove trihalomethane (THM) precursors, taste substances, and odor substances

If the pilot tests were successful, the utility would be able to avoid the expensive option of using ozone treatment to control taste and odor.

The tests were performed during September and October of 1986 and verified by a design consultant. There were no major algae blooms during the test period.

The pilot test report indicated that the new plant could remove a high percentage of turbidity. Furthermore, low concentrations of potassium permanganate could be used to control taste and odor substances. In addition, the powdered activated carbon assisted in the removal of THM precursors. The report indicated that the adsorption clarifiers were effective at dosages of up to 25 mg/L, with no filter breakthroughs or short filter cycles.

Based on the pilot test results, an adsorption clarifier package plant and chemical feed system were recommended. The adsorption clarifier will have a design loading of 6.79 L/sec/m^2 (10 GPM/sq ft), a filtration rate of 3.4 L/sec/m^2 (5 GPM/sq ft), and a tri-mixed filter media. The chemical feed system will use the following chemical dosages:

- 1 to 2 mg/L of ozone
- 10 to 20 mg/L of alum
- 0.2 to 1.0 mg/L of potassium permanganate
- 1 to 10 mg/L of cationic polymer
- 10 to 20 mg/L of powdered activated carbon
- 0.1 to 0.5 mg/L of nonionic polymer
- 20 to 40 mg/L of sodium hydroxide

Table A-12. Clear Lake Water Quality at DWR Sampling Station No. 1 at Lakeport

Date Sampled	Turbidity (NTU)	Sp. Cond. (µMHOS/cm)	pH	DO (mg/L)	Sodium (mg/L)	Alkalinity ($CaCO_3$)	Chloride (mg/L)	Nitrates (mg/L)	Ortho Po_4 (mg/L)
1/7/77	21	307	8.3	9.9	13	142	6.8	0.25	0.05
2/3/77	9	308	8.1	11.8	13	144	7.0	0.14	0.00
3/10/77	25	308	7.7	9.2	12	144	6.6	0.12	0.01
4/7/77	18	316	8.1	7.7	12	147	6.4	0.69	0.00
5/5/77	14	321	8.0	9.1	14	156	6.5	0.01	0.00
6/16/77	35	337	7.8	4.5	14	154	7.9	0.05	0.00
7/14/77	14	346	8.0	5.0	14	160	8.8		
8/11/77	7	302	8.0	4.3	16	138	9.6	0.04	0.09
9/22/77	60	364	8.2	2.6	17	165	8.8	0.05	0.08
10/14/77	50	385	8.1	3.6	17	170	10	0.01	0.03
11/4/77	33	373	8.3	5.7	16	174	7.7	0.14	0.02
12/9/77	36	366	8.2	6.8	16	169	8.5	0.17	0.02
1/6/78	67	193	7.3	9.7					
2/10/78	49	197	7.6	10.0					
3/9/78	20	216	7.6	9.1					
4/6/78	13	223	7.6	9.1					
5/11/78	6	227	8.2	10.4					
6/8/78	3	225	8.2	6.9					
7/13/78	3	238	8.1	8.1					
8/10/78	7	258	7.8	5.9					
9/14/78	18	271	7.8	7.0					
10/5/78	12	275	8.4	10.6					
11/9/78	22	290	7.5	4.3					
12/7/78	20	294	7.7	9.5					
1/5/79	21	299	7.8	10.4	13	138	8.1		
2/16/79	18	255	7.7	10.4					
3/15/79	12	281	7.2	6.7					
4/5/79	20	260	8.1	10.2					
5/10/79	8	259	8.2	9.9					
6/8/79	5	284	7.8	7.0					
7/12/79	6	291	8.1	7.0	12	130	6.0		
8/10/79	4	304	8.0	5.8					
9/7/79	13	311	8.1	5.3	13	143	6.0		0.08
10/4/79	17	320	8.6	7.0					
11/8/79	28	299	7.7	8.5					
12/7/79	19	299	7.5	7.1					
1/11/80	28	260	8.3	10.2					
2/6/80	17	245	7.9	10.3					
3/6/80	3	228	7.7	9.8					
4/10/80	19	235	7.7	9.6					
5/8/80	3	236	8.1	8.8					
6/6/80	6	246	8.1	8.5					
7/11/80	5	263	8.1	7.7	10				N.D.
8/22/80	5	268	7.9	6.7					
9/18/80	12	280	8.3	4.8					
10/15/80	9	185	8.3	7.1					
11/13/80	13	256	8.4	9.1					

Appendix B

Case Histories of Emerging Disinfection Technologies

Several case histories of water supply systems that have used disinfectants other than chlorine are presented in this appendix. The descriptions of each case include the modifications in disinfection/oxidation techniques necessary to (1) attain the desired degree of disinfection while minimizing the formation of disinfection by-products; and (2) cope with other common problems such as taste, odor, color, and algae.

The case histories cover ozone, ultraviolet radiation, chlorine dioxide, and chloramine. Section B.1 presents case histories involving the use of ozone as a primary disinfectant and as a preoxidant for lowering THM levels. Section B.2 illustrates the use of ultraviolet radiation as a primary disinfectant for surface water followed by chlorination. Three case studies involving the uses of chlorine dioxide as a preoxidant for THM control, as a primary and secondary disinfectant, and in combination with chloramines to lower THM production are presented in Section B.3. Lastly, chloramine case histories are presented in Section B.4.

B.1 Ozone Case Histories

B.1.1 Primary Disinfection with Ozone: North Andover, Massachusetts

Under normal circumstances, total coliforms in North Andover's raw water supply, Lake Cochichewick, are between 50 and 500 per 100 mL and turbidity is between 1 and 2 NTU. However, in early 1986, 18 cases of *Giardia*sis were reported. Residents were instructed to boil tap water for at least 5 minutes before use, while public health officials sought to locate and correct the problem.

The raw water was found to contain *Giardia* cysts in samples taken in April 1986. In addition to the concern that residential septic systems were affecting the water supply, local officials found that the lake's watershed had a large muskrat population.

At the time of the *Giardia*sis outbreak, treatment of North Andover water included chlorination without filtration. Two pumping stations distributed the water through the piping system, each station handling 0.12 to 0.13m^3/sec (2.5 to 3 MGD). Over the years, the distribution pipes have formed substantial biofilm and scale, due to high humic concentrations in the water.

Increased chlorination overcame the immediate *Giardia* problem, but subsequent descaling of the distribution pipes released coliform organisms to the water. The State Department of Environmental Quality Engineering (DEQE) had to continue the "boil water" notice. In addition, THM levels rose to above 200/µg/L due to the increased chlorination, and residents began complaining about the strong chlorine concentrations. Thus, ozone was investigated as an alternative primary disinfectant.

Interim Solution

In October 1986, ozonation systems, rented from an equipment supplier, began operating at both Lake Cochichewick pumping stations. One system had a 22.7-kg/day (50-lb/day) ozone generating capacity and the second had a 68.04-kg/day (150-lb/day) capacity. When the efficacy of ozonation was proven, the town purchased and installed two 68.04-kg/day (150-lb/day) ozonation systems, one at each pumping station. At four points in the distribution system, chlorine was added to provide a residual disinfectant.

These two ozonation systems were installed as a stop-gap measure to control *Giardia* cysts until a proposed $10.5 million 0.53 m^3/sec (12-MGD) plant was designed and constructed. The new plant, to be on line in 1991, will provide complete treatment, including ozonation, filtration, and postfiltration GAC adsorption with residual chlorination.

The State Department of Environmental Quality Engineering (DEQE) provided emergency funding of \$2.5 million for installing the interim ozonation systems, connecting pipelines to the three neighboring communities, relining pipes with cement, and replacing water mains. Annual rental fees for the two ozonation systems were approximately \$90,000. The two 68.04-kg/day (150 lb/day) ozonation systems were purchased for \$325,000 (total). This price included the air preparation systems, ozone generators, diffuser contactors, and ozone destruction equipment; plus the appropriate instrumentation, including a residual ozone monitor for the outlet of the contact chamber.

Each of the purchased contacting chambers is 3 m (10 ft) wide, 6 m (20 ft) long, and 4.8 m (16 ft) deep. The contactors have five baffled sections to which equal amounts of ozone are applied. Plug flow is maintained throughout the ozone contactors. Applied ozone dosages are 5 mg/L and the dissolved ozone concentration at the outlet of each contact chamber is between 0.9 and 1.0 mg/L. For design purposes, the average ozone concentration in each contactor was assumed to be 0.5 mg/L. Total residence time of water in the contactors is 10 minutes during the summer at full pumping rate. During the winter, when pumping rates are reduced by 50 percent, the residence time is 20 minutes.

Water temperatures of the lake vary from about 5°C (41°F) in winter to just under 20°C (68°F) during the summer. Therefore, the appropriate CT values for 99.9 percent inactivation of *Giardia lamblia* range from 1.9 to 0.72 mg/L-min. During the summer, the period when the shortest contact time is experienced, at an average dissolved ozone concentration of 0.5 mg/L, a T_{10} contact time of 5 minutes (50 percent of the peak-flow hydraulic detention time) would attain a CT value of 2.5 mg/L-min, which is more than adequate. During winter (T_{10} = 10 min), the CT will be 5 mg/L-min.

The Results

After approximately 90 days of ozone treatment, the State DEQE unconditionally lifted the "boil water" order, which had been in effect for 9 months. Both *Giardia* cysts and coliform organisms were eliminated from the North Andover water supply.

In addition to the microorganism control, several other benefits resulted from use of ozonation. Prior to the outbreak problem, THM values ranged from 8 to 120 µg/L. Since ozonation, measured THM levels range from 1.1 to 2 µg/L. In addition, the color of the treated water has improved significantly (65 to 95 percent lower) and taste and odor of the finished water greatly improved.

Permanent Solution

The planned 0.53 m^3/sec (12-MGD) treatment plant will include ozonation, filtration, and postfiltration GAC adsorption, followed by secondary disinfection. Preozonation will be applied before rapid mix and filtration. After dual media filtration, GAC adsorption is incorporated, followed by chlorination for secondary disinfection. In addition to providing additional removal of contaminants, the GAC step will allow mineralization of much of the biodegradable organic fractions of the water.

B.1.2 Preozonation for THM Control: Kennewick, Washington (Cryer, 1986)

Prior to 1977, the City of Kennewick drew its water supply from a system of five Ranney collectors located in the Columbia River. The water was chlorinated before distribution. When installed, this system was capable of producing approximately 0.876 m^3/sec (20 MGD); however, its output deteriorated to about 0.657 m^3/sec (15 MGD) by 1977. By 1978, peak water demand reached the capacity of the system. It was determined that direct utilization of the Columbia River would be the only long-term reliable water source.

Since the raw water would be drawn directly from the river, additional treatment was necessary to maintain finished water quality. A pilot plant study was undertaken to test alternative water treatment processing steps. This study included the use of preozonation and conventional postfiltration GAC adsorption, in addition to conventional and direct filtration procedures.

Average values of the raw water quality parameters of the Columbia River are:

- TTHM - 1 µg/L
- TTHM Formation Potential - 136 µg/L (7 days chlorine contact time)
- TOC - 2.4 mg/L
- No. Particles - 11,650/mL
- Particle Volume - 160,700 nL/L
- Turbidity - 1.7 NTU

In the pilot study, the preozonation and coagulation/filtration steps each provided approximately 30 percent reduction in TTHM formation potential (TTHMFP) levels and 10 percent reduction in TOC levels. The combined processes gave approximately 60 percent reduction in levels of TTHMFP and 20 percent reduction in TOC levels. The preozonation, coagulation, and filtration steps combined were determined to be operationally equivalent to activated carbon adsorption for the removal of THM precursors; the process was also determined to be the most cost-effective method of treating organics in the Columbia River water supply.

Solution

A 1.314-m^3/sec (30-MGD) water treatment plant was constructed in four stages, adding 0.329m^3/sec (7.5 MGD) capacity in each stage. The new plant includes preozonation, flash mix, coagulation, flocculation, filtration, and postchlorination. Design criteria for the preozonation facilities are given in Table B-1.

Table B-1. Design Criteria for Kennewick Water Treatment Plant Preozonation Facilities

Item	Units	Design Criteria Initial	Ultimate
Plant Capacity			
Design	MGD	7.5	30.0
	cfs	11.6	46.41
	GPM	5,200	20,800
Ozone Contact Basins			
No. of basins		4	16
Detention time at design flow	min	10	10
Basin dimensions – inside	ft	14 × 8	14 × 8
Average water depth	ft	16	16
Basin volume	ft^3	1,792	1,792
	gal	13,400	13,400
Total basin volume	ft^3	7,168	28,672
Chemical Feed Rate (max. dosage at design flow)	lb/day	250	1,000
Chemical Feeders	lb/day	2 × 125	3 × 250 2 × 125
Chemical Storage Capacity		250	1,000
Number of Ozone Generators		2	5

1 pound = 0.4536 kilograms; 1 MGD = 0.044 m^3/sec; 1 GPM = 0.063 L/sec; 1 foot = 0.3 meters; 1 ft^3 = 0.0283 m^3.

At the point of application, ozone dosages averaged 1.5 mg/L and peaked at 4.0 mg/L. The contactors provide 10 minutes of detention time. Raw water total coliform levels are consistently less than 50 per 100 mL. Raw water turbidities range from 1.5 to 2.0 NTU.

The new 0.329-m^3/sec (7.5-MGD) direct filtration treatment plant operates from May through October, when system demand exceeds the capacity of the Ranney collector system. The collector system is still used because it operates at lower cost than the new treatment plant. Generally, the service area customers have been satisfied with the quality of water provided. Judging from the limited number of complaints, the treated river water has a higher customer acceptance than the water from the Ranney collector. Water quality from the collector system is not very different from the raw water from the Columbia River.

THM analyses indicate that the treated water from the new plant has average TTHM concentrations of approximately 14 µg/L, while the water from the Ranney collector averages approximately 107 µg/L TTHM.

Applied Ozone Dosages

Applied ozone dosage rates have ranged from 1.7 to 2.5 mg/L. Until 1983, ozone residual levels at the contactor outlet were maintained at approximately 0.5 mg/L. In 1983, the city installed a dissolved ozone analyzer to control the ozone dosage, which has lowered the dissolved ozone residual concentrations to 0.1 mg/L, thus saving ozone, and still controlling biological growth in the filters and basins prior to chlorination.

Assuming an ozone concentration of 0.1 mg/L is present throughout the 10-minute hydraulic contact time (T_{10} = 5 min), a CT value of 0.5 mg/L-min is obtained. Since the treatment process includes filtration, which provides a 2-log reduction of *Giardia* cysts, only 1 additional log reduction is required to meet the disinfection requirements of the Surface Water Treatment Rule. The CT value attained by ozonation (0.5 mg/L-min) is less than the 1-log reduction requirement at 5°C (0.63 mg/L-min) but provides at least 80 percent of the required disinfection. In addition, the 60-minute chlorine contact time (at pH 8.0) for secondary disinfection, provides an additional 36 percent of the required disinfection. Consequently, the combined disinfection with ozone and free chlorine provides at least 116 percent of the primary disinfection required by the Surface Water Treatment Rule.

System Operation

Operationally, the ozone generation equipment has performed very well. The compressors have required only preventive maintenance. The ozone generators required the replacement of only three burned out tubes during the first 6 years of operation. The major maintenance problem appears to be tube fouling caused by excessive moisture in the feed gas. This situation was caused by two factors. First, after several years of operation, the refrigerant air dryer unit had developed a small leak, which reduced the effectiveness of the air preparation system. Second, the absorptive medium in the desiccant drier should have been replaced when its regeneration capacity was reduced to 40 percent of its original capacity. Cleaning of dielectric tubes has become an annual maintenance procedure.

The only other significant operational problem concerned the ozone contactors. Excess foaming and scum production can occur during spring and late summer, primarily caused by algae destruction. This problem was remedied by installing surface skimmers and froth spray equipment. In addition, the stainless steel tubes holding the ceramic diffusers corrode after about 2 to 3 years of use, and must be checked and occasionally replaced when the diffusers are cleaned.

System Costs

Ozone generation efficiency has averaged approximately 35.3 to 37.5 kWh/kg (16 to 17 kWh/lb) of ozone produced. Based on qualitative observation, the addition of ozone prior to flocculation and filtration results in a 10 percent reduction in the amount of flocculants. In 1987, the treatment plant was producing finished water at a cost of approximately \$3.73/1,000 m^3 (\$14.10/mil gal).

B.2 UV Radiation Case Histories

B.2.1 Ultraviolet Radiation for Primary Disinfection: Fort Benton, Montana

The city of Fort Benton, Montana, obtains drinking water from the Missouri River. The then current filtration plant (20 to 30 years old) was in need of upgrading. Rather than building a new filtration plant, the city built a new 0.088m^3/sec (2-MGD) treatment plant in 1987. Water is drawn through Ranney collectors installed 6 to 7.5 m (20 to 25 ft) below the river bed, a system that allows the river bed to naturally filter the raw water. Turbidities of water entering the treatment plant average 0.08 NTU. No *Giardia* cysts have been found in the water from the Ranney collectors.

The water is treated with UV radiation for primary disinfection, then chlorinated for secondary disinfection. An applied chlorine dosage of only about 1 mg/L is necessary. The entire water treatment system is housed in a 2.97-m^2 (32-sq ft) building.

The UV disinfection system consists of six irradiation chambers, two control cabinets with alarms, chart recorders, relays, hour-run meters, lamp and power on-lights, six thermostats, electrical door interlocks, mimic diagrams, and six UV intensity monitors measuring total UV output. Each irradiation chamber contains one 2.5-kW mercury vapor, medium-pressure arc tube, generating UV radiation at 253.7 nm.

The initial UV dosage is 41,000 μW sec/cm^2 at maximum water flow 104 L/sec (1,650 gal/min) through each irradiation unit. Expected arc tube life is 4,500 operating hours, providing a minimum UV dosage of 25,000 μW sec/cm^2. These conditions are designed to reduce concentrations of *Escherichia coli* organisms by a minimum of 5 logs (10^5 reduction).

The system is equipped with a telemetry control system and fully automated backup system. Each bank of three irradiation chambers has two units on line at all times, with the third unit serving as backup. In the event that the UV intensity drops below acceptable limits (20,000 μW sec/cm^2) in any of the chambers, the automatic butterfly valve will close, stopping flow through the chamber; at that time, the automatic butterfly valve on the standby unit will open. The alarm system also is activated if UV intensity drops below acceptable limits in any of the chambers. The UV alarm system is interfaced with the automatic dialer and alarm system.

In 1987, total equipment costs for the six-unit UV irradiation system with butterfly valves was \$74,587.

EPA's latest Draft Guidance Manual for compliance with disinfection requirements (EPA, 1989a) contains "CT Values" for inactivation of viruses by UV radiation independent of temperature:

Log Virus Inactivation	CT Values by UV (mW-sec/cm^2)
2.0	21
3.0	36

For the UV facility at Ft. Benton, the initial UV dosage of 41 mW-sec/cm^2 provides well in excess of 3 logs inactivation of viruses. However, after 4,500 hours of UV tube operation, the anticipated decrease in UV dosage (to 25 mW-sec/cm^2) will provide only 2 logs of viral inactivation.

B.3 Chlorine Dioxide Case Histories

B.3.1 Predisinfection for THM Control: Evansville, Indiana (Lykins and Griese, 1986)

In order to comply with the November 1979 Amendments to the National Interim Primary Drinking Water Regulations, the Evansville Water and Sewer Utility had to reduce THM levels in drinking water. At the time, raw water was chlorinated, treated with alum prior to primary settling, treated with lime to raise the pH to 8, passed through secondary settling, fluoridated, filtered through rapid sand, and finally postchlorinated. These processes were conducted in two 1.31-m^3/sec (30-MGD) treatment trains. THM levels exceeded the THM standard of 100 μg/L. Prechlorination doses averaged 6 mg/L and distribution system residence time averaged 3 days.

Pilot Plant Study

With EPA's assistance, research evaluating the use of chlorine dioxide as an alternate disinfectant was initiated in a 6.3-L/sec (100-gal/min) pilot plant. One train of the adjacent full-scale treatment plant served as the control for the pilot plant study.

In the first phase of the study, chlorine dioxide was substituted for postchlorination. The water treated with chlorine dioxide was stored 3 days to simulate the distribution system residence time. TTHM concentrations were much lower using chlorine dioxide postdisinfection, 1.2 µg/L (1.9 mg/L residual) as compared with 141 µg/L (2.5 mg/L residual) using chlorine postdisinfection.

Predisinfection with chlorine dioxide to maintain a residual through the pilot plant did not increase the THM concentration and provided adequate disinfection. The chlorine dioxide residual decreased from 4 to 0.3 mg/L through the pilot plant.

In the second phase of the study, the pilot plant procedures were evaluated throughout the year to determine the extent of seasonal effects. In this phase, predisinfection with chlorine dioxide (1.1 mg/L average applied dosage) reduced the amount of THMs formed during postchlorination by approximately 60 percent. The idea of postdisinfecting with chlorine dioxide was abandoned because of the difficulty in maintaining an adequate residual in the distribution system while meeting the current EPA recommendation for total oxidant concentrations from chlorine dioxide (1 mg/L maximum). The average chlorine residual concentration in the clearwell was 2.1 mg/L.

Plant Modifications

Based on data obtained from the pilot plant study, changing the primary disinfectant from chlorine to chlorine dioxide was judged to be the most effective, lowest cost procedure for meeting the THM standard.

A separate building was constructed to house the chlorine dioxide generator and two 907.2-kg (1-ton) cylinders of chlorine. The installation is capable of generating 6.46 kg (14.24 lb) of chlorine dioxide per hour, which can be divided in any proportion between the two halves of the treatment plant. Gaseous chlorine and 25 percent sodium chlorite solution are delivered to the chlorine dioxide reactor under partial vacuum generated by an eductor. Both reagent flows are controlled by flow rate meters, and the system is designed to shut down if the eductor water supply fails or if chemical feed lines break.

Operating Experience

During the first 5 months of use, various chlorine dioxide dosages were used to determine the optimum dosage for maximum reductions in THMs and the portion of the total dosage that would appear as an oxidant residual in the finished water. A general review of system operations was conducted at the same time.

During this period, operators encountered only one major problem. The concentrated disinfectant oxidized the brass corporation cocks used to connect the PVC chlorine dioxide feed lines to the water influent piping. This oxidation and subsequent leaking of chlorine dioxide solution temporarily disrupted operations. The problem was resolved by sliding a section of PVC pipe through new corporation cocks into the main stream of the raw water lines. This modification permitted the PVC piping to serve as an inductor, while preventing direct contact of the brass corporation cocks with concentrated chlorine dioxide solution.

Since the implementation of the new chlorine dioxide system, total oxidant levels from chlorine dioxide have been maintained consistently below the 1.0 mg/L recommended by EPA. With an average applied chlorine dioxide pretreatment dosage of 1.2 mg/L, total oxidant concentrations in the finished water have averaged 0.5 mg/L. These data show that approximately 42 percent of the chlorine dioxide dosage remains as total oxidant. In addition, the new system maintains TTHM levels in the distribution system between 50 and 80 µg/L.

B.3.2 Primary and Secondary Disinfection with Chlorine Dioxide: Hamilton, Ohio (Augenstein, 1974; Miller et al., 1978; U.S. EPA, 1983)

In 1956, a 0.657-m^3/sec (15-MGD) lime-softening ground water treatment plant began operating in Hamilton, Ohio (18 wells, 60-m [200-ft] deep). Chlorine was used as the sole disinfectant when the plant opened. Because of customer complaints of chlorinous tastes and odors, however, chlorine dioxide was tested and then substituted for chlorine in 1972 as a primary and secondary disinfectant.

Hamilton's treatment process now includes aeration, lime addition, flash mixing, sedimentation, recarbonation (with food-grade carbon dioxide), filtration, fluoridation (sodium silicofluoride), chlorine dioxide for primary and secondary disinfection, and clearwell storage. Raw water turbidities are below 1 NTU, and raw water total coliforms are less than 1 per 100 mL.

Generation of Chlorine Dioxide

Chlorine dioxide is generated by mixing 37 percent aqueous sodium chlorite and aqueous chlorine in a ratio of 1:1 by weight, 0.24 g of each reagent/m^3 of water to be disinfected (2 lb of each reagent/mil gal water). This provides an applied chlorine dioxide dosage of 0.25 mg/L. The residual leaving the clearwell is approximately 0.15 mg/L, and is 0.10 mg/L at the extremities of the distribution system.

The generation system consists of one plant-fabricated reactor vessel for chlorine dioxide production, one peristaltic pump for sodium chlorite solution, and two chlorinators (one serves as a standby). Aqueous sodium chlorite solution (37 percent) is delivered in 90.72-kg (200-lb) drums. Chlorine gas is delivered to the site in 68-kg (150-lb) cylinders, and two 68 kg (150-lb) liquid chlorine cylinders are positioned next to the chlorinators. The weight of the cylinder contents is measured by a scale. Switching over from one cylinder to the other is conducted manually. PVC tubing connects the chlorinator and chlorine dioxide reactor vessel; heavy Tygon tubing transports the sodium chlorite solution from the drum to a small plastic day-tank and the reactor. After about 1 month, the Tygon tubing loses its rigidity and must be replaced. The semi-transparent day-tank allows visual inspection of the sodium chlorite level, thus enabling the operator to maintain an acceptable suction head on the peristaltic pump.

The chlorine dioxide reactor vessel is constructed of Schedule 80 PVC piping, 45.72 cm (18 in) high and approximately 15.24 cm (6 in) in diameter. The vessel is filled with PVC rings, 2.5 cm (1 in) in diameter. The chamber is opaque except for the sight glass mounted in-line on the discharge piping. A white card is positioned behind the sight glass for better observation of the chlorine dioxide color.

Other Benefits of the Chlorine Dioxide System

Prior to the installation of the chlorine dioxide system, customers complained about brown iron stains on laundry articles. The switch to chlorine dioxide treatment loosened the brown slimes from the mains. The distribution system was flushed, and shortly thereafter the staining problems were resolved. Plant personnel attributed the source of the problem to *crenothrix* and *leptothrix* bacteria (iron bacteria) that had been present in the extremities of the distribution system before chlorine dioxide was introduced.

Costs of Chlorine Dioxide

In 1977, the costs of chlorine dioxide disinfection for Hamilton were about 3.6¢/capita/year. Total chemical costs averaged \$0.05/m^3 (\$0.19/1,000 gal), chlorine dioxide accounting for only a fraction of the total. The operating and maintenance costs were less than \$50 annually. The plant-fabricated reactor, piping, hardware, and installation, which was performed by plant personnel, cost roughly \$400 in 1977. The peristaltic pump for sodium chlorite solution cost less than \$200. The two chlorinators, each worth \$600, were already on line at the plant.

Implications of the SWTR CT Values for Hamilton

Although the Hamilton raw water is ground water, and therefore probably will not be subject to the requirements of the Surface Water Treatment Rule, it is interesting to consider the effects if such disinfection requirements as listed in the latest EPA Guidance Manual (U.S. EPA, 1989a) were to be levied on this water supply system.

Chlorine dioxide is added to the Hamilton water in applied doses of 0.25 mg/L as it enters the clearwell. The water temperature is about 20°C (68°F) year round, and the pH is 9.4 to 9.5. Hamilton's first customer is located about 0.5 miles from the plant. Thus there is very little contact time in the distribution system.

The plant filters efficiently, and therefore only 1-log additional inactivation of *Giardia* cysts and 2-logs inactivation of viruses need to be provided by the chlorine dioxide. The latest EPA Guidance Manual shows that at 15°C (59°F) CT values of 5 mg/L-min and 2.8 mg/L-min will provide the required degrees of disinfection for *Giardia* cysts and viruses, respectively.

The Hamilton clearwell holds 1,892,500 L (500,000 gal). During periods of high water use, water is produced at the rate of 0.83 m^3/sec (19 MGD). In periods of low water use, only 0.35 m^3/sec (8 MGD) are produced. Thus the contact time in the Hamilton clearwell ranges from 30 to 90 minutes, at the high and low production rates, respectively. Assuming that the average concentration of chlorine dioxide in the clearwell is 0.2 mg/L, then the CT values provided are 6 and 18 mg/L-min, respectively.

Thus the current disinfection conditions using chlorine dioxide meet the CT requirements for both *Giardia* cysts and viruses.

Secondary Disinfection with Chlorine Dioxide at Hamilton

The recently promulgated Surface Water Treatment Rule requires that a minimum secondary disinfectant concentration of 0.20 mg/L be present as the water enters the distribution system or that the HPC level be less than 500 mL in the distribution system. Since Hamilton's chlorine dioxide concentration entering the distribution system is 0.15 mg/L, the city will either have to increase this residual to 0.20 mg/L or rely on HPC analyses to comply.

B.3.3 Preoxidation with Chlorine Dioxide, Postchlorination with Chlorine Dioxide and Chloramine: Galveston, Texas (Myers et al., 1986)

The Galveston County Water Authority (GCWA) owns and operates an 18-MGD water treatment plant in Texas City, Texas. In 1986, the plant produced 0.526 m^3/sec (12 MGD) from the Brazos River. Raw water had high color; variable turbidities (68 to 111 NTU); high organic content; high iron (2.7 to 3.8 mg/L); and seasonally high algae content, sometimes reaching levels of 5,000 blue-green algae per mL. Total THM formation potentials for Brazos River water between September 1983 and April 1984 ranged from 400 to 650 µg/L. Finished water THM concentrations during the same period ranged from 180 to 350 µg/L. During periods of intermittent raw water algae blooms and associated high organic carbon contents, numerous taste and odor complaints were received. These problems prompted an investigation of alternative disinfectants.

The original treatment process included the addition of cationic polymers for primary coagulation, lime for pH adjustment, prechlorination for taste and odor control, and ferric sulfate as a flocculent aid prior to upflow reactors and clarifiers, which provided flocculation and sedimentation. Dual media filtration, chlorine disinfection (2.4 to 5.0 mg/L total available chlorine residual), and fluoridation followed.

Pilot Study

Chloramines and chloride dioxide were selected for further study and pilot plant testing. Several preoxidation/postdisinfection combinations involving chlorine, chloramine, and chlorine dioxide were studied. The studies indicated that preoxidation with chlorine dioxide and postdisinfection with chlorine dioxide in combination with either chlorine or chloramines provided effective taste and odor control, maintained an active disinfectant residual, and minimized THM formation (to a level of about 68 µg/L). A brief summary of the results obtained from the combinations studied follows:

- *Chlorine/chlorine* - Finished water exhibited intermittent algae-related tastes and odors; THM levels were in excess of Federal standards; bacterial quality was excellent.

- *Chloramine/chloramine* - Chlorine-to-ammonia weight ratios of 3:1 and 7:1 were used. With the 3:1 ratio, THM levels were lowered to about 60 µg/L, but bacterial counts for coliforms indicated an inadequate residual in the distribution system. The experiments were repeated using a 7:1 chlorine-to-ammonia ratio. Acceptable bacterial quality was achieved, but numerous taste and odor complaints were received during this period.

- *Chlorine dioxide/chlorine* - Chlorine dioxide was installed in May 1984, using a generator with a conversion efficiency (chlorite ion to chlorine dioxide) of approximately 80 percent. After preoxidation, the clarified water showed no traces of THMs. No taste and odor complaints were received, despite very high raw water algae counts (up to 5,000 blue-green algae/mL). However, finished water THM levels sometimes persisted above 100 µg/L. In November 1984, a chlorine dioxide generator with a greater than 98 percent conversion efficiency was installed. THM levels averaged 102 µg/L, and no taste and odor complaints were received.

- *Chlorine dioxide/chlorine dioxide* - Chlorine dioxide for preoxidation and postdisinfection was tested in March 1985. The Texas Department of Health required that a maximum chlorine dioxide residual of 1.0 mg/L be maintained and that finished water quality be monitored throughout the distribution system. During the test period, finished water THM concentrations averaged 60 µg/L and turbidities were the lowest of any of the alternative disinfectant scenarios. Bacterial counts generally were excellent, but intermittent elevated counts were noted at the clearwell and at two locations 3.2 and 8 km (2 and 5 mi) from the plant. Additionally, bacterial counts displayed a shift from orange to yellow-staining gram negative (-) rods to white-staining gram positive (+) rods, similar to slime-forming *Bacillus* sp.

- *Chlorine dioxide/chlorine and chlorine dioxide* - The above test was repeated, adding chlorine in conjunction with chlorine dioxide. Excellent bacterial quality was obtained with plate counts at or below the 500 colonies per 100 mL for all monitoring locations. The shift in bacterial species distribution continued as the plate counts decreased, so that over 95 percent of all colonies examined were either yellow gram negative rods or white gram positive rods. THM levels of the finished water rose to an average of 81 µg/L.

- *Chlorine dioxide/chloramine and chlorine dioxide* - These tests were conducted in December 1985. THM levels of the finished water averaged 68 µg/L, and the bacterial quality remained excellent. No coliforms were found in the clearwell or in the distribution system and bacterial counts had ranged from <1 to 30 colonies per 100 mL. Of those colonies identified, over 95 percent were white-staining, gram positive *Bacillus* sp. with less than 1 percent belonging to the orange-staining gram negative type.

Chlorine dioxide preoxidation has proved an excellent algicide and biocide. The effectiveness of chlorine dioxide in removing algae in flocculation and sedimentation has resulted in a decrease in filter fouling and improved finished water turbidities. Odors experienced at the flocculating clarifiers, and taste and odor complaints were reduced.

Bacterial counts, using the membrane filter method for coliforms, have greatly improved as well. With the new system, all bacterial counts obtained from samples collected at the distribution point have been below 30 colonies per 100 mL, and all but two samples have been below 10 colonies per 100 mL. Samples from the distribution system are continually below the guideline of 500 colonies per mL, and often are below 5 colonies per 100 mL. These counts indicate an adequate and stable residual.

B.4 Chloramine Case Histories

B.4.1 Prechlorination, Postchloramination: Bloomington, Indiana (Singer, 1986)

Bloomington obtains raw water from a lake with TOC levels of 4 to 6 mg/L. The water is treated with alum coagulation, flocculation, sedimentation, and filtration through pressure filters. Prior to September 1984, chlorine was applied to the raw water and just before the pressure filters. Average chlorine dosages were 1.8 and 1.0 mg/L at each point, respectively. According to quarterly compliance monitoring reports, average TTHM concentrations were exceeding the 100 µg/L limit a majority of the time.

In September 1984, the Bloomington water utility changed from post chlorination to postchloramination. Before the pressure filters, an average 0.54 mg/L of ammonia was applied along with 1.5 mg/L of chlorine. The desired residual chlorine concentration leaving the plant of 1.0 mg/L of free chlorine was changed to 1.5 mg/L of combined chlorine. After the change, quarterly THM levels ranged from 24 to 57 µg/L.

Table B-2 summarizes THM and TOX (total organic halide) data for samples collected at points in the treatment train when chlorine was used for both pre- and posttreatment. The data show that TOX levels increase with TTHM levels. Table B-3 summarizes similar data after postchloramination was instituted. These data show that although the TTHM formation ceases after the addition of ammonia, the production of TOX continues, but at a greatly reduced rate.

Thus, as MCLs for halogenated organic materials other than THMs are promulgated, utilities using postchloramination should plan to determine the makeup of their TOX fraction.

Table B-2. Summary of THM Data at Bloomington, Indiana, with Free Chlorination, August 16, 1984

Sampling Point	Chlorine Residual (mg/L)	Toc (mg/L)	Tthm (µg/L)	Tox (µg/L)
Raw water	–	4.3	1	23
Settled water	0.25	3.6	48	127
Filtered water	1.0	2.4	81	205
Dist. system #1	1.8	–	110	291
Dist. system #2	0.65	–	151	363

Source: Singer (1986).

Table B-3. Summary of THM Data at Bloomington, Indiana, with Postchloramination, August 26, 1984

Sampling Point	Chlorine Residual (mg/L)	Toc (mg/L)	Tthm (µg/L)	Tox (µg/L)
Raw water	–	4.1	0	17
Settled water	Trace, free	2.8	53	94
Filtered water	1.2 combined	2.8	55	91
Dist. system #1	1.0 combined	–	52	115
Dist. system #2	0.9 combined	–	57	116

Source: Singer (1986).

Since switching to postchloramination, the utility has experienced no adverse effects in operations or in finished water quality. According to distribution system monitoring records, the microbiological quality of the water has been maintained.

B.4.2 Prechlorine Dioxide, Prechlorination, and Postchloramination: Philadelphia, Pennsylvania (McKeon et al., 1986)

The Baxter Water Treatment Plant, a 12.35-m^3/sec (282-MGD) conventional treatment plant built in 1960, supplies drinking water from the Delaware River to a population of over 800,000. Chemicals used in treatment include chlorine, ferric chloride or ferrous sulfate, lime, fluoride, and ammonia. Powdered activated carbon is used on demand for control of taste and odor, and chloride dioxide is used for control of THMs, tastes, and odors. The chlorine dioxide system was left over from the previous water treatment plant on that site. In the 1950s, it was used to oxidize phenolic compounds found in the watershed, which have since been eliminated.

Prior to 1976, the Baxter plant practiced breakpoint chlorination at the raw water basin and maintained free chlorine in the distribution system. A total of 96 hours of free chlorine contact time was typically achieved.

In 1978, analyses of THMs showed peak concentrations above 300 µg/L with an annual average of 140 µg/L. In light of these results, the Philadelphia Water Department began to reevaluate its disinfection strategies. Factors that affected the selection of an alternative included THM, bacterial, taste, odor, algae, and corrosion control; residual duration; and economics.

Process Modifications

Modifications were implemented between 1976 and 1983. Chloramination of the finished water was introduced in 1976 to reduce free chlorine contact times and THM levels. Ammonia was added to convert free chlorine to monochloramine. The monochloramine reduced contact time from 96 to 24 hours, provided a stable residual in the distribution system, improved the organoleptic properties of the water, and reduced the corrosion rates associated with the use of free chlorine. Adequate disinfection was assured by maintaining a free chlorine residual of 2 to 3 mg/L throughout the plant treatment process. THM formation potential within the treatment process was reduced by 40 percent (from 231 to 174 µg/L with the 96-hour contact time) under summer conditions.

Between 1976 and 1979, the addition of chlorine at the raw water basin inlet was gradually phased down from 7.19 g (60 lb) in 1975 to between 2.4 and 3.6 g chlorine/m^3 (20 to 30 lb/mil gal) in 1979. While algae blooms, taste, and odor problems were controlled, THM levels were still too high (200 g/L in the summer, and 140 µg/L annual average).

In early 1980, routine use of chlorine at the raw water intake basin was abandoned. In its place, chlorine dioxide was added in dosages between 0.5 and 1.0 mg/L. Free chlorine contact times were reduced to 5 hours. Summertime THM values were reduced from 200 µg/L to 140 µg/L. This treatment was sufficient to control taste and odor problems from algae at all times except during the spring algae bloom. For that period of time, breakpoint chlorination of the intake water and/or 12 to 24 g powdered activated carbon/m^3 (100 to 200 lb/mil gal) were added to eliminate vegetative tastes and odors.

In the fall of 1980, a chlorine application point was installed in the "applied to filters" channel which allowed for increased flexibility in the use of chlorine. Free chlorine contact time was reduced from 5 hours to 1. Chlorine was added at the rapid mix to barely achieve breakpoint and provide a residual, which dissipated within a few minutes. Chloramines were carried across the flocculation and sedimentation basins. Sufficient chlorine was then added at the new application point to achieve a free chlorine residual of 1.5 to 2.0 mg/L. This residual was converted to chloramines 1 hour later as the water left the filter building.

This treatment regimen gave adequate control of taste and odor, again, except during spring algae blooms, which forced a reversion back to free chlorine at the intake. THM levels, with only 1 hour of free chlorine contact time, resulted in summer values averaging 100 µg/L with an annual average of 60 µg/L.

In November 1982, a 10-minute chlorine contact time was tested. Results indicated that satisfactory disinfection could be achieved with only chloramines carried through the flocculation/sedimentation basins when the water temperature was below 15.6°C (60°F). This strategy was initiated on a plant scale in December 1982. Adequate disinfection was achieved, but periodic taste and odor problems persisted, especially after storms. Average annual THM levels were reduced from 60 to 50 µg/L.

The 10-minute chlorine contact time trial was terminated in December 1983, because the disinfection scheme did not adequately address taste and odor problems. The treatment regimen returned to a 1-hour free chlorine contact time.

Costs

Over the 10-year period, a 70 percent reduction in THM concentrations was realized. In 1978, disinfection cost $1.32/1,000 m^3 ($5.01/mil gal); in 1986, the cost was $1.46 ($5.52) (1977 dollars). Cost increases were minimized because the reduced chlorine contact times resulted in less evaporative losses of chlorine, which netted a 20 percent decrease in the amount of chlorine needed.

Future Considerations

The near-term goal of the Philadelphia Water Department is to reduce the annual average THM concentrations to below 50 µg/L, a 15 percent reduction from 1986 levels. This can be achieved at minimum expenditure by installing a pH adjustment point at posttreatment. The existing treatment scheme calls for raising the raw water pH to 8.4 during rapid mix and carrying this high pH through the distribution system for corrosion protection. Addition of a pH adjustment point at posttreatment will allow a pH of 7.5 to be used through the flocculation/sedimentation basins and filters, moving adjustment to a 8.4 pH to just after chloramination. Plant-scale trials of this strategy yielded a 20 percent reduction in THM formation (to about 40 µg/L).

If the THM MCL is reduced to below 50 μg/L, ozone and/or GAC become the likely alternatives at the Baxter plant. Extensive laboratory and pilot plant evaluations have developed conceptual full-scale plant designs incorporating these two treatment techniques.

Estimated annual amortized capital and operating costs for ozone at the Baxter plant are estimated to be about $13.21/1,000 m^3 ($50/mil gal). The associated costs for GAC postfiltration (15-minute EBCT) with a 75-day regeneration frequency would be about $56/1,000 m^3 ($212/mil gal). This design configuration (ozone plus postfiltration GAC) is capable of producing THM concentrations of less than 10 μg/L.

Appendix C

Experience with Treatment Technologies for Organic Contaminants

This appendix is a collection of field experiences with organics treatment technologies. Section C.1 includes descriptions of GAC systems in use, C.2 describes packed tower aerators, and C.3 describes powdered activated carbon.

C.1 Experience with Granular Activated Carbon

C.1.1 GAC for VOC Removal: Washington, New Jersey (Chrobak et al., 1985)

A well with a 0.04 m^3/sec (0.8-MGD) capacity in this community was contaminated with four VOCs. The VOC levels fluctuated throughout the well's 9-hour daily operating cycle according to the following concentration ranges:

- Tetrachloroethylene (PCE): 50 to 500 µg/L
- Trichloroethylene (TCE): 1 to 10 µg/L
- 1,1,1-Trichloroethane: 1 to 20 µg/L
- Carbon tetrachloride: 1 to 5 µg/L

The town selected a GAC system with two downflow pressure contactors in parallel. The filters have 2.1-m (7-ft) diameters, 3-m (10-ft) carbon depths, hydraulic loadings of 4.8 L/sec/m^2 (7.1 GPM/ft^2), and EBCTs of 10.5 minutes. Filter backwashing is performed once a month due to solids in the influent. Wash water is filtered with sand media and recycled. Figure C-1 shows a schematic diagram of the plant.

Carbon usage rates for meeting the PCE standard averaged about 10.8 g GAC/m^3 (90 lb/mil gal) of treated water. Meeting the 1,1,1-trichloroethane standard required carbon usage rates of 25.2 g GAC/m^3 (210 lb/mil gal) of treated water. GAC consumption for the two compounds before breakthrough (any contamination of the effluent, as opposed to effluent concentrations over the regulated standards) was about 12 and 32.4 g/m^3 (100 and 270 lb/mil gal) for PCE and 1,1,1-trichloroethane, respectively.

The capital cost for the system was $508,500 in 1981 dollars, while the operating costs have been about $15,000 annually based on 9 hours of operation per day.

C.1.2 GAC for Contaminant Control: Cincinnati, Ohio (DeMarco, 1983) (Westerhoff and Miller, 1985)

This case involves a 9.6-m^3/sec (220-MGD) plant that uses Ohio River water. The proposed addition of GAC to the original process train is shown in Figure C-2. Figure C-3 shows TOC removal during a GAC pilot study conducted at the plant. The chief goals of adding GAC to this plant were to (1) reduce TOC in the effluent to less than 1.0 mg/L, (2) maximize the use of existing plant facilities, (3) maximize the flexibility to accommodate future requirements, and (4) keep costs reasonable.

The primary design elements for the addition were a postfiltration GAC adsorption unit that used downflow deep-bed contactors, post-GAC chlorination, and onsite carbon regeneration using a fluidized-bed furnace. In addition, the design was intended to keep carbon losses to a minimum.

The addition of the GAC process resulted in the following system capacities:

- 7.7 m^3/sec (175 MGD) maximum daily plant flow rate
- 5.4 m^3/sec (124 MGD) average daily average plant flow rate
- 15-minute EBCT- 3-m (10-ft) GAC bed depth
- 3-meter (10-foot) GAC bed depth
- 24,494.4 kg/day (54,000 lb/day) average carbon usage

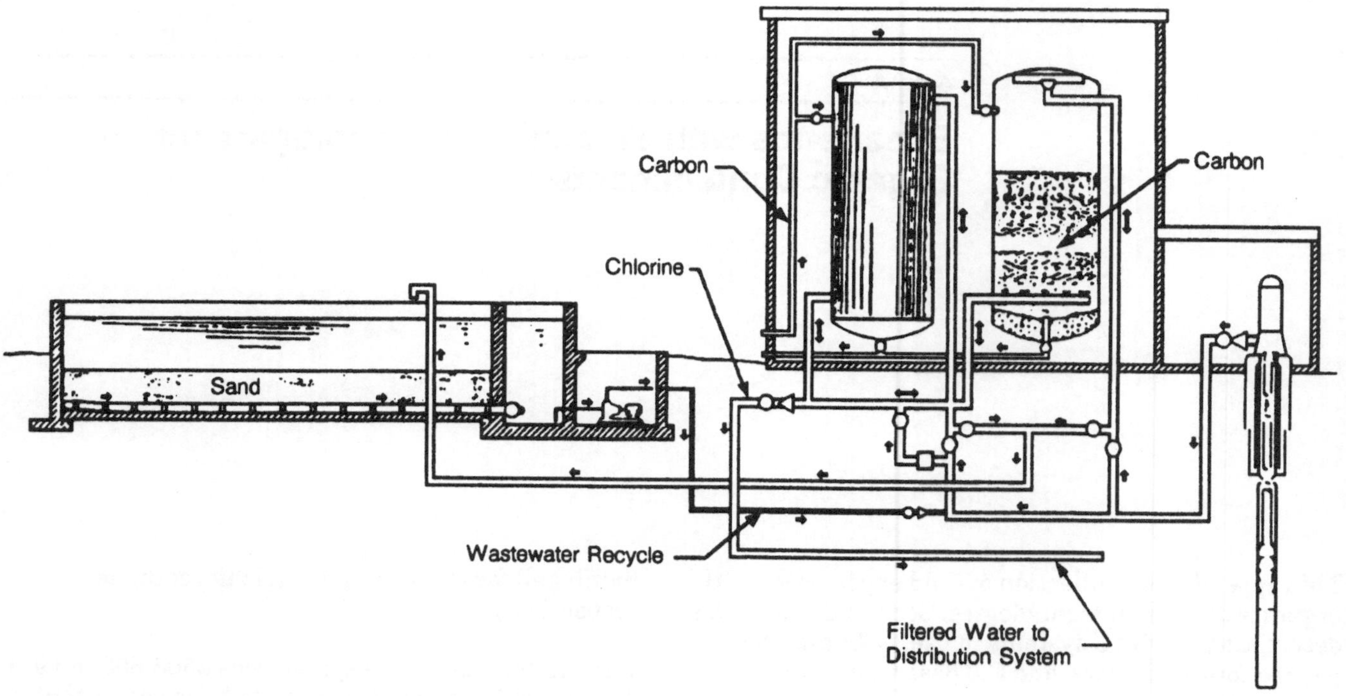

Figure C-1. GAC treatment plant schematic Vannatta Street Station.

- 41,731.2 kg/day (92,000 lb/day) peak carbon usage rate

The carbon transportation system employed for regeneration uses schedule 10 316L stainless steel pipe and maintains water velocities of 91.4 to 152.4 cm/sec (3 to 5 ft/sec). The layout design required that pipe bends in the transportation system have minimum radii to allow the free flow of the carbon and minimize abrasion. As a result, pipe bends for 7.6-cm (3-in) diameter pipe used 60.96 cm (24-in) minimum radii, 10.16-cm (4-in) pipe used 91.4-cm (36-in) radii, and 20.32-cm (8-in) pipe used 121.9-cm (48-in) radii. Figure C-4 shows the regeneration system.

The capital cost of adding the GAC system was $57.7 million (1988 dollars). This included the GAC contactors, regeneration equipment, intermediate pumping equipment, outside piping, and modifications to the existing plant. The operating and maintenance costs are estimated to average $3 to $4 million annually, including costs for labor, power, natural gas, and replacement carbon. These combined costs resulted in a 30 to 40 percent increase in the consumers' average annual water charge from $80 to about $110 per year.

C.1.3 EPA Health Advisory Example

This case involves a system serving 30,000 persons. The plant had a capacity of 0.22 m^3/sec (5.1 MGD), an average demand of 0.13 m^3/sec (3.0 MGD), and peak demands of 0.18 m^3/sec (4.2 MGD). Table C-1 presents a profile of this 1957 system.

The influent for the system is characterized in Table C-2. The presence of vinyl chloride required use of PTA because GAC is ineffective for that compound. However, GAC was also needed to remove trichloroethylene and aldicarb.

Costs for three technological alternatives were developed. The alternatives included GAC and PTA individually and GAC and PTA together. PTA proved to be almost one fourth as expensive as GAC. However, combining the two treatments proved less expensive than adding both treatments separately.

C.2 Experience with PTA: Scottsdale, Arizona (Cline et al., 1985)

The only application of the PTA system described in this appendix is located in Scottsdale, Arizona. In the Scottsdale system, PTA was added to a system with 24 wells and the combined capacity of 1.75 m^3/sec (40 MGD). Two of the wells were contaminated with TCE at levels of 18 to 200 µg/L and 5 to 43 µg/L, respectively.

Both PTA and GAC were considered as potential treatment solutions; PTA was chosen for its cost effectiveness. For this application, GAC was estimated to cost from $0.04 to $0.10/$m^3$ ($0.17 to $0.38/1,000 gal), while PTA was estimated to cost only $0.02/$m^3$ ($0.07/1,000 gal).

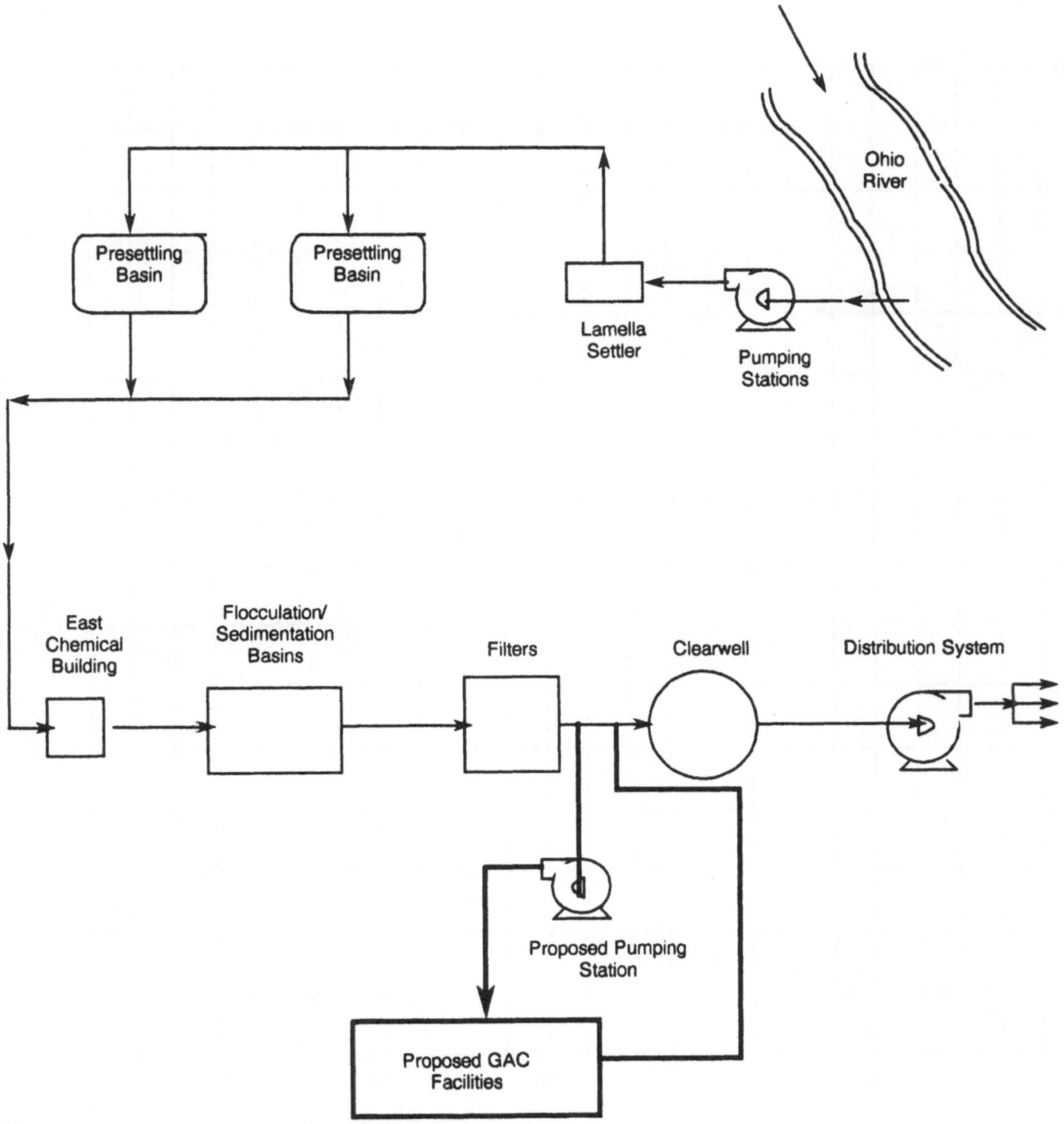

Figure C-2. Cincinnati treatment train with addition of GAC.

The PTA unit was designed to manage a 75.7 L/sec (1,200 GPM) flow rate with 3.6 m (12 ft) of packing material, a 50:1 air-to-water ratio, and a column diameter of 3 m (10 ft). Figure C-5 is a schematic diagram of this facility.

The PTA achieved over 98 percent reductions from the initial influent levels, producing effluent levels of TCE ranging from 0.5 to 1.2 µg/L. The system's capital cost was $300,000, and its operation and maintenance costs average $25,000 annually.

C.3 Experience with PAC

Currently only pilot-scale units operate in the United States. There are, however, full-scale Roberts-Haberer units operating in Africa and Europe, including a 0.42 m^3/sec (9.5-MGD) plant in Wiesbaden, West Germany, that has operated since 1970.

In Newport News, Virginia, a Roberts-Haberer pilot plant reduced TOC by 90 percent during initial operations. Within a 24-hour period, TOC removal decreased to only 20 percent. During that same period, this unit significantly reduced THM precursors.

In Shreveport, Louisiana, a Roberts-Haberer unit was used as a roughing filter to reduce THM formation potential. The unit succeeded in reducing THM formation potential from about 265 µg/L by 7 to

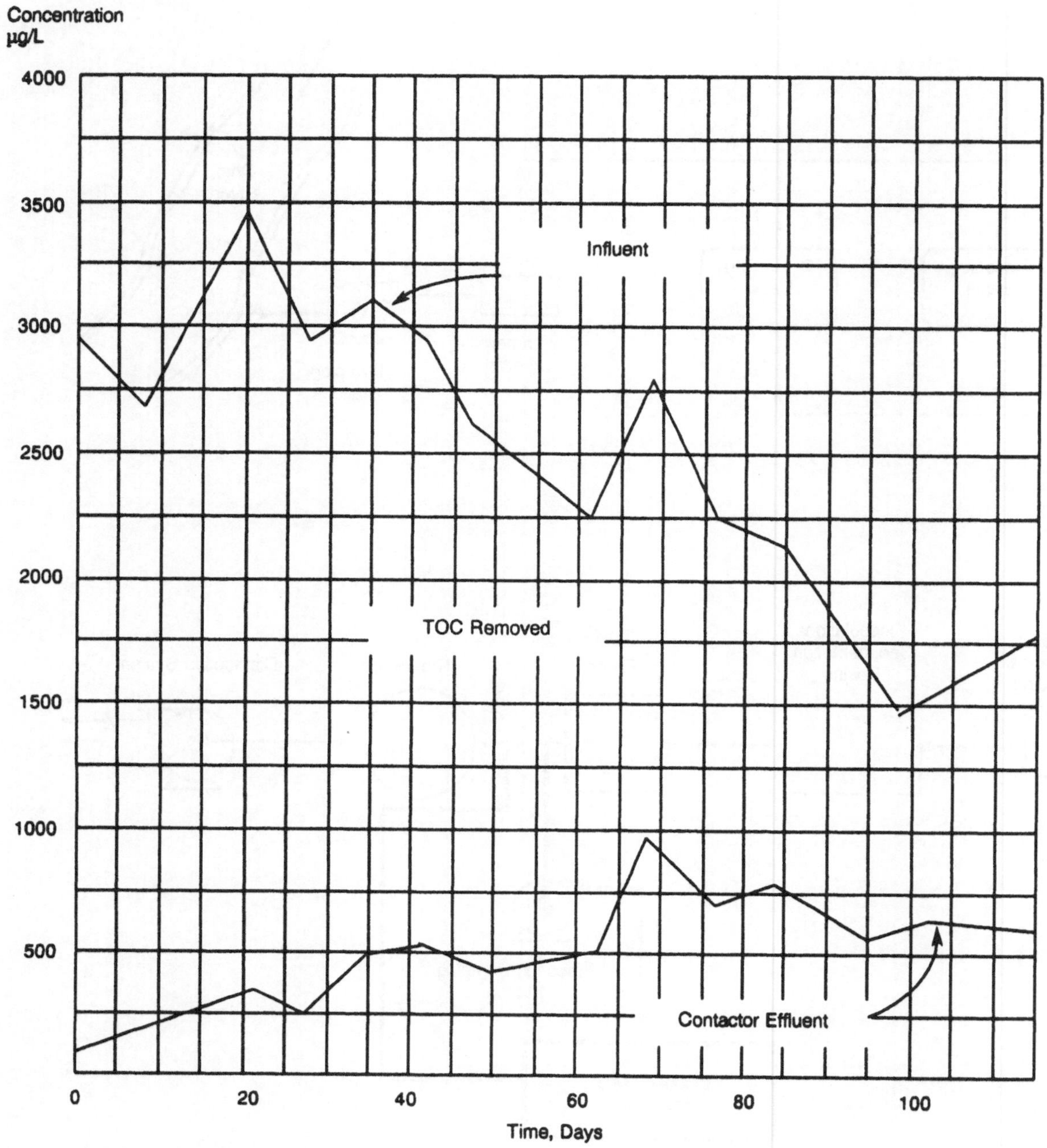

Figure C-3. Typical TOC reduction curve during pilot study.

31 µg/L with PAC dosage of 26 to 29 mg/L. This operation, however, did not use the optimum PAC dosage.

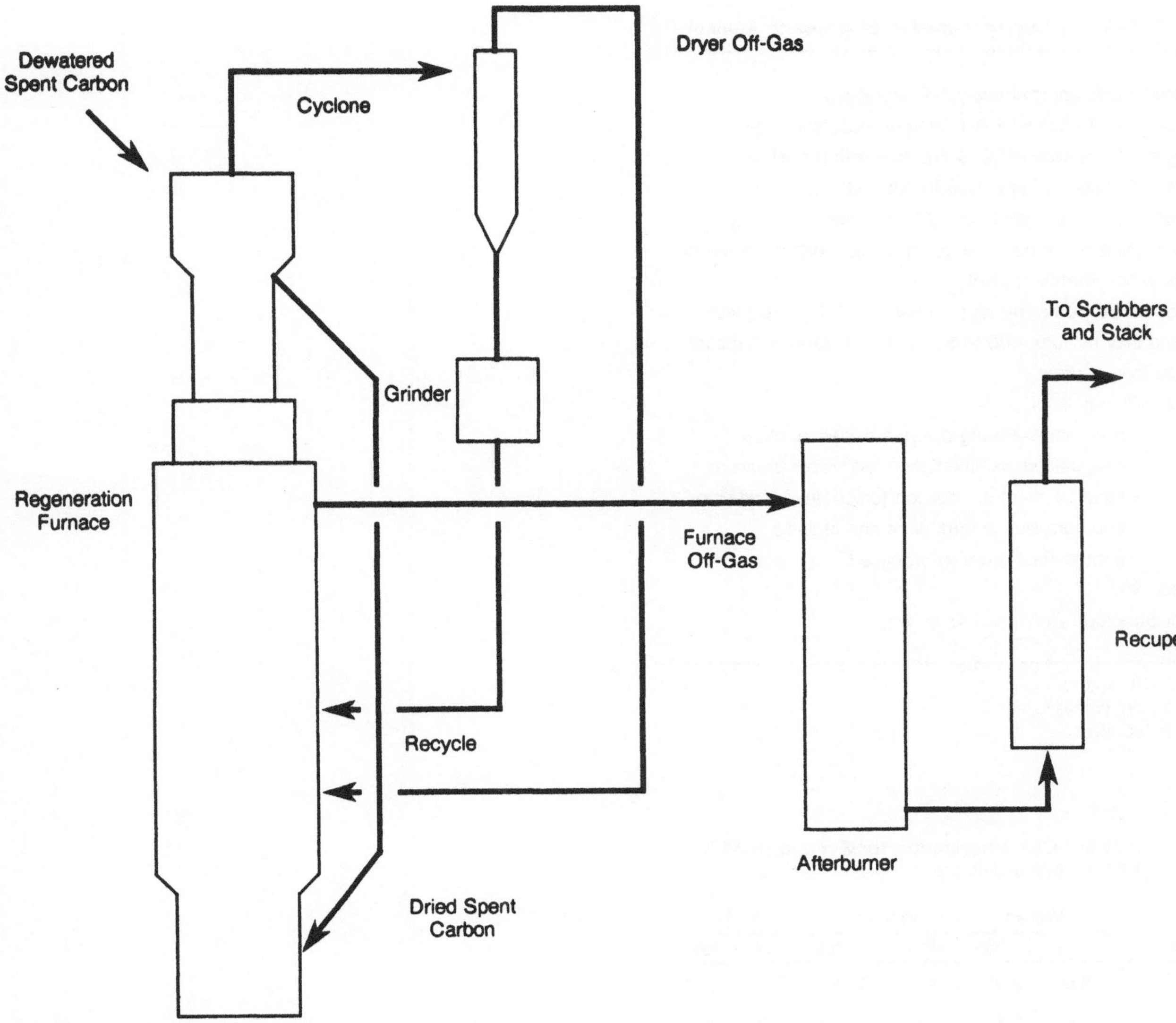

Figure C-4. Regeneration system schematic.

Table C-1. Profile of System used in EPA Health Advisory

Source:

- Three wells approximately 500 feet deep
- Capacity of each well is 1.8 million gallons per day
- Screened between 400 to 500 feet with gravel pack
- 18-inch steel casing from 0 to 400 feet
- Portland cement grout from 0 to 200 feet
- All wells are pumped to a common manifold that flows to the water treatment plant
- Soil profile: 0 to 100 feet, sandy soil; 100 to 400 feet, sand clay mixture; 400 to 500 feet, wet sand and gravel; 500 feet, bedrock

Storage:3.5 million gallons

Treatment: Iron removal using chlorine oxidation, alum coagulation, sedimentation, and rapid pressure sand filtration; disinfection (chlorine), fluoridation, and corrosion control (lime and metallic phosphates) are also practiced.

Constructed:1957

Mechanical Structural Condition: Excellent

1 foot = 0.3048 meters.
1 inch = 2.54 centimeters.
1 gallon = 3.785 liters.

Table C-2. Influent Characterization for System Used in EPA Health Advisory

	Well #1		Well #2		Well #3	
Parameter	Raw	Treated	Raw	Treated	Raw	Treated
Iron (mg/L)	3.0	0.05	2.2	0.05	2.0	0.05
pH	6.0	7.8	5.9	7.8	6.2	7.8
Alkalinity (mg/L)	10	110	14	110	12	110
Vinyl chloride (μg/L)	40	20	14	20	6	20
Trichloroethylene (μg/L)	50	60	30	60	100	60
Aldicarb (total) (μg/L)	30	30	30	30	30	30
Total organic carbon (mg/L)	3.0	1.0	2.1	1.0	1.0	1.0

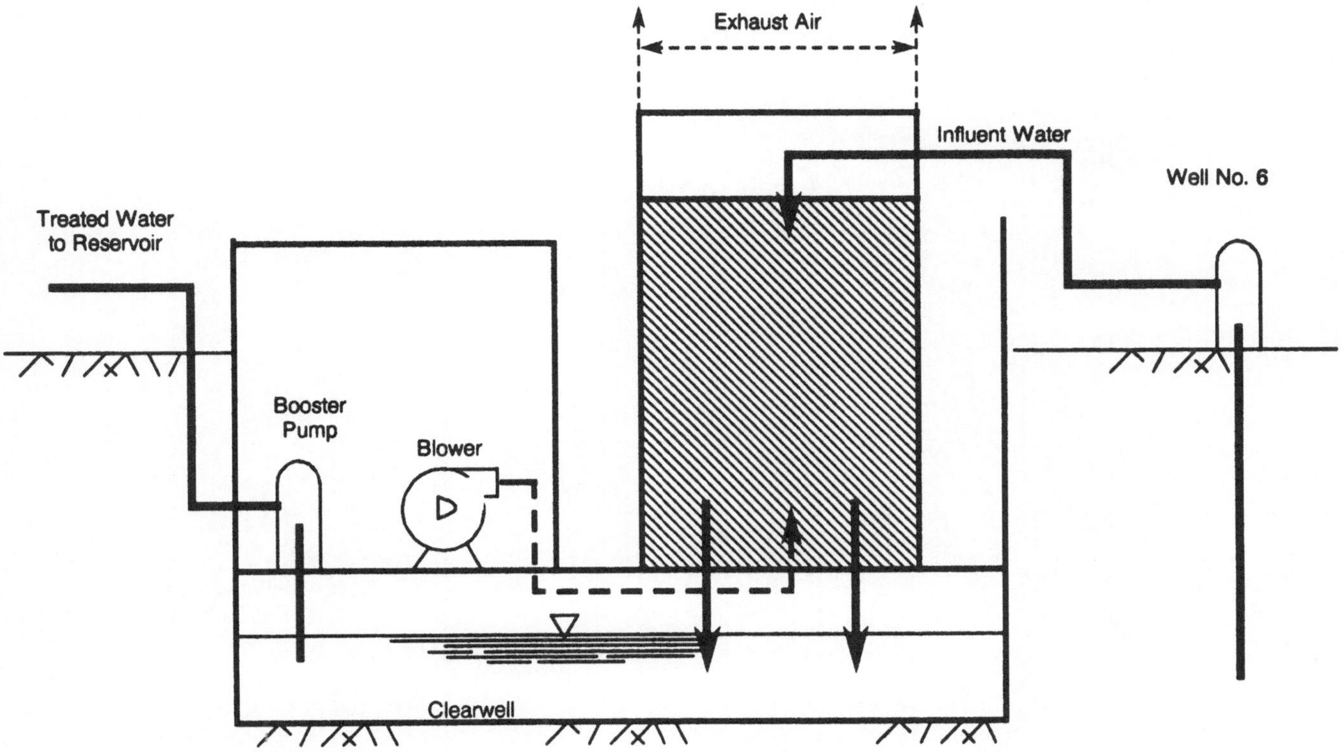

Figure C-5. Schematic diagram of a Scottsdale packed column.

Appendix D

Experience with Treatment Technologies for Inorganic Contaminants

This appendix describes a number of field experiences with inorganics treatment technologies. Section D.1 covers two cases of corrosion control; Section D.2, the use of coagulants; Section D.3, reverse osmosis; Section D.4, ion exchange; and Section D.5, activated alumina.

D.1 Corrosion Control

D.1.1 Controlling Lead: Seattle, Washington

Seattle, Washington, has instituted a successful corrosion control program. The program has a component that assists residents in materials selection, in addition to a water treatment modification program. The city encourages the use of plastic piping, types K and L copper piping, and solder with less than 0.2 percent lead. In addition, the city requires dielectric insulators to mitigate corrosion from joining dissimilar metals.

D.1.2 Controlling Lead with pH Adjustment: Boston, Massachusetts

A 1975 survey found significant levels of lead in tap water supplied to Boston residents by the Metropolitan District Commission (MDC). The protected watersheds of the MDC's main water sources precluded contamination of the raw water. Instead, the lead contamination came from a combination of corrosive water and extensive use of lead in service lines and plumbing. The corrosive water was characterized as acidic and low in hardness and alkalinity. Table D-1 presents water quality characteristics for the MDC's raw and finished water. Prior to discovery of the lead problem, the finished water underwent only chlorination and ammoniation treatments, as shown in the column for "raw water."

The MDC, in conjunction with U.S. EPA, developed a water monitoring program to ascertain the optimal

Table D-1. Metropolitan District Commission Water Quality Data

Parameters	Shaft 4 (Southborough, MA) Raw Water	Norumbega Reservoir (Weston, MA) Finished Water
Hardness (as $CaCO_3$)	12	12
Alkalinity (as $CaCO_3$)	8	12
Total Dissolved Solids	37	46
Calcium	3.2	3.4
Sodium	5.5	9.7
Sulfate	<15	<15
Chloride	<10	<10
Specific conductance (micromhos)	59	78
pH (units)	6.7	8.5
Copper	<0.02	<0.02
Iron	<0.10	<0.10
Zinc	<0.02	<0.02
Lead	<0.005	<0.005

Note: All values in mg/L unless otherwise specified.
Source: U.S. EPA (1984).

solution for corrosion control. The monitoring program was also designed to measure progress in reducing corrosion with the chosen solution. The MDC selected two treatment alternatives.

The first treatment, plan adding zinc orthophosphate, was implemented from June through December 1976. While reductions in lead contamination were attained, levels were still above the MCL.

The second treatment plan data shown in the second column in Table D-1, which proved more successful, consisted of adding sodium hydroxide to adjust the pH. During 1979, pH levels dropped below 8.0 and then were increased to the target level of 9.0. Lead levels in the water rose and then dropped with the resumption of the higher pH level. The pH increase also reduced copper contamination levels, but had little effect on iron corrosion.

The treatment consisted of applying a dosage of 14 mg/L of a 50 percent sodium hydroxide solution to about 13.18 m^3/sec (301 MGD). This dosage, in 1981, cost $2.54/1,000 m^3 ($9.64/mil gal of treated water). Chemical costs for that year were $900,000 and operation and maintenance costs were $161,000.

D.2 Coagulation to Control Barium: Illinois

This case involves a conventional treatment plant in northeastern Illinois with high barium concentrations in its source groundwater. Barium concentrations in the plant's influent ranged between 0.4 and 8.5 mg/L; the MCL for barium is 1 mg/L. Contamination was found in small areas of the water utility's watershed and the contaminated water was drawn through Cambrian/Ordovician soils. No barium contamination occurred in wells with sulfate levels greater than 50 mg/L, because of the low solubility of barium sulfate.

Plant operators conducted many jar tests to evaluate alternative chemical additives to address the barium contamination. Influent samples with 7.4- mg/L levels of barium were used. The tested additives and their operational purposes include:

- Alum as a coagulant-precipitant
- Ferrous sulfate as a coagulant-precipitant
- Calcium hydroxide as a precipitant-pH adjuster
- Sulfuric acid as a pH adjuster-precipitant
- Hydrochloric acid as a pH adjuster
- Sodium hydroxide as a pH adjuster
- Potassium hydroxide as a pH adjuster
- Calcium sulfate (gypsum) as a precipitant
- Commercial gypsum as a precipitant
- Sodium bisulfate as a precipitant
- Anionic polymer as a flocculent-filter aid
- Diatomaceous earth as a filter precoat

The laboratory jar tests determined that the most effective additive was calcium sulfate. Optimum dosages ranged between 75 and 175 mg/L with a pH of 11.0. A 0.066-m^3/sec (1.5 MGD) pilot plant provided precipitation, direct filtration, and polymer additions for a variety of barium concentrations, which confirmed the laboratory approximation.

At full-scale operation, the plant uses dosages of 100 mg/L of gypsum and 0.25 mg/L of polymer to achieve a 91 percent reduction in barium from 6 mg/L to 0.5 mg/L. The plant's filter operates at 1.0 L/sec/m^2 (1.5 GPM/ft^2) at a pH level of 11.0.

Total capital costs were $2,366,000 (1980 dollars) to address the barium contamination. The costs to address the barium contamination covered 10 capital components, including aerator, rapid mix tank, flocculation basin, gravity filter, recarbonation system, transfer pumps, potassium hydroxide system, gypsum system, polymer system, and appurtenances. Total construction costs were $1,068,100. Annual operating and maintenance costs were $155,900.

D.3 Reverse Osmosis: Sarasota, Florida

This case involves a study of eight existing groundwater supplies in Sarasota County, Florida. Water treatment plants ranged in capacity up to 3.03 m^3/day to 0.044 m^3/sec (from 800 to 1 MGD) and served five mobile home/trailer parks, one school, and two communities. The study was performed cooperatively by U.S. EPA's Division of Water Supply Research and the Sarasota County Board of Health between January and June 1977.

The influent for the eight plants was naturally contaminated with concentrations of radium-226 ranging from 3.4 to 20.2 pCi/L due to the presence of phosphatic limestone. In addition, many supplies had high levels of total dissolved solids.

Design parameters for the eight treatment plants are summarized in Table D-2. These plants added reverse osmosis units with either hollow-fiber or spiral-wound membrane designs from six different manufacturers. Their pretreatment processes included cartridge filtration, pH adjustment, and ion sequestration. Posttreatment processes included pH adjustment, degassification, and chlorination. Table D-3 lists the pre- and posttreatment processes for each plant.

All eight plants achieved compliance with the 5 pCi/L standard for radium-226, with operating pressures ranging from 14.1 to 29.9 kg/cm^2 (200 to 425 psi). Finished or product water consisted of 28 to 54 percent of the influent volume, and finished water concentrations ranged from 0.14 to 2.0 pCi/L. Waste or reject brine water contained from 7.8 to 37.8 pCi/L of radium-226.

Table D-4 presents chemical analyses for the eight plants. The table contains analyses of the raw water and product water collected after at least 1 hour of operation. Analytical data were examined at several different points during operation to assess the effect of operating duration on removal efficiency. Although the eight systems attained 76 percent efficiency upon initiation of operation, two systems examined attained peak operational efficiency after about 5 minutes of operation.

Table D-2. Design Criteria of Reverse Osmosis Equipment

System	Population	Product Water Design Capacity kL/day	Product Water Design Capacity GPD x 10^3	Reverse Osmosis System Manufacturer	Membrane Type	Pumps kw	Pumps hp	Operating Pressure kPa	Operating Pressure psi	Stages	Recovery Percent Design	Recovery Percent Actual
Venice	15,000	3,800	1,000	Polymetrics, Santa Clara, CA	Hollow fiber[a]	4,900	500	2,800	400	2	50	54
Sorrento Shores	1,300	780	200	Permutit, Paramus, NJ	Spiral wound[b]	1,470	150	2,900	425	1	75	39
Spanish Lakes MHP	800	265	70	Universal Oil Products (UOP), San Diego, CA	Spiral wound[b]	490	50	2,800	400	2	66	31
Bay Lakes Estates MHP	340	150	40	Purification Techniques, Avon-by-the-Sea, NJ	Hollow fiber[a]	290	30	2,800	400	2	55	NA
Kings Gate TP	800	115	30	Purification Techniques, Avon-by-the-Sea, NJ	Hollow fiber[a]	390	40	2,800	400	2	70	NA
Sarasota Bay MHP	135	19	5	Polymetrics, Santa Clara, CA	Hollow fiber[a]	50	5	2,800	400	1	50	50
Bay Front TP	39	6	1.6	Continental Water Conditioning, El Paso, TX	Hollow fiber[a]	7	3/4	1,400	200	1	28	28
Nokomis school	800	3	0.8	Basic Technologies, West Palm Beach, FL	Spiral wound[c]	5	1/2	1,400	200	1	35	NA

[a]DuPont, Wilmington, Delaware.
[b]Universal Oil Products (UOP), San Diego, California.
[c]Basic Technologies, West Palm Beach, Florida.
NA = not available.
1,000 gallon - 3.78 m^3
1 psi = 0.0703 kg/cm^2
Source: Sorg (1980a).

The operating costs for these units ranged from $0.16 to $0.41/m³ ($0.60 to $1.54/1,000 gal) of treated water (see Table D-5). These costs include chemicals, electrical power, filter cartridge replacement, and labor. However, these data are not complete or comparable because they were derived from interviews with operators and owners, and do not necessarily use the same basis for each cost element.

D.4 Ion Exchange: McFarland, California

This case involves a 0.044 m³/sec (1-MGD) plant with four ground-water wells contaminated with nitrates from agricultural application of fertilizers and manure. Influent concentrations of nitrate ranged from 6.8 to 22.1 mg/L and averaged 16 mg/L.

To address this contamination problem, ion exchange units were selected for wellhead application because they are effective and easy to operate. The treatment process included:

- Anion exchange with A-101-D, Duolite resin
- Sodium chloride regeneration with slow rinse and declassification
- Aerated lagoons and spray irrigation for brine waste treatment

The ion exchange units used three reaction basins, each measuring 1.8 m (6 ft) in diameter and 3 m (10 ft) in height. The standard operational height of the reaction basin is only 0.9 m (3 ft), with operational maximums of 1.5 m (5 ft). According to system design, one of the basins undergoes regeneration, while the other two operate. The plant uses a 2.5-minute EBCT; the treated water flow rate was 15.77 L/sec (250 GPM) with surface loading rates of 6.13 L/sec/m² (9.03 GPM/ft²). Treated water was blended with raw water in a 7-to-3 ratio.

The regeneration process used a 6 percent sodium chloride brine. Regeneration involved quick rinse, slow rinse, and resin reclassification procedures that required 981 kg (2,162 lb) of salt daily during periods of continuous operation. The process produced saturated brine at a rate of 2.27 L/sec (36 GPM) and diluted brine at 12 L/sec (190.5 GPM). Brine was discharged to a municipal wastewater treatment plant, where it was diluted by the other waste streams and then placed in aeration lagoons. This

Table D-3. Pre- and Posttreatment Reverse Osmosis Processes

System	Pretreatment: pH Adjustment Chemical	Pretreatment: Sequestering Agent	Pretreatment: Filters µm	Posttreatment: Degassification or Aeration	Posttreatment: pH Adjustment Chemical	Posttreatment: Chlorination	Blending Raw and Treated Water	Reject Water Disposal Method
Venice	H_2SO_4	$Na_3(PO_3)_6$	5	Yes	a	Yes	a	Creek
Sorrento Shores	H_2SO_4	$Na_3(PO_3)_6$	10	No	Na_3CO_3	Yes	75% reverse osmosis	Bay
Spanish Lakes MHP	H_2SO_4	$Na_3(PO_3)_6$	25	Yes	None	Yes	No	Pond
Bay Lakes Estates MHP	H_2SO_4	$Na_3(PO_3)_6$	10	Yes	None	Yes	84% reverse osmosis	Storm sewer
Kings Gate TP	H_2SO_4	$Na_3(PO_3)_6$	10	Yes	None	Yes	No	Ditch
Sarasota Bay MHP	HCl	$Na_3(PO_3)_6$	5	Yes	None	Yes	No	Subsurface drainfield
Bay Front TP	None	poly-stabilizer A-5	10	Yes	None	Yes	55% reverse osmosis	Subsurface drainfield
Nokomis school	CO_2	None	5	Yes	None	Yes	No	Subsurface drainfield

[a] Blended with lime-softened water in approximate ratio of 2 parts lime-softened water to 1 part reverse osmosis water.
Source: Sorg (1980a).

aerated solution was spray- irrigated onto animal feed crops and cotton.

The treated water had nitrate levels of 2 to 5 mg/L. The blended finished water nitrate levels ranged from 6 to 10 mg/L with a 7 mg/L average.

The total construction costs for the ion exchange units were $354,638 in 1983 dollars. The operating and maintenance costs totaled $0.03/m³ (12.8 cents/1,000 gal) of treated water. Table D-6 shows the components of both types of costs. These costs include the annual loss of 20 percent of the resin.

D.5 Activated Alumina: Gila Bend, Arizona

This case involves a ground-water supply with undesirable levels of fluoride ranging from 4 to 6 mg/L, with a 5 mg/L average. The plant, with an average capacity of 37.8 L/sec (600 GPM) and a maximum capacity of 56.77 L/sec (900 GPM), was equipped with the following elements to manage the fluoride contamination:

- Activated alumina: Alcoa activated alumina, grade F-1, -28 to +48 mesh
- Caustic regeneration
- Acid neutralization
- Evaporation pond for regenerant waste treatment

The treatment process produces 90 percent treated finished water and 10 percent waste water. The finished water averages 0.7 mg/L of fluoride, with a maximum of 1.4 mg/L.

The alumina medium was placed in two vessels 3 m (10 ft) in diameter and height. The alumina takes up 1.5 m (5 ft) of vessel height, and expands about 50 percent of its original height during backwash operations. Approximately 15 cm (6 in) of basin freeboard is provided. The water's superficial residency time is 5 minutes. The maximum operational flow rate in the basins is 4.75 L/sec/m² (7 GPM/ft²), while the backwash rate is 7.47 L/sec/m² (11 GPM/ft²).

For every 13.2 to 15.1 thousand cubic meters (3.5 to 4 million gallons) of water treated, a 10-hour regeneration cycle is required. Annual losses of alumina due to regeneration range from 10 to 12 percent. Regeneration of the alumina medium is accomplished with a 1 percent solution of sodium hydroxide. The 1 percent solution flows through the basins at a maximum rate of 1.70 L/sec/m² (2.5 GPM/ft²), with a detention time of 24 minutes. The regeneration process uses 757 L (200 gal) of sodium hydroxide solution/lb fluoride removed.

The caustic water from the regeneration process requires a 0.04 percent solution of sulfuric acid for neutralization. The acid solution is derived by diluting a bulk 93 percent acid solution. (The neutralization process flow rate is 4.75 L/sec/m² [7 GPM/ft²] at most.) The goal of the neutralization process is to produce acceptable pH levels from 6.5 to 8.5 for disposal. The backwash and neutralization rinse water wastes are discharged to the sewer. The

Table D-4. Chemical Analyses of Sarasota County Reverse Osmosis Systems

Parameter[a]	Bay Lakes Estates MHP Water		Kings Gate TP Water		Spanish Lakes MHP Water: Defective Membranes (3/77)		Spanish Lakes MHP Water: New Membranes (3/78)		
	Raw	Product	Raw	Product	Raw	Product	Raw	Product	Reject
TDS	2,532	113	1,620	256	1,194	496	1,327	158	3,380
Specific conductance-µS	2,525	196	1,822	385	1,401	630	1,580	255	3,224
Turbidity - NTU	0.9	0.12	0.45	0.15	0.23	0.11	0.24	0.11	0.70
Color - *color units*	8	3	5	4	8	4	10	4	25
pH - *pH units*	7.3	6.0	7.5	6.35	7.55	7.3	8.0	6.1	6.7
Alkalinity (as $CaCO_1$)	114	20	144	30	164	68	176	10	24
Hardness (as $CaCO_3$)	1,620	68	1,020	134	750	286	865	98	2,130
Calcium	385	15.1	244	32.2	214	75.0	222	23.0	570
Magnesium	153	6.6	90	12.6	56.4	21.4	63.8	6.59	153
Chloride	105	13	85	27	71	32	66	14	135
Sulfate	1,460	59	840	118	580	210	670	80	2,000
Sodium	48	8.6	51	21.3	40	19	39	7.5	91
Lithium	0.07	<0.01	0.02	<0.01	<0.01	<0.01	<0.01	<0.01	.11
Silica	8.4	<1.0	12.9	2.8	9.2	4.9	12.7	3.2	30.6
Arsenic	0.021	<0.005	0.01	<0.005	0.008	<0.005	0.03	<0.005	0.075
Selenium	0.019	<0.005	0.014	<0.005	0.01	<0.005	0.017	<0.005	0.03
Fluoride	2.0	0.4	0.7	0.3	0.5	0.3	0.4	0.2	0.7
Radium-226 - pCi/L ±2Θ[b]	3.2 ±0.07	0.14 ±0.02	15.74 ±0.26	2.01 ±0.04	11.41 ±0.15	3.97 ±0.06	10.49 ±0.11	1.18 ±0.03	20.48 ±0.02

Table D-4. Continued

Parameter[a]	Sorrento Shores Water			Venice Water			Bay Front Water		
	Raw	Product	Reject	Raw	Product	Reject	Raw	Product	Reject
TDS	3,373	404	5,330	2,412	129	5,238	895	66	1,578
Specific conductance-µS	3,900	406	5,505	2,781	216	5,040	1,215	123	1,967
Turbidity - NTU	1.5	0.06	0.4	0.75	0.09	0.4	0.54	0.11	0.50
Color - *color units*	6	3	5	5	3	6	50	3	70
pH - *pH units*	7.15	5.25	5.1	7.25	5.9	5.45	7.45	6.75	7.5
Alkalinity (as $CaCO_1$)	114	8	6	110	20	14	294	36	540
Hardness (as $CaCO_3$)	1,980	70	3,100	1,425	44	3,050	570	32	980
Calcium	488	16.9	780	326	9.4	700	171	9.4	292
Magnesium	171	5.7	263	144	4.5	313	28.2	1.6	47.5
Chloride	520	95	690	300	33	500	95	<10	150
Sulfate	1,880	42	2,800	1,200	43	2,800	240	10	480
Sodium	258	45	366	140	25.4	249	59	13	100
Lithium	0.12	<0.01	0.26	0.04	<0.01	0.36	<0.01	<0.01	<0.01
Silica	7.6	1.9	11.1	7.2	<1.0	16.8	8.1	<1	14.8
Arsenic	0.095	0.008	0.190	0.015	<0.005	0.025	0.008	<0.005	0.016
Selenium	0.025	<0.005	0.035	0.015	<0.005	0.045	<0.005	<0.005	0.018
Fluoride	2.2	0.8	2.8	2.2	0.8	4.0	0.4	0.1	0.5
Radium-226 - pCi/L ±2Θ[b]	4.59 ±0.08	0.21 ±0.02	7.86 ±0.11	3.37 ±0.08	0.26 ±0.02	7.84 ±0.09	12.10 ±0.19	0.62 ±0.03	19.38 ±0.2

[a]All units are reported as mg/L except as noted.
[b]±2Θ represents instrument counting error of Ra-226 analysis.
Source: Sorg (1980a).

Table D-5. Reverse Osmosis Treatment System Costs

System	Year Con-structed	Capital Cost[a] $/1,000 gal	Estimated Operating Cost[b] $/kL	$/1,000 gal
Venice	1975	1,000	0.16	0.60
Sorrento Shores	1975	250	0.29	1.08
Spanish Lakes	1974	150	0.41	1.54
Bay Lakes Estates MHP	1973	25	0.33	1.06
Kings Gate MPH	1974	30	c	c
Sarasota Bay MPH	1975	14	0.32	1.22
Bay Front TP	1976	11.2	0.28	1.06
Nokomis school	1976	5.5	c	c

[a]Capital costs include reverse osmosis units and pumps, pre- and post- treatment equipment, filters, flushing equipment, booster pumps, and shelter.
[b]Operating costs include chemicals, power (0.4/kwhr), filled cartridge replacement, and labor. Excluded were amortization costs and membrane replacement. Blending was not taken into account.
[c]Insufficient data.
1,000 gallons = 3.78 m^3
Source: Sorg (1980a).

Table D-6. Ion Exchange Cost Components for McFarland, California ($1983)

Construction Costs Components in (expressed in$):	
Ion exchange unit vessels	111,741
Onsite Construction	81,154
Resin	56,610
Engineering	46,388
Brine tank	18,700
Other	40,045
Total	354,638
Operation and Maintenance Costs Components (expressed in cents/1,000 gallons):	
Salt	3.4
Resin replacement	3.2
Power	2.2
Normal operating and maintenance	1.9
Operating labor	1.3
Miscelleneous	0.8
Total	12.8

1,000 gallons = 3.78 m^3

regenerant waste is discharged to a lined evaporation pond which is 73 m (240 ft) by 134 m (440 ft) by 2.7 m (9 ft).

The construction costs for the unit totaled $285,000 in 1978 dollars, and included treatment facility, well, 1,892,500-L (500,000-gal) steel tank, evaporation pond, booster pumps, standby generator, and chlorination facilities. The plant's operating costs were 7.1 to 7.4 cents/m^3 (27 to 28 cents/1,000 gal) of treated water. These costs included salaries, power, chemicals, and media replacement.

Appendix E

Summary of Corrosion Indices

Index	Equation	Parameters	Meaning
Langelier Saturation Index (LSI)	$LSI = pH - pH_s$	Total alkalinity, mg/L as $CaCO_3$ Calcium, mg/L as $CaCO_3$ Hardness, mg/L as $CaCO_3$ Total dissolved solids, mg/L Onsite pH Onsite temperature	LSI > 0 = Water is supersaturated; tends to precipitate $CaCO_3$ LSI = 0 = Water is saturated (in equilibrium); $CaCO_3$ scale is neither dissolved nor deposited LSI < 0 = Water is undersaturated; tends to dissolve solid $CaCO_3$
Aggressive Index (AI) (for use with asbestos cement)	$AI = pH + \log[(A)(H)]$	Total alkalinity, mg/L as $CaCO_3$ Hardness, mg/L as $CaCO_3$ Onsite pH	AI < 10 = Very aggressive AI = 10 -12 = Moderately aggressive AI > 12 = Nonaggressive
Ryznar Stability Index (RSI)	$RSI = 2pH_s - pH$	Total alkalinity, mg/L as $CaCO_3$ Calcium, mg/L as $CaCO_3$ Hardness, mg/L as $CaCO_3$ Total dissolved solids, mg/L Onsite pH Onsite temperature	RSI < 6.5 = Water is supersaturated; tends to precipitate $CaCO_3$ 6.5 < RSI < 7.0 = Water is saturated (in equilibrium); $CaCO_3$ scale is neither dissolved nor deposited RSI > 7.0 = Water is undersaturated; tends to dissolve solid $CaCO_3$
Riddick's Corrosion Index (CI)	$\frac{75}{Alk}\left[CO_2 + 1/2\left(Hardness - Alk\right) + Cl^- + 2N\right] \times \left(\frac{10}{SiO_2}\right)\left(\frac{DO + 2}{Sa + DO}\right)$	CO_2, mg/L Hardness, mg/L as $CaCO_3$ Alkalinity, mg/L as $CaCO_3$ Cl^-, mg/L N, mg/L DO, mg/L Saturation DO[a] (value for oxygen saturation), mg/L	CI = 0-5 Scale forming 6-25 Noncorrosive 26-50 Moderately corrosive 51-75 Corrosive 76-100 Very corrosive 101 + Extremely corrosive
Driving Force Index (DFI) $CaCO_3$	$\frac{Ca^{++}(ppm) \times CO_3^{=}(ppm)}{K_{SO} \times 10^{10}}$	Calcium, mg/L as $CaCO_3$ $CO_3^{=}$ = mg/L as $CaCO_3$ K_{SO} = solubility product of $CaCO_3$	DFI > 1 = Water supersaturated; tends to precipitate DFI = 1 = Water saturated (in equilibrium); $CaCO_3$ scale is neither dissolved nor deposited DFI < 1 = Water undersaturated; tends to dissolve $CaCO_3$

[a]DO = dissolved oxygen
Source: U.S. EPA (1984).